ERGEBNISSE DER BIOLOGIE
ADVANCES IN BIOLOGY

HERAUSGEBER · EDITORIAL BOARD

H. AUTRUM · E. BÜNNING · K. v. FRISCH
E. HADORN · A. KÜHN · E. MAYR · A. PIRSON
J. STRAUB · H. STUBBE · W. WEIDEL

REDIGIERT VON · EDITED BY
HANSJOCHEM AUTRUM

BAND/VOLUME 26

SPRINGER-VERLAG
BERLIN · GÖTTINGEN · HEIDELBERG
1963

ORIENTIERUNG DER TIERE
ANIMAL ORIENTATION

SYMPOSIUM IN GARMISCH-PARTENKIRCHEN

17.– 21. 9. 1962

MIT 137 ABBILDUNGEN

WITH 137 FIGURES

SPRINGER-VERLAG

BERLIN · GÖTTINGEN · HEIDELBERG

1963

ISBN-13: 978-3-642-99874-4   e-ISBN-13: 978-3-642-99872-0
DOI: 10.1007/978-3-642-99872-0

© by Springer-Verlag OHG. Berlin · Göttingen · Heidelberg 1963
Softcover reprint of the hardcover 1st edition 1963
Library of Congress Catalog Card Number 26—11246

Vorwort

Vom 17. bis 21. September 1962 fand in der Vogelwarte Garmisch-Partenkirchen ein Symposium über die Fragen der Orientierung der Tiere statt. Die Anregung zu diesem Treffen ging von Professor HENRI PIÉRON, Paris, aus.

Die Analyse der Orientierung der Tiere im Raum und in der Zeit beginnt mit der — keineswegs auch nur annähernd vollständigen — phänomenologischen Bestandsaufnahme. Allein diese Bestandsaufnahme stößt auf erhebliche Schwierigkeiten, die nur mit kritischen experimentellen und einwandfreien statistischen Methoden überwunden werden können. Ein weiterer Schritt ist die Bestimmung der sinnesphysiologischen Leistungen, die an den Orientierungsvorgängen beteiligt sind. Die Kenntnis der Phänomene und der receptorischen Voraussetzungen der Orientierung stellen aber erst den Anfang dar. In erster Linie sind es zentralnervöse Vorgänge, die eine Orientierung ermöglichen. Hier spielt die zentrale Verrechnung der sinnesphysiologischen Daten, hier spielen Regelvorgänge und es spielen — bisher kaum bekannt, noch weniger analysiert — autonome zentralnervöse Prozesse eine entscheidende Rolle. Daher sind Sinnesphysiologie, Verhaltensforschung, Neurophysiologie, Kybernetik, Biophysik und mathematische Statistik in gleicher Weise an der Erforschung der Orientierungsleistungen beteiligt. Diese Arbeitsrichtungen zusammenzubringen und Anregungen auszutauschen, war eine der Hauptaufgaben des Symposiums.

Die Phänomene und Probleme der Orientierung in der Zeit wurden ausgeklammert. Sie sind Gegenstand eines Cold Spring Harbor-Symposiums im Jahre 1960 gewesen. Ihre Behandlung hätte den Rahmen gesprengt und die für eine fruchtbare Diskussion notwendigerweise kleine Teilnehmerzahl allzu stark vermehrt.

Unser herzlicher Dank gilt all denen, die das Zustandekommen des Symposiums ermöglichten: Die technischen und organisatorischen Vorarbeiten hat J. SCHWARTZKOPFF übernommen und mit dem ihm eigenen

Impetus durchgeführt. The International Council of Scientific Unions stellte Mittel für Reisekosten und Organisation zur Verfügung. Der Springer-Verlag sorgte für die rasche Publikation. Die Vogelwarte Garmisch-Partenkirchen überließ den Teilnehmern ihre Räume in groß-zügiger Weise. So war es möglich, intensive wissenschaftliche Arbeit fern vom Lärm und den Ablenkungen der Großstadt mit Erholung in der herrlichen Natur zu verbinden.

Februar 1963   H. Autrum

Inhaltsverzeichnis

Teilnehmerliste

Symposium
Orientierung der Tiere

In Garmisch-Partenkirchen vom 17.—21. September 1962

Adler, H. E., Dr. (New York, U.S.A.)
Autrum, H., Prof. Dr. (München, Deutschland)
Birukow, G., Prof. Dr. (Göttingen, Deutschland)
Braemer, Frau H., Dr. (Seewiesen, Deutschland)
Bückmann, D., Dr. (Göttingen, Deutschland)
Carmichael, L., Prof. Dr. (Washington, U.S.A.)
Carr, A., Prof. Dr. (Gainesville, U.S.A.)
Creutzberg, F., Dr. (Den Helder, Holland)
Dijkgraaf, S., Prof. Dr. (Utrecht, Holland)
Ferguson, D. E., Prof. Dr. (Mississippi, U.S.A.)
Gunning, G. E., Dr. (New Orleans, U.S.A.)
Jacobs, W., Prof. Dr. (München, Deutschland)
Klingler, J., Dr. (Wädenswil, Schweiz)
Kunze, P., Dr. (New Haven, U.S.A.)
Lindauer, M., Prof. Dr. (München, Deutschland)
Médioni, J., Dr. (Toulouse, Frankreich)
Mittelstaedt, H., Dr. (Seewiesen, Deutschland)
Moulton, J. M., Dr. (Brunswick, U.S.A.)
Novick, A., Dr. (New Haven, U.S.A.)
Paillard, J., Prof. Dr. (Marseille, Frankreich)
Papi, F., Dr. (Pisa, Italien)
Pennycuick, C. J., Dr. (Cambridge, England)
Perttunen, V., Dr. (Helsinki, Finnland)
Pringle, J. W. S., Prof. Dr. (Oxford, England)
Pye, J. D., Dr. (London, England)
Renner, M., Dr. (München, Deutschland)
Robert, P., Dr. (Colmar, Frankreich)
Sauer, E. G. F., Prof. Dr. (Gainesville, U.S.A.)
Schaller, F., Prof. Dr. (Braunschweig, Deutschland)
Schmidt-König, K., Dr. (Durham, U.S.A.)
Schneider, D., Dr. (München, Deutschland)
Schneider, F., Dr. (Wädenswil, Schweiz)
Schöne, H., Dr. (Seewiesen, Deutschland)
Schwartzkopff, J., Prof. Dr. (München, Deutschland)
v. St. Paul, U., Frl. Dr. (Seewiesen, Deutschland)
Susec-Michieli, S., Dr. (Ljubljana, Jugoslawien)
Teichmann, H., Dr. (Gießen, Deutschland)
Wallraff, H. G., Dr. (Wilhelmshaven, Deutschland)
Waterman, T. H., Prof. Dr. (New Haven, U.S.A.)
Wells, M. J., Dr. (Cambridge, England)
v. Zwehl, V., Frl. Dr. (München, Deutschland)

The Proprioceptive Background to Mechanisms of Orientation

By J. W. S. Pringle

Department of Zoology, Oxford University (England)

Contents

I. Introduction

The sensory physiology of orientation, with which this first part of the symposium is concerned, is a large subject which cannot be adequately reviewed in a single lecture. One may define orientation as the process whereby animals establish or maintain their body attitude in relation to the external environment, and many different types of sensory mechanisms are used by animals to this end. The subject matter in the study of the sensory physiology of orientation is, however, definable as a particular part of the overall problem; orientation involves a reaction on the part of the organism and, at least conceptually, a distinction can be made between the processes by which the relationship with the environment is perceived and the response processes, in this case a movement of some sort, by means of which the organism brings about or restores its correct angular relationship. The contributions of sensory physiology to the study of orientation can be well illustrated by considering the proprioceptive senses and it is with these that this lecture will be concerned.

II. The evolution of proprioception

It is at first necessary to consider the evolution of animal orientation in general terms. If, as is commonly supposed, the peculiarly animal (holophytic) type of organisation appeared first in an aqueous medium, the planes of space had little importance for the most primitive organisms. Movement is an almost necessary characteristic of a holophytic organism, and directed movement in relation to chemical and other features of the environment must early have been evolved in order to ensure that the animal remained, on average, in the optimum conditions. But orientation of position in direct response to external stimuli only begins to be important at a certain level of structural and behavioural complexity. Although, therefore, orientation responses undoubtedly have their origin in kineses and taxes and the evolutionary study of sense organs must be started at this level, its experimental study can be commenced only high up the scale of evolution. As with other branches of comparative physiology, it is the misfortune of zoologists that the systems in higher animals are easier to study than those in lower.

This difficulty is emphasized in the study of the sensory physiology of proprioception and of its role in orientation phenomena. Very little indeed is known about proprioceptive senses in lower animals. There may be a good reason why do they not exist. It was G. H. PARKER (1918) who first pointed out that the most primitive effectors in the animal kingdom were probably directly sensitive to stimuli from the environment, so that the whole machinery of regulation of movement was initially contained in the single cell or cell complex. Contractile cells responding directly to stretch form the only effector system of sponges (PROSSER, 1962), and the property is retained in various ways by muscles as diverse as the short-fibred visceral muscles of vertebrates (BURNSTOCK and PROSSER, 1960) and the fibrillar flight muscles of higher insects (PRINGLE, 1949). Sense organs reacting to the results of movement are necessary only when the effector itself has lost the ability to respond directly. In the effector system for movement, this differentiated condition is normally a considerable advance since, with integrative actions in the nervous system, compensatory responses can then be made by the animal as a whole and not merely by the part receiving the stimulus; a true integration of movement thus becomes possible, rather than a series of local responses. It is, however, important to appreciate this functional origin of the proprioceptive sense, since it follows that the sense organs appeared initially in relation to local needs and that their pattern of reflex connexions will initially have been local. Integration of the proprioceptive information into a *system* for the body as a whole is essentially a later evolutionary development.

The point is made all the more clearly because there are other senses whose organs must, at a much earlier stage, have had nervous connexions affecting the total pattern of behaviour. Receptors for visible light probably fall in this category. Sensitivity to damaging ultra-violet radiation may be a valuable local sensibility, but visible light is a stimulus whose importance for animals is always indirect, in the correlation which normally exists in nature between light intensity and other, directly significant, environmental features. There are thus two ways in which a sensory system can evolve: either by the gradual integration of local sensibility, as for proprioception, or with an influence from a very early stage on the behaviour of the whole animal.

At the level of the lower metazoan phyla, the relationship of proprioception to orientation reactions is already well defined. Local reflexes mediated through proprioceptors have become at least partially integrated into a central nervous machinery which achieves co-ordination of relative position of different parts of the body and the correct sharing of motor effort in promoting movement, and the orientation of the body as a whole in relation to the environment, both statically and dynamically, is controlled, through exteroceptors, by an effector system of which the proprioceptive machinery is an integral part. The proprioceptors do not play a direct role in the sensory mechanisms of orientation. This is a perfectly satisfactory state of affairs for a swimming animal or for one which lives in mud; we should not expect to find in such an animal, or in one which has only recently (in evolutionary terms) adopted a different mode of life, sensory mechanisms of orientation mediated directly through the proprioceptive system.

The change comes with life on the hard bottom of the sea or on land. Provided that the exteroceptive tactile sense is sufficiently well developed to provide information about the nature of the contact with the ground, an integration of proprioceptive information about the relative position of parts of the body could now form the basis for an orientation sense. Since there is also a need under these conditions for a more refined co-ordination of the effector system for movement, one might expect that there would develop central nervous mechanisms for the integration of the primitive proprioceptive information and for its correlation with that from tactile receptors. But the evidence is that this does not occur, or occurs only to a limited extent. Instead, a new system of proprioceptive sense organs evolves from exteroceptors to provide the required information, leaving the primitive proprioceptors to perform their original role. The reason for this is perhaps to be found in the elaborate nature of the computation which must be performed in order to derive orientation information; for it to come into existence, there had first to be an elaborate system of sense organs with some other (exteroceptive) function and at least the basis of a central nervous machinery required to perform the computation. Enough is now known about the pattern of proprioceptors in four animal groups to see how this new function has taken shape. They are the vertebrates, insects, aquatic arthropods and cephalopods, and it is instructive to look at the organs, reflexes and behaviour of these four groups from this point of view.

III. Types of proprioceptive sense organs

1. Vertebrata

It is first necessary to know the structure and location of the proprioceptive sense organs and, so far as possible, their central connexions. In the vertebrates, the primitive proprioceptors are the muscle receptors: initially, in fish, endings in the connective tissue on the surface of muscle fibres (KIRSCHE, 1948) and then becoming differentiated into the encapsulated endings of the muscle spindles and the Golgi tendon organs. These two main types lie, respectively, in the middle region of the muscle fibres, functionally in parallel with the main mass of the muscle, and at the ends of the fibres or in the tendons, functionally in series with the main muscle. The sensory and reflex mechanisms of the muscle spindles are becoming fully understood (COOPER, 1960; JANSEN and MATTHEWS, 1962); their role in behaviour is to maintain the relative position of parts of the body and to produce the mechanical stability required for precise movements. The role of the tendon organs in series with the muscle fibres is less clear; electrophysiological evidence suggests an inhibitory reflex action on synergic muscles (ECCLES, ECCLES and LUNDBERG, 1957) but cybernetic considerations suggest a more important role in active movement which may not be apparent in spinal animals (PRINGLE, 1961). Between them these two types of muscle sense organ comprise the afferent component of the movement coordination system which is the effector mechanism for orientation responses in the vertebrates. These may be the only types of proprioceptor found in the lower vertebrates. But in birds and mammals there is another system superimposed on the first: a system of connective tissue capsules (Pacinian corpuscles in mammals, Herbst corpuscles in birds), of a type generally reacting to deep pressure on the skin, but now, through corpuscles located in joints, tendons and elsewhere, giving effectively new information about the relative position of different parts of the body and of the forces acting on them. It is known that in mammals and probably in birds reactions of orientation can be mediated directly by the proprioceptive system and there is medical evidence that it is the joint sense and not the muscle receptors which is responsible. There is, indeed, no evidence that, in man, information from the muscle spindles or Golgi tendon organs ever reaches consciousness, and it is probable that it is solely the corpuscular system — a new development in the highest vertebrates — that is responsible for the sensory role of the proprioceptive sense in orientation.

2. Arthropoda

Turning now to the Arthropods, where knowledge of the structures involved is much more recent, it is necessary to give a more detailed review (see PRINGLE, 1961). Unlike the vertebrates where muscle endings appear to be the only primitive type, in arthropods there are always two distinct types of primitive proprioceptor, which have been categorized by their histology and possible ontogeny (Table 1). Of the

Table 1. *Classification of arthropod proprioceptors*
(modified from Table 1 of PRINGLE, 1961)

Characteristics	Type I	Type II	Type III
Cell origin	Differentiation from hypodermis	Migration from C.N.S. (?)	Differentiation from hypodermis
Histology	Unbranched distal process containing scolopales	Branching distal processes without chromophilic bodies	Unbranched distal process
Association	Cuticle	Connective tissue	Cuticle
Function	Co-ordination of movement	Co-ordination of movement	Orientation
Examples			
Crustacea	Chordotonal organs	Muscle receptor organs: innervated elastic strands	Joint endings
Chelicerata	Slit sensilla (lyriform organs)	Deeper-lying endings	Joint hairs
Insecta	Chordotonal organs: campaniform sensilla	Stretch receptors	Hair plates

Type I endings, the chordotonal organs of crustacea and insects are strands stretching across joints and attached at each end by modified hypodermal cells, from which the whole organ differentiates in ontogeny (SCHÖN, 1911). The best-known type of crustacean chordotonal organ is that described by BURKE (1954) from the propodite-dactylopodite joint of *Carcinus* and now found by WHITEAR (1962) to be present at most of the leg joints. In insects, chordotonal organs often serve as vibration or sound receptors which do not concern us here, but they may also be proprioceptors (BECHT, 1958) and are widely distributed both in the legs and between the joints of the body (DEBAISIEUX, 1938; EGGERS, 1928). They have not been described from the Chelicerata. Another, quite different Type I sense organ differentiated from the hypodermal cells (WIGGLESWORTH, 1953) is in the insect campaniform

sensillum and arachnid slit sensillum or lyriform organ (Pringle, 1955);
this is absent from crustacea, probably owing to the greater rigidity
of the calcified cuticle, since the strains produced by muscular contraction are the adequate stimulus. Type II endings occur in all the
arthropod groups and are multipolar sense cells associated with the
connective tissue of muscles or elastic strands. Alexandrowicz (1951,
1958) has described various types in crustacea, Pringle (1956) those of
Limulus and Finlayson and Lowenstein (1958) the abdominal stretch
receptors of insects.

When the attempt is made to determine the function of these various
arthropod proprioceptors and to classify them as position-indicating,
movement-indicating and force-indicating sense organs, there appears
to be no simple correlation with the histology. As with the vertebrate
muscle endings, it seems probable that in the primitive condition all
three functions are served by the same organ; only with further differentiation (which has not always occurred) does the functional division
become clear. Thus the sensilla of the crab PD organ (Burke's organ)
are a diverse population and include some tonic endings signalling joint
position and some phasic endings signalling joint movement (Wiersma
and Boettiger, 1959; Wiersma, 1959); the reflex effects of these sensory
discharges confirm their mixed function (Bush, 1962). In the Type II
stretch receptor of the silk-moth pupa there is no sign of morphological
differentiation between position and movement reception (Lowenstein
and Finlayson, 1960), but the similar organ in the crayfish is clearly
differentiated to distinguish these two types of stimulus (Wiersma,
Furshpan and Florey, 1953). The cuticular endings show stages in the
differentiation of force reception. In scorpions the slit sensilla signal
joint position, but are also excited by the strains produced by resistance
to muscular contraction (Pringle, 1955); in insects, many of the groups
of campaniform sensilla are not excited by free joint movement, but
only by the forces produced as the result of muscular activity (Pringle,
1938a, b). The general point which emerges from the study of the physiology of arthropod proprioceptors is that it cannot be assumed, because
senses of relative position, movement and force are known from behavioural studies to be present, that the sense organs have differentiated in
any particular way. The whole system of sense organs and reflexes
evolves in relation to the animal's needs, and the conceptual distinctions
which the investigator can make between these three roles of proprioception are not necessarily distinctions made in the physiological
machinery of the animal. This is a difficulty which arises in any attempt
to analyse animal behaviour and it must be borne in mind when considering the sensory physiology of orientation. The investigator has to
set a conceptual framework for his analysis. The animal may not conform

to this; the evolution of its organisation has produced responsiveness to the environment so as to ensure survival, and, of this process, the gathering of information is only one part.

As yet, one important system of arthropod proprioceptors has been omitted from the review; these are the cuticular hairs at the joints, present in crustacea and chelicerata and reaching their greatest refinement in the hair plates of insects (PRINGLE, 1938c). These endings come in a different category and are probably no part of the proprioceptive machinery concerned with the precise co-ordination of movement. In the same way that the Pacinian corpuscle system of mammals has been superimposed on the more primitive system of vertebrate muscle receptors, so the hair plate system has been evolved from originally tactile hairs and superimposed on the more primitive arthropod proprioceptive system in those animals where the relative position of parts of the body is especially important. It is known through the work of MITTELSTAEDT (1950, 1957), LINDAUER and NEDEL (1959), WENDLER (1961) and MARKL (1962) that the hair plate system is of direct importance in orientation and, indeed, gives insects their main gravity sense. As with the Pacinian corpuscle system of mammals, the joint position information from the hair plates needs to be integrated over the whole or large parts of the body in order to be useful, and the necessary processes of central integration are more similar to those required for an exteroceptive than for a proprioceptive sense. In its function and probably in its nervous connexions, the hair plate system is quite distinct from the primitive proprioceptors, whose function is the local co-ordination of movement and whose role in orientation is more correctly regarded and studied as part of the effector machinery.

3. Relation to special organs of mechanical sense

The special organs of mechanical sense (the inner ear of vertebrates and the statolith organs of invertebrates) are often classified as proprioceptors, because they are excited by the stimulus of movement. These sense organs will not be discussed in detail in this review, since they are to be described by other contributors to the symposium, but it is important to consider their relationship to the proprioceptors already described. In many animals, they have differentiated to mediate not only a static gravity sense but also dynamic responses to linear and angular accelerations. They are clearly important in orientation and, in that the gravity field is one of the physical characteristics of the environment, they may, from this point of view, be regarded as exteroceptors. Like other organs of special sense, they control the responses of the whole animal, but are situated on one particular part of the body, and for this to be possible the prior existence of a good proprioceptive sense of the relative position of the parts of the body is necessary for their proper functioning. The role of the neck proprioceptors in righting reactions is one of the classic investigations of human physiology (MAGNUS, 1924). But this does not imply that the primitive system of diffuse proprioceptors must play a direct role in the sensory mechanism of orientation. Information from the diffuse proprioceptors does not need to be integrated with information from a statocyst in producing what MITTELSTAEDT (1962) calls the "command order" for the orientation response; the diffuse proprioceptors may be more correctly regarded as part of the effector than of the sensory system. This may be more than just a conceptual distinction, since it could determine the pattern of nervous connexions in the central nervous system.

4. Cephalopoda

The cephalopods illustrate this point, and their orientation behaviour can only be understood if it is borne in mind. Little is known about the structure and sensory physiology of cephalopod proprioceptors. ALEXANDROWICZ (1960) has described endings in the pallial and branchial musculature of *Eledone*, but was unable from their structure to suggest their mode of action. From behavioural studies by BOYCOTT and YOUNG (1950) it is known that proprioception must be important in the coordination of movement by the arms, but until proprioceptive sense organs have been found it remains a possibility that this may be an example of the retention of the primitive sensitivity to stretch of the muscles themselves or of elements in the peripheral nervous system. If proprioceptive sense organs exist, there are several good pieces of evidence that, in *Octopus*, the information from them is not used as a sensory system for orientation or for other types of higher nervous behaviour. BOYCOTT (1960) has shown that octopuses do not learn to use proprioceptive information from the arms to orientate in relation to the ground after statocyst removal. WELLS and WELLS (1957) established that octopuses cannot learn to discriminate objects differing merely in qualities for the perception of which proprioceptive sensory integration would be needed, and further (1960) that even the statocyst is subservient to the eyes and the optic centres in orientation responses. Statocyst information is used to control the orientation of the retina in relation to gravity, and neither statocyst nor diffuse proprioceptive information is integrated centrally with retinal input. This is an entirely different state of affairs to that found in the mammals or insects. In mammals, information from the inner ear is integrated with information from the eyes in producing the "command order" for orientation. In insects, where there is no statocyst, MITTELSTAEDT (1957) has shown in the *Mantis* how information from the hair plate system is integrated with that from the eyes in determining the unmonitored command order for the grasping response. One may speculate on the evolutionary explanation of this difference. The octopus has evolved late in evolutionary history from purely nektonic swimming cephalopods whose whole behavioural orientation machinery centres round the optic information. Both mammals and insects have had a longer history of walking on a solid substratum; in the vertebrates, which were swimming animals up to the fish stage, perfection of the sensory system of the inner ear preceded the perfection of pattern vision and probably conditioned the development of the nervous centres used in orientation; but there has been a long history of locomotion by walking. Insects and other arthropods have been in contact with the ground for an even longer period and the sensory and central machinery for orientation has developed from an integrated tactile sense, successive parts of which have become functional as proprioceptors. It would be hard to find a clearer case of the impact of evolutionary history on the form of a physiological system.

IV. Gyroscopic sense organs

Detection of the results of angular movement can be achieved in two different ways: by means of a static, loosely suspended mass whose movement lags behind that of the body, or by sensing the Coriolis forces acting on a mass which is actively moved in relation to the body. The majority of animals use the former method, which is capable, by refinement of the sense organs associated with the statolith, of giving good indication of angular position and angular acceleration. A rapid measure of angular velocity is provided in the animal kingdom only by the gyroscopic mechanism of the halteres of Diptera, which are known to indicate rotation of the body of the fly in all three planes of space (FAUST, 1952; PRINGLE, 1948, 1957). This complex sensory system has evolved in particular relation to the need for maintenance of equilibrium in flight, and since the reaction time of the fly to angular disturbances is of the order of fractions of a second, equilibration comes in rather a different category from most of the orientation responses to be considered in this symposium. It is, however, comparable in speed to the acoustic orienting reactions of bats and cannot be omitted from the review. The theoretical possibility exists that long-term maintenance of direction could be achieved by integration of the haltere information, but this has never been demonstrated; in most circumstances the visual system is better suited to this purpose.

V. Summary

(1) The role of proprioceptive senses in orientation reactions is different from that of exteroceptors which give a direct indication of the animal's relationship with the environment. Proprioceptive sense organs

first evolved as part of the mechanism of co-ordination of movement, and are functionally part of the effector rather than the receptor system.

(2) In the higher vertebrates and arthropods, a secondary proprioceptive system (Pacinian corpuscles and hair plates) is present, information from which is treated differently in the central nervous system. Through this system, but probably not through the primitive proprioceptors, the animal obtains an integrated picture of the relative position of parts of its body. By correlation with information from the exteroceptive tactile system a new type of orientation sense emerges.

(3) The special organs of mechanical sense (statoliths and associated endings) often provide an alternative type of proprioceptive orientation sense, but observations of *Octopus*, where the statocysts do not directly control orientation reactions, show that caution must be exercised in interpreting their function.

(4) Comparative studies of the role of proprioceptors in orientation suggest that the evolutionary history of a group has a marked influence on the detailed mechanisms involved.

References

ALEXANDROWICZ, J. S.: Muscle receptor organs in the abdomen of *Homarus vulgaris* and *Palinurus vulgaris*. Quart. J. micr. Sci. **92**, 163—199 (1951).
— Further observations on proprioceptors in Crustacea and a hypothesis about their function. J. mar. biol. Ass. U.K. **37**, 379—396 (1958).
— A muscle receptor organ in *Eledone cirrhosa*. J. mar. biol. Ass. **39**, 419—431 (1960).
BECHT, G.: Influence of DDT and lindane on chordotonal organs in the cockroach. Nature (Lond.) **181**, 777—779 (1958).
BOYCOTT, B. B.: The function of the statocysts of *Octopus vulgaris*. Proc. roy. Soc. B **152**, 78—87 (1960).
— and J. Z. YOUNG: The comparative study of learning. Symp. Soc. exp. Biol. **4**, 432—453 (1950).
BURKE, W.: An organ for proprioception and vibration sense in *Carcinus maenas*. J. exp. Biol. **31**, 127—138 (1954).
BURNSTOCK, G., and C. L. PROSSER: Responses of smooth muscles to quick stretch; relation of stretch to conduction. Amer. J. Physiol. **198**, 921—925 (1960).
BUSH, B. M.: Proprioceptive reflexes in the legs of *Carcinus maenas* (L.). J. exp. Biol. **39**, 89—105 (1962).
COOPER, S.: Muscle spindles and other muscle receptors. In: The Structure and Function of Muscle, Vol. 1, Ed. BOURNE, G. H. New York and London: Academic Press 1960.
DEBAISIEUX, P.: Organes scolopidiaux des pattes d'insectes. Cellule **47**, 78—202 (1938).
ECCLES, J. C., R. M. ECCLES and A. LUNDBERG: Synaptic actions on motoneurones caused by impulses in Golgi tendon organ afferents. J. Physiol. (Lond.) **138**, 227—252 (1957).
EGGERS, F.: Die stiftführenden Sinnesorgane. Zool. Baust. **2**, 353 (1928).
FAUST, R.: Untersuchungen zum Halterenproblem. Zool. Jb. (Abt. allg. Zool. u. Physiol.) **63**, 325—366 (1952).
FINLAYSON, L. H., and O. LOWENSTEIN: The structure and function of abdominal stretch receptors in insects. Proc. roy. Soc. B **148**, 433—449 (1958).

JANSEN, J. K., and P. B. C. MATTHEWS: The central control of the dynamic response of muscle spindle receptors. J. Physiol. (Lond.) **161**, 357—378 (1962).

KIRSCHE, W.: Histologische Untersuchungen über das peripherische Nervensystem der Teleostier. Anat. Anz. **96**, 419—454 (1948).

LINDAUER, M., u. J. O. NEDEL: Ein Schweresinnesorgan der Honigbiene. Z. vergl. Physiol. **42**, 334—364 (1959).

LOWENSTEIN,O., and L. H. FINLAYSON: The response of the abdominal stretch receptor of an insect to phasic stimulation. Comp. Biochem. Physiol. **1**, 56—61 (1960).

MAGNUS, R.: Körperstellung. Berlin: J. Springer 1924.

MARKL, H.: Borstenfelder an den Gelenken als Schweresinnesorgane bei Ameisen und anderen Hymenopteren. Z. vergl. Physiol. **45**, 475—569 (1962).

MITTELSTAEDT, H.: Physiologie des Gleichgewichtsinnes bei fliegenden Libellen. Z. vergl. Physiol. **32**, 422—463 (1950).

— Prey capture in mantids. In: Recent advances in invertebrate physiology. Ed. SCHEER, B. T. University of Oregon Press 1957.

— Control systems of orientation in insects. Ann. Rev. Entomol. **7**, 177—198 (1962).

PARKER,G.H.: The elementary nervous system. Philadelphia: J.B.Lippincott Co.1918.

PRINGLE, J. W. S.: Proprioception in insects. I. A new type of mechanical receptor from the palps of the cockroach. J. exp. Biol. **15**, 101—113 (1938a).

— Proprioception in insects. II. The action of the campaniform sensilla on the legs. J. exp. Biol. **15**, 114—131 (1938b).

— Proprioception in insects. III. The function of the hair sensilla at the joints. J. exp. Biol. **15**, 467—473 (1938c).

— The gyroscopic mechanism of the halteres of Diptera. Philos. Trans. roy. Soc. B **233**, 347—384 (1948).

— The excitation and contraction of the flight muscles of insects. J. Physiol. (Lond.) **108**, 226—232 (1949).

— The function of the lyriform organs of arachnids. J. exp. Biol. **3**, 270—278 (1955).

— Proprioception in *Limulus*. J. exp. Biol. **33**, 658—667 (1956).

— Proprioception in arthropods. In: The cell and the organism. Ed. RAMSAY, J. A., and V. B. WIGGLESWORTH, Cambridge University Press 1961.

PROSSER, C. L.: Unpublished results reported to International Congress of Comparative Neurophysiology. Leiden 1962.

SCHÖN, A.: Bau und Entwicklung des tibialen Chordotonalorgans bei der Honigbiene und bei Ameisen. Zool. Jb. (Abt. Anat.) **31**, 439—472 (1911).

WELLS, M. J.: Proprioception and visual discrimination of orientation in *Octopus*. J. exp. Biol. **37**, 489—499 (1960).

— and J. WELLS: The function of the brain of *Octopus* in tactile discrimination. J. exp. Biol. **34**, 131—142 (1957).

WENDLER, G.: Die Regelung der Körperhaltung bei Stabheuschrecken *(Carausius morosus)*. Naturwissenschaften **48**, 676—677 (1961).

WHITEAR, M.: The fine structure of crustacean proprioceptors. I. The chordotonal organs in the legs of the shore crab, *Carcinus maenas*. Philos. Trans. roy. Soc. Lond. B. **245**, 291—325 (1962).

WIERSMA, C. A. G.: Movement receptors in decapod Crustacea. J. mar. biol. Ass. U. K. **38**, 143—152 (1959).

— and E. G. BOETTIGER: Unidirectional movement fibres from a proprioceptive organ of the crab, *Carcinus maenas*. J. exp. Biol. **30**, 136—150 (1959).

— E. FURSHPAN and E. FLOREY: Physiological and pharmacological observations on muscle receptor organs of the crayfish, *Cambarus clarkii* GIRARD. J. exp. Biol. **30**, 136—150 (1953).

WIGGLESWORTH, V. B.: The origin of sensory neurons in an insect, *Rhodnius prolixus* (Hemiptera). Quart. J. micr. Sci. **94**, 93—112 (1953).

Mechanisms of Echolocation *

By J. D. PYE

Institute of Laryngology and Otology, London (England)

With 7 Figures

The cries of bats, although unperceived by our unaided ears, may be made audible by simple electronic apparatus. With such a detector, it is fascinating to observe a bat navigating, negotiating obstacles and catching its prey in the dusk, and to remember that this performance is achieved by acoustic means. The question always arises "If the bat can do this with two ears and a larynx, why can man achieve only a very imperfect obstacle sense with the same organs?" The answer is partly that the bat both produces and hears very much higher sound frequencies which permit finer discrimination, and partly that its ear is much more sensitive to short-delay echoes. The physiological mechanisms behind all these effects are of fundamental interest but here only the last aspect will be considered — that of echo detection and interpretation.

The principle problem the bat faces is that it has just made, or is still making, a very loud sound when it has to detect a very faint and precise copy of that sound. The small size of the head and the relationship of aperture sizes to wavelength would seem to preclude much acoustic isolation of transmitter and receiver so that masking must be expected. Also the problem is seldom one of a single echo. Bats hunt quite successfully among rain-drops, foliage, or close to large surfaces and can select single insects from a swarm. A single target with dimensions larger than a wavelength will return a complex composite echo rather than a simple copy of the emission. The system is also able to cope with the presence of large numbers of other active bats, as when colonies emerge from roost and even hunt together without interference. Any theory of the mechanisms involved must embrace all these situations.

A major clue appears to lie in the structure of the signals employed for orientation. Most, if not all, bats produce simple pulses of ultrasound with a rapid frequency sweep, although other components may be

* *Acknowledgement:* The research reported in this document has been sponsored by the Air Force Office of Scientific Research, OAR through the European Office, Aerospace Research, United States Air Force.

present. The extremes of this variation appear to be represented by the Vespertilionidae with frequency modulation throughout the pulse, and the Rhinolophidae with a variable, but usually long, constant-frequency phase before the sweep. Fortunately the information potential of frequency-modulated pulses has been calculated for Radar purposes by KLAUDER et al. (1960). Their formulae can be applied to the information on bats given by GRIFFIN (1958) with the results shown in Table 1. These figures make it clear that the full bandwidth must be taken into account in deriving a satisfactory explanation of bat performance.

Table 1

Species	Minimum range differences detectable by unmodulated pulses[1] cm	Minimum range differences detectable by frequency-modulated pulses[2] cm
Myotis lucifugus	39.1	0.43
M. keenii septentrionalis . .	15.3	0.35
Antrozous pallidus	54.4	0.65
Plecotus rafinesqui	28.9	1.42
Tadarida yucatana	90.1	1.21

[1] $\Delta R = \dfrac{c \cdot T}{2}$, where c is the velocity of conduction, T is pulse duration.

[2] $\Delta R = \dfrac{c}{2 \cdot \Delta f}$, where Δf is the width of the frequency sweep in cycles per second.

The theories put forward so far fall into four categories. The first is exemplified by the work and views of GRINNELL (1962) and supposes that each echo is individually detected by the cochlea for normal representation by impulse patterns in the acoustic nerve. The greatly developed acoustic centres of the brain are then wholly responsible for the interpretation of echo information from acoustic delay, frequency and binaural differences. By means of extensive and elegant recording of central neural responses, GRINNELL has produced considerable evidence that the acoustic sensory system of the bat is indeed capable of performing these feats, at least for the single echo situation. Simultaneous treatment of each component of the frequency sweep could utilise the full bandwidth and achieve the necessary discrimination, although the evolution of such a complex and specialised neural arrangement is difficult to imagine.

MILLS (1958) and others have shown that the minimum difference of binaural arrival time that can be detected by man is 10 microseconds. A similar performance by a small bat with an interaural distance of 1 cm would allow binaural comparison to within about 20° of the heading.

One may expect the animal to resolve rather finer time differences, say 2 microseconds, permitting directional measurement to within 4° of the heading. Otherwise the indefinable region directly ahead might be embarrassingly wide.

Further work on these lines could well show that this direct approach with greatly refined organisation has been adopted by all bats with a simple frequency sweep such as the Vespertilionids. But it does not seem adequate at present to explain the presence of other components such as the constant frequency of Rhinolophids.

The second theory is that of NORDMARK (1960) which proposes that bats rely on the time-difference tone first described by THURLOW and SMALL (1955). This is a faint tone heard when pairs of waves or sound pulses are presented to the ear, and its period is equal to the temporal separation of the pulses of each pair. The pitch of this tone for pulse-and-echo pairs is therefore a measure of target distance. If THURLOW and SMALL are correct in arguing that this tone is produced centrally, then this is not a new theory, but evidence that even the human brain is capable of registering small time delays. But other views on the origin of the effect are possible (vide THURLOW, 1957). GRINNELL has criticised its application here on the grounds that a train of pulse-pairs is required whereas the bat appears to make an initial discrimination with very few pulses, possibly a single one.

Thirdly STROTHER (1961) has suggested that bats could operate on the principle of Chirp Radar as described by KLAUDER et al., involving a frequency-sensitive delay which temporally compresses a frequency-modulated pulse. There is no evidence for such a system in mammals, indeed the only frequency-sensitive delays known are in the wrong sense, leading to pulse expansion rather than compression. All the advantages of the Chirp system for both bat and Radar appear to lie in the band-width of a frequency-modulated pulse, and the pulse compression receiver merely facilitates standard visual display and electronic timing circuits. The bat does not use visual display and it seems more pertinent to look for a mechanism suited to auditory display.

Such a system has been proposed independently by KAY (1961, 1962) and PYE (1960, 1961) and is applicable in conditions where the echoes return to the ear before pulse production is complete. If beat-notes are then produced by a multiplicative process, their frequencies (for which the cochlea appears to be an admirable analyser) represent all the relevant echo information. Frequency-modulated pulses give a beat frequency which is directly proportional to target distance and sufficiently sensitive to allow fine binaural comparison over the whole azimuth. A constant frequency pulse gives relative velocity by Doppler shift and, if coupled with rapid ear movement, permits a high degree of directional reception.

If beat-notes are produced artificially, the human ear immediately becomes capable of fine discrimination as an echolocation receiver, as can be demonstrated by models of these systems.

The theoretical attractiveness of this scheme is fourfold. First, as shown in Fig. 1, it avoids problems of masking and congestion of the auditory nerve. Eighth-nerve fibres may be classified both according to effective stimulus frequency and to dynamic range to give a population probably like that within the outer dashed line. The outgoing pulse,

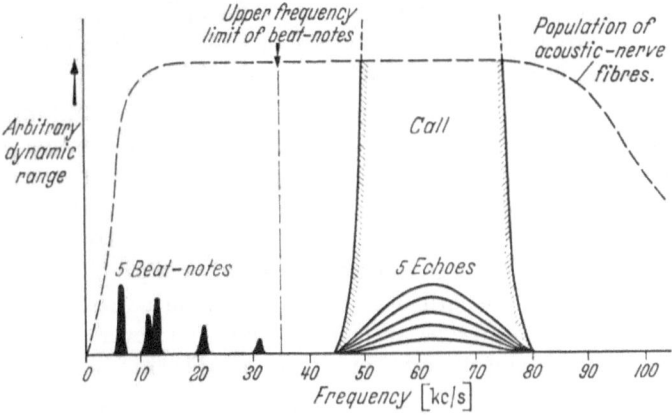

Fig. 1. A hypothetical scheme of the stimulation of acoustic-nerve fibres by a frequency-modulated pulse, five echoes and their beat-notes. No account is taken of time relations in this diagram

being very loud, might activate most or all of the fibres within its spectrum. Now a central discrimination theory demands that the most sensitive fibres in this same band respond again, within a very short time, to every echo and every component of a complex echo. It is, of course, possible to postulate neural mechanisms which could ignore the first signal, but the beat-note transformation shifts the onus to an entirely different part of the population. Multiple and complex-echo responses become much less difficult to comprehend. Fig. 1 is drawn for a frequency-modulated system but its application to constant-frequency pulses, where masking is more acute, gives an even greater separation.

Secondly this information transformation is extremely simple. It could be effected by non-linear transmission and a distorting ear is not difficult to imagine. The result then depends only on the pulse structure which is already known. GRINNELL has pointed out that the conversion implies a loss of energy when minimum effective echo energies are already small. But all the remaining energy is concentrated in a much smaller bandwidth and, even with low conversion efficiency, temporal integration could recover and even increase the energy available to each receptor cell.

Thirdly the system is economical since, by transferring some of the responsibility to a peripheral level, it reduces the redundancy of information which the eighth nerve is required to transmit. Echo frequency is probably of little importance since it is a close copy of the transmission, and it is rejected although Doppler shifts remain as bandspreading of the beat-note; echo delay is coded as beat frequency and echo intensity remains as beat intensity.

Finally this is a unifying hypothesis applicable to a wide range of bats, including the forms which produce pulses intermediate in form between those of Vespertilionids and Rhinolophids. It also provides ready explanations for variations in the behaviour of different bats such as the dependence on binaural reception.

Despite these encouraging features there is at present no direct evidence for the formation or use of beat-notes by bats, and for Vespertilionids one piece of contrary evidence is emerging. GRIFFIN has shown that during the approach to a target, pulse repetition rate rises and pulse duration is reduced. Now WEBSTER and CAHLANDER (private communication) have indicated by stero-cinematography and simultaneous sound recording that pulse length may be steadily reduced in proportion to range so that temporal overlap of pulse and echo never occurs. If this is so, beat-notes could not arise at a time when their advantages would seem greatest.

But for Rhinolophids the theory is more promising since long pulse duration ensures that overlap must nearly always occur. Sensitivity to Doppler shifts is deemed very likely because of the dramatic response these bats give to small quick movements even at several metres range; heterodyning is one of the most sensitive methods of measuring Doppler shifts. Some recent information also has direct bearing on these speculations.

Reference has been made to the possible function of rapid ear movement in Rhinolophids for the detection of echoes at constant range and for improving directional reception. SCHNEIDER and MÖHRES (1960) have described the musculature responsible for these movements and have shown that denervation of the muscles results in a severe reduction of orientation ability. Recently GRIFFIN et al. (1962) discovered by high-speed cinematography that the ear movements of *Rhinolophus ferrumequinum* tend to be synchronised with the emission of ultrasound pulses. That is, there is a forward movement of one ear and a backward movement of the other associated with each probing sound up to the highest rate of sound production — over 60 per second.

This finding has been confirmed by PYE, FLINN and PYE, using different methods. These turned the tables by using echolocation on the bat as shown in Fig. 2 and 3. The ears of a hand-held bat were placed in a steady high-frequency sound field and the echoes were heterodyned to produce a Doppler beat whose frequency indicated the speed of movement. At the same time sounds emitted by the bat were separately heterodyned to about 1 kc/s. and both signals were recorded by a two-channel tape-recorder or with an oscilloscope and camera.

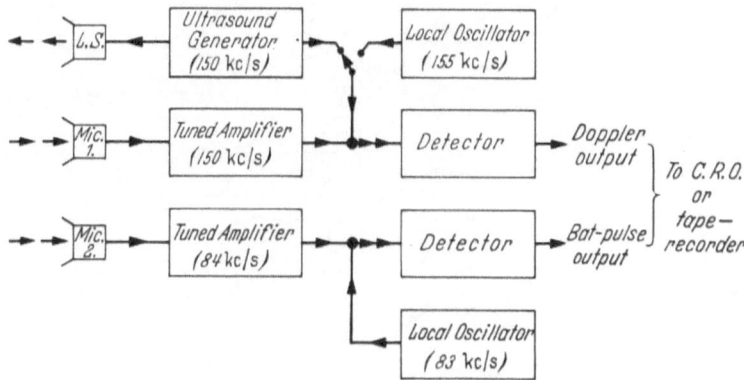

Fig. 2. A plan of the apparatus used for observing the ear movements of *Rhinolophus ferrumequinum* by Doppler echolocation

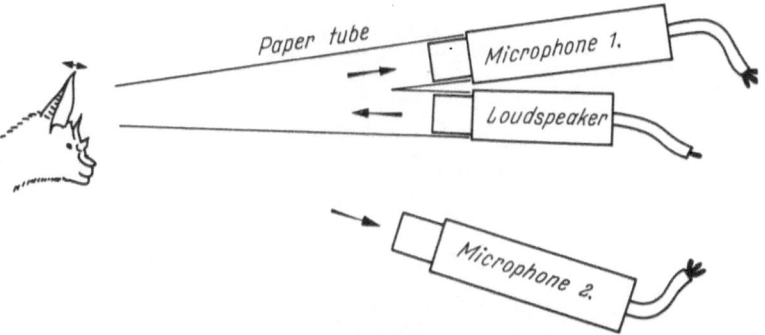

Fig. 3. The arrangement of the transducers used in Fig. 2. Those of the Doppler system were masked in order to restrict the effective field to a single ear

A full account of the results has been published along-side the paper by GRIFFIN et al. and further examples of the recordings are shown in Fig. 4 and 5. Synchronism of movement and pulse is often very precise although not so consistent in the restrained animal as in the free-ranging specimens observed by GRIFFIN. Short bursts of pulses at rates exceeding 50 per second with perfectly correlated pinna movements are not uncommon.

A fine electromyograph needle, 300 microns in diameter, has been introduced below the scalp. Some of the recorded potential spikes were clearly associated with pulse production as shown in Fig. 6. The genesis of tetanus has also been investigated in anaesthetised bats. The superficial cervicoauricularis muscle was stimulated electrically at various

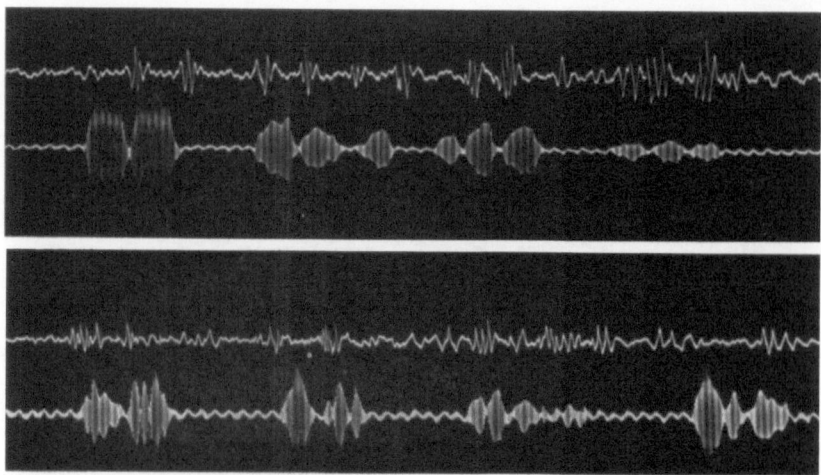

Fig. 4

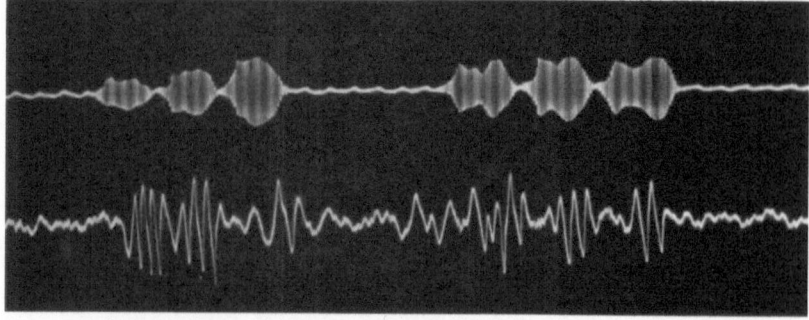

Fig. 5

Fig. 4 and 5. Upper traces: Doppler beats from ear movement using a sensing signal of 95 kc/s. Lower traces: probing pulses from the bat heterodyned from 85 kc/s. to 1 kc/s., with 100 c/s. ripple for time marking. These and all subsequent traces read from left to right

rates while ear movement was recorded by a light-weight liquid potentiometer. As shown in Fig. 7, the twitches are quite separate at the rates normally employed by the bat and considerable ear vibration can be induced at much higher rates. Of course these experiments do not prove that beat-notes are invoked by Rhinolophid bats but they do permit that possibility.

Finally, as this symposium is not concerned solely with bats, other echolocating animals must be mentioned. Briefly it is considered that insufficient is yet known of the behaviour and accuracy of these groups

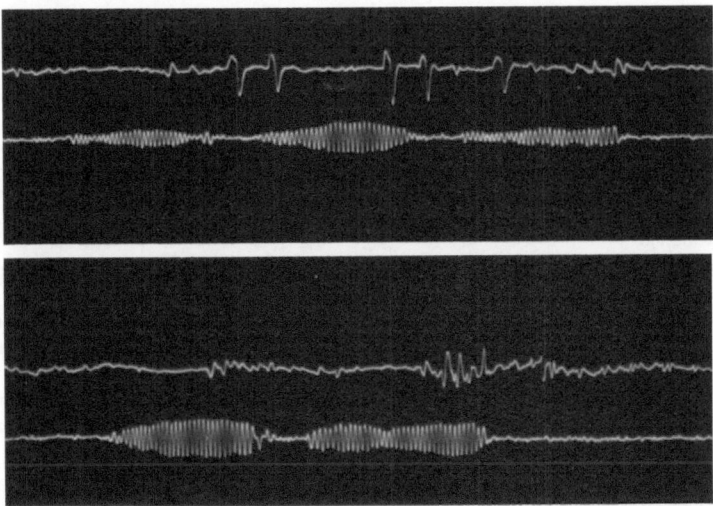

Fig. 6. Upper: traces potentials recorded from *Rhinolophus ferrumequinum* by an electromyograph needle, possibly arising from the superficial cervicoauricularis muscle. Lower traces: bat pulses simultaneously heterodyned to 1 kc/s

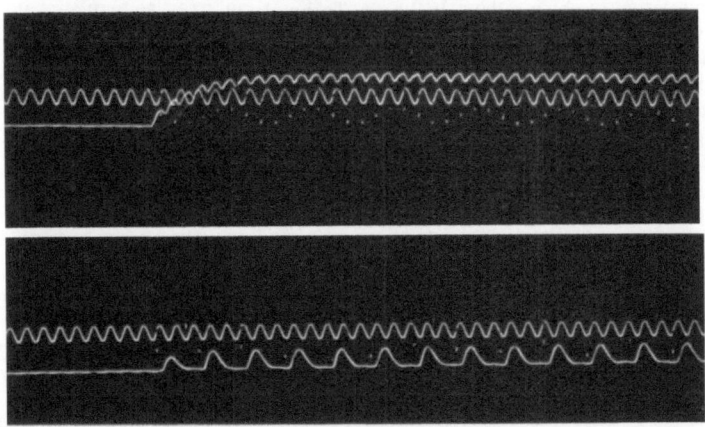

Fig. 7. Upper traces: 50 c/s. time marker with superimposed stimulation pulses of 1 msec. duration at 20 c/s. and 60 c/s. Lower traces: movements of the ear of *Rhinolophus ferrumequinum* as recorded by a liquid potentiometer

for speculation to be well founded. The birds *Steatornis* and *Collocalia* may not achieve fine discrimination with their click-like pulses although the Megachiropteran bats *Rousettus*, with apparently similar sounds, are able to avoid fine obstacle-wires (GRIFFIN et al., 1958).

2*

A clue to the Cetacea may lie in the great diameter of the eighth nerve and of its individual fibres. By increasing conduction rate, this feature renders the idea of central information handling rather more feasible than in the relatively tiny bats. But it is already clear that a beat-note theory finds no application outside the Microchiroptera, since no other known group produces sounds with a simple frequency structure.

References

GRIFFIN, D. R.: Listening in the dark. 413 pp. Yale Univ. Press 1958.
— D. C. DUNNING, D. A. CAHLANDER and F. A. WEBSTER: Correlated orientation sounds and ear movements of Horseshoe Bats. Nature (Lond.) **196**, 1185—7 (1962).
— A. NOVICK and M. KORNFIELD: The sensitivity of echolocation in the fruit bat *Rousettus*. Biol. Bull. **115**, 107—113 (1958).
GRINNELL, A. D.: Neurophysiological correlates of echolocation in bats. Techn. Rep. 30, Office of Naval Res. U. S. N. 1962 and Ph. D. Thesis Harvard University 1962.
— The neurophysiology of audition in bats. (4 papers in press) J. Physiol. (Lond.).
KAY, L.: Orientation of bats and men by ultrasonic echo location. Brit. Comm. Electronics **8**, 582—586 (1961).
— Perception of distance in animal echolocation. Nature (Lond.) **190**, 361—362 (1961).
— A plausible explanation of the bat's echo-location acuity. Anim. Behav. **10**, 34—41 (1962).
KLAUDER, J. R., A. C. PRICE, S. DARLINGTON and W. J. ALBERSHEIM: The theory and design of chirp radars. Bell Syst. Techn. J. **39**, 745—808 (1960).
MILLS, A. W.: On the minimum audible angle. J. acoust. Soc. Amer. **30**, 237—246 (1958).
NORDMARK, J.: Perception of distance in animal echo-location. Nature (Lond.) **188**, 1009—1010 (1960).
— Perception of distance in animal echolocation. Nature (Lond.) **190**, 363—364 (1961).
PYE, J. D.: A theory of echolocation by bats. J. Laryng. **74**, 718—729 (1960).
— Echolocation by bats. Endeavour **20**, 101—111 (1961).
— Perception of distance in animal echolocation. Nature (Lond.) **190**, 362—363 (1961).
— M. FLINN and A. PYE: Correlated orientation sounds and ear movements of Horseshoe Bats. Nature (Lond.) **196,** 1186—8 (1962).
SCHNEIDER, H., u. F. P. MÖHRES: Die Ohrbewegungen der Hufeisenfledermäuse (Chiroptera, Rhinolophidae) und der Mechanismus des Bildhörens. Z. vergl. Physiol. **44**, 1—40 (1960).
STROTHER, G. K.: Notes on the possible use of ultrasonic pulse compression by bats. J. acoust. Soc. Amer. **33**, 696—697 (1961).
THURLOW, W. R.: Further observation on pitch associated with a time difference between two pulse trains. J. acoust. Soc. Amer. **29**, 1310—1311 (1957).
— and A. M. SMALL: Pitch perception for certain periodic auditory stimuli. J. acoust. Soc. Amer. **27**, 132—137 (1955).

Pulse Duration in the Echolocation of Insects by the Bat, *Pteronotus* *

By Alvin Novick

Department of Biology, Yale University, New Haven, Conn. (U.S.A.)

With 4 Figures

The pulse durations of the orientation sounds of the bat, *Pteronotus*, are progressively shortened during the pursuit of a fruit-fly. If the position of the bat is calculated for each pulse, the echo from the fly is found to overlap the pulse by about 1.5 msec, in the approach phase, and by about 1.0 msec, in the terminal phase of each pursuit. Such pulse-echo overlaps has not previously been demonstrated in bats.

The orientation pulses of 13 families of bats have been recorded and analyzed but only a few genera of the family Vespertilionidae have been observed and recorded while hunting insects or avoiding obstacles (Griffin, 1953, 1958; Grinnell and Griffin, 1958; Griffin, Webster and Michael, 1960; Webster, 1962). Pye (1960, 1961a, b) and Kay (1961, 1962) have recently advanced hypotheses for the mechanism of echolocation based essentially upon the overlap in time between the orientation pulses and their echoes, the bat perceiving a beat note between the two because of the frequency modulated design of the pulses which bats of many families produce (Griffin, 1958; Novick, 1958, 1963). Depending upon the precision of perception and measurement, such a system could be used for calculating the distance of a detected object. These hypotheses have not been supported by the observations on the vespertilionids, where the pulses, indeed, shorten as the bat and its prey converge but apparently in order to *preclude* overlap (Griffin, 1958; Griffin, Webster and Michael, 1960; Webster, 1962).

The orientation sounds of one individual *Pteronotus davyi* (Phyllostomatidae) have now been recorded while, at a rate of up to 25 per

* I am grateful to Drs. Bernardo Villa-R. and A. Grinnell and Mr. R. Grummon for help in capturing and training this bat, to the Institute of Biology, National University of Mexico for the use of their facilities during part of this work, to Drs. D. R. Griffin and F. A. Webster for technical assistance and the loan of equipment, and to Mr. L. P. Granath for creating a microphone without which these records would not have been possible. This work has been supported in part by a Lalor Foundation Summer Fellowship and by the National Institutes of Health.

minute, the bat pursued and captured common fruit flies, *Drosophila* sp., which had been released in a laboratory flight room. About 60 satisfactory hunting sequences of this bat were recorded with a custom made (Granath) microphone and an Ampex tape recorder (Model 407) at

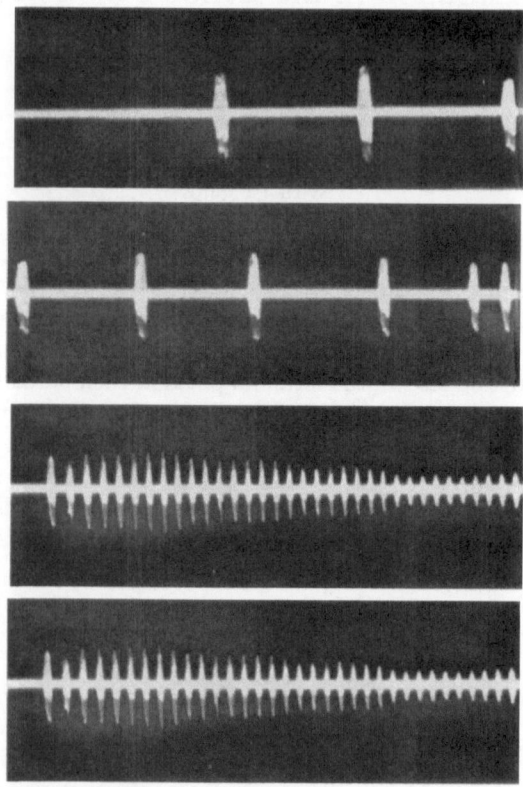

Fig. 1. The approach and terminal phases of the pursuit of a fruit-fly by the bat, *Pteronotus*. Photographs of oscillograph tracings of the bat's recorded orientation sounds are shown. Each sweep about 170 msec in duration. The last pulse of each sweep is reproduced as the first of the next sweep to improve continuity. Pulses next preceeding and following the events shown were about 108 and 84 msec away respectively

60 i. p. s. Eight of these sequences were analyzed in detail from filmed oscillograph tracings. The basic design of the orientation sounds of this bat have been described elsewhere (NOVICK, 1963). All of the pulses are frequency modulated in the latter portion following an initial plateau (of somewhat more than half of the total duration) of constant frequency. All have a prominent second harmonic.

The present analysis deals chiefly with the significance of variations in pulse duration seen in these records. As in the vespertilionids (GRIFFIN, WEBSTER and MICHAEL, 1960), these pursuits may be subdivided as follows. A *search* phase is characterized by pulses of about 3.9 to 5.0 msec in duration, spaced from about 70 to more than 200 msec apart but commonly with pulse-to-pulse intervals of about 75 to 100 msec (Figs. 1 and 2). An *approach* phase, about 3 to 10 pulses covering some 350 msec on the average, when an insect has been detected, is characterized by progressive shortening of the pulse-to-pulse interval, first to about 50 msec and then to about 25 msec, and by progressive shortening of

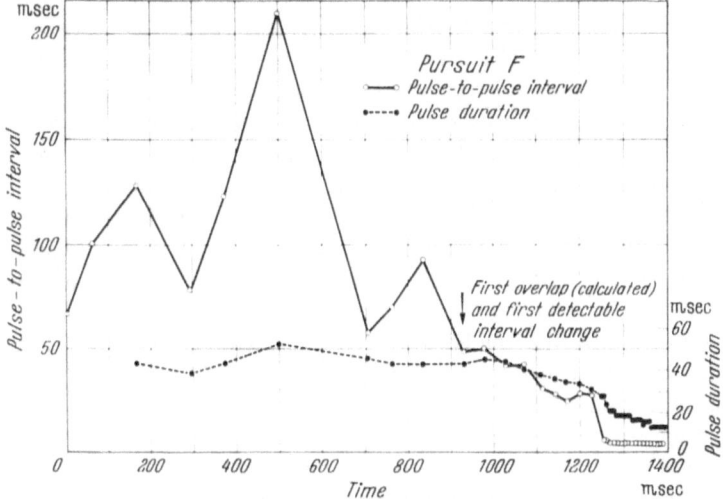

Fig. 2. Plot of bat's position in time vs. the pulse-to-pulse interval and the pulse duration for search, approach, and terminal phases of a typical fruit-fly pursuit (not the one in fig. 1). Note coincidence of point of first distinct interval change and first calculated pulse-echo overlap

the pulse duration as well so that the last pulse of this phase is usually 2.4 to 3.2 msec. Presumably, during this phase, the bat has turned toward its prey.

There then follows a *terminal* phase, leading up to the capture of the insect, which lasts from 150 to 214 msec in 7 analyzed pursuits consisting of from 27 to 39 pulses. The pulse-to-pulse interval drops from an initial value of about 7 msec to about 4.5 msec terminally while the pulse duration drops from about 2.3 to 2.9 msec initially to 1.0 to 1.25 msec terminally (Figs. 1 and 2).

The relative spacing of these events in time is objective. In order to convert these values into distances, however, lacking clear photographic records of the events, three assumptions are convenient and seem reasonable. First, assume that the fly was captured just after the last pulse of the terminal phase. This appears to be compatible with the events concerning hunting vespertilionids recorded by WEBSTER (1962) and GRIFFIN et al. (1960) on film. Second, assume that the bat was closing in on the fly at 1.25 mm/msec. Such a speed is not inconsistent with observed bat flight speeds and leads to interesting calculations below. Third, assume a constant relative speed for the bat vs. the fly. This, of course, is an approximation. The bat's approach to the fly often involves turns. Though the bat may conceivably maintain a constant speed in hunting in order to eliminate variations in its own speed from its navigational calculations, it has no control, as far as we know, over the speed or direction of the fly. In any event, the fly's speed is small relative to the bat's and to the speed of sound. We shall also ignore changes in distance of the bat from the microphone during these pursuits since we have no way of reconstructing them.

Using these assumptions, the bat's position in space relative to the fly can be plotted for each part of its pursuit. Setting the speed of sound at 344 mm/msec (temperature and pressure in the room were not recorded), the time at which the echo of each pulse returned to the bat may be calculated. During the search phase, the pulses and echoes did not overlap. Given pulses of 3.9 to 5.0 msec, overlap will occur only when the bat is within about 680 to 860 mm of the fly. In the three best cases, the first evidence that the bat had detected an insect (a change in pulse-to-pulse interval which has been defined as the initiation of the approach phase) occurred at 687—790 mm. This close association between the first recognizable change in pulse-to-pulse interval and the first calculated overlap of pulse and echo suggests that detection, or at least initiation of pursuit, is associated with the occurrence of an overlap of pulse and echo (Figs. 2, 3, and 4). In fact, not uncommonly, at about the time of the first calculated overlap, the bat produced one or two pulses of unusually great duration which suggested that the bat might, upon hearing an echo occurring shortly after its own outgoing pulse was over, lengthen the next pulse deliberately so as to assess the echo more exactly. This possibility can not be substantiated from the present records.

During the approach phase, the pulse durations are progressively shortened as the bat moves closer to the fly so that the pulses and echoes ordinarily overlap by 1.3 to 1.9 msec. After a rapid transition, the pulses and echoes of the terminal phase overlap by about 0.9 to 1.2 msec in the large majority of cases. Furthermore, in any given pursuit, the overlap values are surprisingly uniform considering the three basic assumptions behind these calculations (Fig. 3 and 4).

If the fly's position relative to the bat at the end of the last pulse were, indeed, greater than 0 mm, say up to 170 mm, the magnitude of the overlap would be smaller. If the fly had, on the other hand, already been captured at the time of the last pulse (a possibility because there is almost always a marked amplitude drop which could result from the closing of the lips), the magnitude of the overlap would be greater. But only if the last pulse occurred while the bat was more than 170 mm from the fly would there fail to be overlap. That the bat would be silent when separated by such a distance from its prey seems most unlikely.

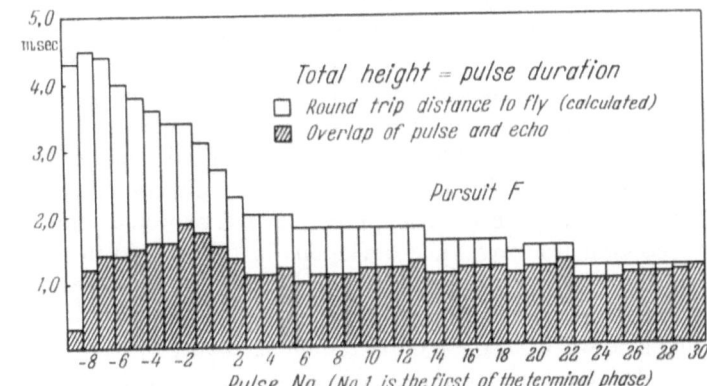

Fig. 3. Plot of pulse duration relative to the calculated round trip distance separating bat and prey (for the same pursuit shown in fig. 2) showing the remarkable constancy of the pulse-echo overlap. The pulses are numbered negatively and positively from No. 1, the first of the terminal phase. Note shift of magnitude from approach to terminal phase

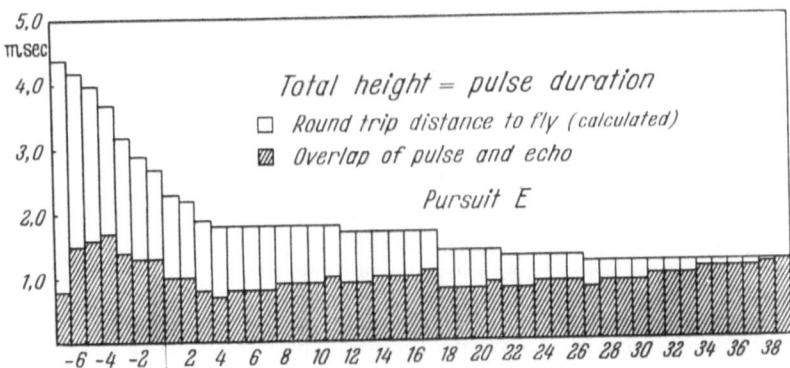

Fig. 4. Same as fig. 3 but showing data for the longest analyzed pursuit

If the bat's speed relative to the fly were much deviant from 1.25 mm per msec, the uniformity of overlap would be lost. The striking uniformity within each pursuit as well as from one to another strongly suggests that we have not introduced a fictitious value. The assumption of constant speed may be responsible for the fluctuation in apparent amplitude of the overlap.

Apparently, therefore, when *Pteronotus* is pursuing fruit-flies, there is a constant temporal association between the outgoing pulse and the returning echo such that the approach phase is initiated about when the first overlap occurs, the overlap equals about 1.5 msec during the approach phase, and the overlap equals about 1.0 msec during the terminal phase of a pursuit. Possibly, during the search phase, the bat's world is divided into two simple categories — objects which echo after the pulse is over (beyond about 800 mm) and those which echo before. Perhaps, only those echoes which overlap, indicating proximity are then carefully assessed. A distance of 800 mm should not involve much danger of collision before a decision could be made to change direction since the bat would ordinarily have travelled only about 125 mm before its next pulse and echoes. So, the first overlap of an appropriate echo may lead to the approach phase (to locate the fly directionally, identify it, and to line up on it). Thereafter, appropriate overlap would immediately reveal successful tracking while deviations in time would indicate deviation in the relative positions of the bat and fly from the expected. The mechanism of perceiving the overlap, whether by beat note or otherwise, cannot be assessed from these data. Assuming the fly to be at 0 mm at the last pulse, the echo overlaps well before the frequency modulated portion of the pulse. But if the bat were actually at a distance of say 25 or 50 mm. at the time of the last pulse — a distance supported by the argument that later information could not be processed anyway before the insect had been captured — then the overlap of the constant frequency portion of the echo would be more closely associated with the FM portion of the pulse. These data are compatible with the hypotheses of PYE and KAY even though the use of beat notes cannot be implied.

References

GRIFFIN, D. R.: Bat sounds under natural conditions, with evidence for echolocation of insect prey. J. exp. Zool. 123, 435—466 (1953).
— Listening in the dark. New Haven, Conn.: Yale Univ. Press. 1958.
— F. A. WEBSTER and C. R. MICHAEL: The echolocation of flying insects by bats. Animal Behav. 8, 141—151 (1960).
GRINNELL, A., and D. R. GRIFFIN: The sensitivity of echolocation in bats. Biol. Bull. 114, 10—22 (1958).
KAY, L.: Perception of distance in animal echolocation. Nature (Lond.) 190, 361 (1961).
— A plausible explanation of the bat's echo-location acuity. Anim. Behav. 10, 34—41 (1962).
NOVICK, A.: Orientation in paleotropical bats. I. Microchiroptera. J. exp. Zool. 138, 81—154 (1958).
— Orientation in neotropical bats. II. Phyllostomatidae and Desmodontidae. J. Mammal. 44, 44—56 (1963).
PYE, J. D.: A theory of echolocation by bats. J. Laryng. Otol. 74, 718—729 (1960).
— Perception of distance in animal echolocation. Nature (Lond.) 190, 362—363 (1961 a).
— Echolocation by bats. Endeavour 20, 101—111 (1961 b).
WEBSTER, F. A.: Chapter 25 in: Human factors in technology. (E. M. BENNETT et al., Eds.) New York: McGraw-Hill (1962, in press).

Acoustic Orientation of Marine Fishes and Invertebrates *, **

By James M. Moulton

Department of Biology, Bowdoin College, Brunswick, Maine (U.S.A.)

With 3 Figures

The obvious problems surrounding a complete understanding of the acoustic orientation of marine fishes and invertebrates are great enough so that at this time it seems only possible to point out (1) some concrete evidences that acoustic orientation is of some importance in the lives of these marine animals, (2) some discussion of the acoustic behaviour of a few species, and (3) some indirect evidences which seem to relate the migrations, movements and behavioural patterns of some of these marine animals to their acoustic environment.

The acoustic environment of a marine organism — that is, the portion of the sound spectrum in the water to which it is sensitive — is of as much significance potentially as a controller of the organism's orientation as is light, tidal effect, food or salinity. The acoustic environment varies geographically, often within relatively narrow limits; it varies in the course of the day and year as do other features of the environment, and in the same place from year to year (Busnel and Dziedzic, 1962; Moulton, 1958a, b, c; Winn and Marshall, 1960).

Therefore, it has seemed of significance to many marine biologists and oceanographers that we should sort out the kinds of noise in the water; to anatomists and behavioural scientists that attention should be paid to the mechanisms by which those sounds stemming from marine or aquatic organisms are produced, and the circumstances surrounding their production; and finally, particularly to physiologists within whose province problems in orientation most properly lie, that we learn those portions of the total sound spectrum which can be heard or otherwise detected by marine organisms — the only sound or resultant of sound

* Part of the work upon which this paper is based was performed under aid of the Bowdoin College Faculty Research Fund established by the Class of 1928, of the National Science Foundation, of a John Simon Guggenheim Memorial Foundation Fellowship, and of the Woods Hole Oceanographic Institution.

** Contribution No. 1313 from the Woods Hole Oceanographic Institution.

vibrations which can have any direct biological significance in the normal sense, and then that we learn how that spectrum is perceived — whether through the general integument, through a lateral line, through ears or statocysts, by substrate vibrations or bristle quivering, or by conbinations of all of these. For all of us, the sound production and perception of marine organisms have presented many unanswered questions as to the orientation of marine animals to sound.

It should also be acknowledged that it has preoccupied the minds of scientists of diverse interests who look to the relationships between science and the welfare of mankind, that sound might somehow be used by fishermen to improve their catches. Finally, ecological and modern taxonomic values of the study of underwater sound are rapidly accruing (FISH and MOWBRAY, 1959; GRAY and WINN, 1961; MOULTON, 1958a, b, c; TAVOLGA, 1958b).

Problems in acoustic orientation that might seem straight-forward have been complicated by a number of factors: by the multiplicity of vibration sensitive pathways in many marine animals, by the variation in response to sound stimuli of marine fishes and invertebrates — both sound production and responses to sound vary with the physiological state of the organism, as well as with species, age and other factors. Also, in fishes which condition so rapidly with continued stimuli of many kinds, experimental results have seemed to suggest a variability of response inconsistent with any real significance of acoustic orientation to their biology. But marine animals may react more subtly to acoustic stimuli than to other kinds of environmental factors; different species may appear to react very differently to the same acoustic stimulus while arriving at the same end result (MOULTON, 1956b).

A further problem of considerable moment lies in the fact that it is very difficult to see well in the ocean, and at the same time to maintain an experimental situation which does not alter or make abnormal the behaviour of organisms under study. There are dangers inherent if the study of schooling, for example, is carried on with a few individuals or a few hundred individuals, when the same species normally schools in very large numbers at sea. Criteria for schooling (for example, the heads of school members all pointed in the same direction) may be very different when one is dealing with a few dozen individuals from what they are when one is considering a school which darkens the sea over some distance, and which is comprised of millions of individual which in their detailed behaviour may not be fulfilling the criteria for schooling in an aquarium. I dwell on schooling here because I believe that it is in part a manifestation of acoustic orientation — that sounds of swimming of a schooling species assist in maintaining the integration of the school (MOULTON, 1960).

The distribution of a species in a given area is the sumtotal of the distributions of all its members; the picture obtained by the behaviour or responses of a few individuals under the experimental conditions frequently demanded may give a false picture of what is going on at sea. This may be more true relative to the acoustic behaviour of marine organisms — their sound production and responses to sound — than to many other kinds of behaviour.

Many species which produce apparently purposeful sounds in nature become silent in captivity (DOBRIN, 1947; FISH, 1948; MOULTON, 1956a); other fishes produce sounds in captivity which in nature during long hours and days of listening over appropriate sea bottom have been heard rarely if at all.

Marine fish sounds produced by specialized means, as distinct from sounds resulting from swimming, feeding, and similar activities, may be roughly divided into three large groups:

(1) Sounds produced in nature but rarely if at all in captivity;

(2) Sounds produced readily in captivity, particularly during molestation, but heard rarely in nature;

(3) Sounds produced equally readily in captivity and in nature.

Although probably all fishes produce some sound when moving violently or rapidly in the water, even a heavy-bodied lobster, *Homarus* or a palinurid, is able to move over either sandy or rocky bottom in a nearly complete absence of any sound. Crustaceans, of course, possess a number of well-described stridulating and snapping mechanisms, and the lobsters *Homarus americanus* and *H. vulgaris* are the sources of grumbling noise, better felt than heard by the investigating scientist; in the case of *H. vulgaris*, it is clearly audible through a wooden rod held to the ear from the lobster's carapace (H. O. BULL, personal communication). Many of the spiny lobsters (Palinuridae) possess a particularly elaborate and highly evolved stridulating mechanism (DIJKGRAAF ,1955; MOULTON, 1957, 1958c).

Sounds as diverse as horn-like sounds, whistles, hammerlike knocks, rattles and buzzes, bell-like notes, clicks and snaps (FISH, 1948, 1954; MOULTON, 1958a) comprise only a part of the variety of sounds of fishes and invertebrates, and are probably cues for the orientation of some organisms. To judge from the sub-tropical North Atlantic and the tropical and sub-tropical Coral Sea, the fish and invertebrate fauna of warm coastal marine waters is likely to include species producing each of these kinds of noise.

In addition to these, one may in coastal waters anticipate also sound stemming from the feeding, swimming and boring of marine organisms, from the movements of barnacle shells (BUSNEL and DZIEDZIC, 1962) and mussel *(Mytilus)* shells (FISH, 1961) and non-biological sounds such as surf or surface wave noise, and noise from moving bottom sediments. Together with the presumably purposeful sounds of fishes and invertebrates, these underwater sounds comprise a spectrum characteristic of a given coast. It was suggested some years ago by Professor Hasler that the characteristic sounds of coastal waters might furnish an orientation clue for migrating fishes (Hasler, 1956).

Nearly all we know of marine fish and invertebrate sounds is related to relatively shallow water; most of the sounds recorded from deeper water have not yet been attached to their proper sources, and we know essentially nothing of their biological significances. A few clues such as the "echo-fish" of Griffin (1955) recorded north of Puerto Rico by the Woods Hole Oceanographic Institution, some other sounds recorded in the deep-sea or along the continental shelf, and sound producing organs found in some deep-sea fishes such as the macrourids suggested by N. B. Marshall (1954) as the source of the "echo-fish" call — all of these suggest that sound may be a significant clue for orientation of marine fishes in waters below the level of light penetration or where visibility is relatively poor.

Fishes of several families are known to develop or to increase sound production with the onset of the breeding season (*Chasmodes bosquianus*, Blennidae — Tavolga, 1958c; *Bathygobius soporator*, Gobiidae — Tavolga, 1956, 1958a; *Notropis analostanus*, Cyprinidae — Winn and Stout, 1960; *Prionotus evolans* and *P. carolinus*, Triglidae — Moulton, 1956a; *Opsanus tau*, Batrachoididae — Gray and Winn, 1961; *Gadus callarias*, Gadidae — Bull and Brawn, 1959; Brawn, 1961a, b; as well as several sciaenids). In males of *N. analostanus*, sound production is stimulated in the male by injections of testosterone (Winn and Stout, 1960).

There are four principal manifestations in fishes of acoustic behaviour relative to sounds developed during the breeding season: (1) stimulation of the female to greater activity making it more likely that she will find a male (Tavolga, 1956, 1958a); (2) stimulation of calling by other individuals of the same species, possibly as an aid to location of the opposite sex in murky waters (Moulton, 1956a); (3) grunting produced by the male *Opsanus* while guarding a nest of eggs (Gray and Winn, 1961); and (4) possible attracting of the male by the female in a cyprinid (Winn). For present purposes, we will only state that these sounds for which biological significances have been reasonably well established are almost certainly of acoustical characteristics such that they can be heard by the species concerned (see review of Lowenstein, 1957).

Most of the significant producers of underwater sound among fishes possess air bladders and it has been shown that the air bladder acts as a resonator of underwater sound, both of sounds produced by a fish, and of sounds reaching the fishes body through the water. Fishes with the greatest hearing ranges and with the most acute hearing couple the resonating air bladder in some way with the inner ears (recently discussed by DIJKGRAAF, 1960). The larger the fish, the larger the air bladder, and the deeper the resonating frequency. The same principle is illustrated by body size of a decapod crustacean, in which increasing carapace size (combined perhaps with other factors) seems to resonate increasingly lowered frequencies with increase in age. In *Panulirus argus*, it should be possible to estimate the size range of populations in various areas of an important habitat by analyzing the principle frequencies of their stridulation; the larger the lobster, the lower the frequency of greatest intensity (MOULTON, 1958c).

However, in fishes the air bladder does not work as a simple resonating system in every case. Many bladders are of complex shape, and the manner in which specific sound producing or resonating organs are used will influence markedly the sounds emitted as in the case of triglids and batrachoidids.

The concomittant use of the air bladder as a hydrostatic organ probably has relatively little effect on sound production within the physiological range of most fishes. In sea robins (Triglidae) the air bladder may undergo a remarkable degree of distortion and collapse without a marked change in the sounds produced by muscles drumming on the bladder walls (MOULTON, 1960); species specific sounds are still produced after considerable deflation.

However, changes in species specific sounds that do occur with increasing size, and seasonal or physiological changes in sound production which may occur within a species must be taken into account in any generalized interpretation of the significance of marine animal sounds to the orientation of these organisms.

Attempts to guide the movements of anadramous and of other fishes with underwater sound have been notoriously unsuccessful, even although primitive fisheries have for centuries used both sound sources and listening methods successfully to improve the catch (BUSNEL, 1959; MOULTON and BACKUS, 1955; PARRY, 1954; WESTENBERG, 1953). It seems likely from the sparse evidence available that sounds thus serving to attract fishes are useful because they are imitative of sounds of prey of the fishes concerned.

Because there is still relatively little comparative information available on the movements of unconditioned fishes in a measured sound field, I should like to describe some work performed at the Woods Hole Oceanographic Institution in 1955, which has hitherto been published only in abstract form (MOULTON, 1956b).

Despite the difficulties inherent in using narrow confines for the study of fish behaviour in relation to sound, visibility requirements and the demands of careful experiment usually dictate such usage. However, certain observations seem valid for both enclosed and free fishes: Firstly, untrained teleosts frequently show quickened swimming movements or diving (MOORHOUSE, 1933; MOULTON and BACKUS, 1955; SHISHKOVA, 1958) when sounds are transmitted into the water; secondly, unconditioned fishes rapidly adapt to a sound signal initially affecting their behaviour.

Most of the information on the hearing capacity of fishes has derived from experiments with conditioned animals and their responses to sound as the conditioned stimulus. With food as the unconditioned stimulus, fishes may be trained to move predictably in a sound field. Of special interest in the present connection are KLEEREKOPER's experiments in which Creek chub, *Semotilus atromaculatus*, trained by feeding to move to a sound source, in KLEEREKOPER's opinion moved along crests of highest sound intensity, and in which when two sound sources of different intensities were presented, the fishes moved to the source of greater intensity (KLEEREKOPER and CHAGNON, 1954).

The experiments to be described now were performed in an attempt to determine whether the initial movements of *untrained* fishes in a suddenly created sound field of measured intensities, might demonstrate a consistent pattern or whether the quickened swimming movements anticipated might be entirely random.

Although there was no attempt in my experiments to determine the frequency or sound pressure sensitivity of the species concerned, one species, the menhaden (*Brevoortia tyrannus* LATROBE — Clupeidae), was selected because of the relation of its air bladder to its inner ear (v. FRISCH, 1936; DIJKGRAAF, 1960). Instead of Weberian ossicles interposed between the two, an arrangement which Professor Dr. AUTRUM and Dr. POGGEN-DORF have shown lends to the ostariophysan, *Amiurus nebulosus*, keen hearing sensitivity (AUTRUM and POGGENDORF, 1951; POGGENDORF, 1952), clupeid fishes possess tubular extensions of the air bladder to membranes separating the gas of these tubes from the perilymph of the inner ears. Judging from available evidence, including the behaviour of the menhaden to be described, clupeids are highly sensitive to sound.

As a contrasting species with no special arrangements between air bladder and inner ear, the butterfish (*Poronotus tricanthus* PECK — Stromateidae) was used.

Adults of both species, in numbers up to 12 in the case of the menhaden, averaging about 30 cm length, and up to 36 in the case of the butterfish, averaging about 18 cm in length, were contained in a wood and screen cage the inner submerged space of which measured 72 in. × × 42 in. × 36 in. in depth (1.8 m × 1.1 m × 0.9 m in depth).

The top of the cage was marked at 6 in. (15.2 cm) intervals on the rim, so that a 6 in. (15 cm) grid measuring 42 in. × 72 in. (1.1 m × 1.8 m) could be visualized. The cage was hung in the center well of a raft over an approximate depth of 70 ft. (21 m) in Great Harbor at Woods Hole, where a tidal current provided a good flow of water through the cage most of the time. All experiments were performed within 36 hours, usually within 6 hours, after delivery of the fish to the cage by boat from the weir in which they were caught. Responses of the fish became less marked in successive experiments, due presumably to adaptation, and after prolonged confinement.

Upon arrival in the live car, the menhaden tended to swim about the cage at the 2-foot depth (0.61 m) in a loose mill, individual fish crossing over and changing direction rather frequently. The fish tended to head into the current during the strongest tidal flow. The butterfish, on the other hand, tended to form a rather compact school within the live car, the majority of the fish with their heads into the tidal current, individuals turning occasionally, and with the mid-depth of the group at the 2-foot level.

Sound was generated through a Hewlett-Packard audio oscillator (Model LAJ), amplified through a Craftsman amplifier (Model C55O), and transmitted into the water through a QBG underwater loudspeaker or transducer with its center at the 2-foot depth. Full gain settings were used in all cases discussed in order to obtain a semi-directional sound field. The direction of facing of the transducer was changed during the experiments.

Sound pressure measurements were taken with a United States Navy OCP-1 monitor of the sound field created in the cage at audio oscillator settings from 2 to 20 kilocycles/sec, at the 1, 2 and 3-foot levels within the cage, and in each square or in several squares of the 6 in. grid. Sound pressure measurements at the 2-foot level, at one of the higher settings used, 15 kilocycles/sec, and at a lower setting used, 2 kilocycles/sec, are shown in Figs. 1 and 2.

The measurements indicated that under the conditions stated, the sound source used was semi-directional at higher frequency settings, as at 15 kilocycles/sec, the regions of diminishing intensities tending to form enlarging cones before the face of the transducer (Fig. 1). At a setting of 2 kilocycles/sec, the sound field was more homogeneous before the transducer than at 15 kilocycles/sec (Fig. 2), but the field still presented a region of highest intensity before the face of the transducer.

With the audio-oscillator sat at 15 kilocycles and during transmission of a steady signal at full gain settings, menhaden moved into the region of higher sound intensities, either forming a tight mill in the center of the cage or moving rapidly to the corner of the cage opposite to the sound source and crowding together with frenzied swimming movements. During experiments with 15 kilocycles/sec, sound pressure levels in the live car varied from 55 to 87 db re 1 microbar. Reactions of the menhaden to the signals involved movement into intensity areas of 75 to 87 db re 1 microbar.

Similar reactions were obtained during transmission of signals with the audio oscillator at various settings from 1 to 20 kilocycles, and

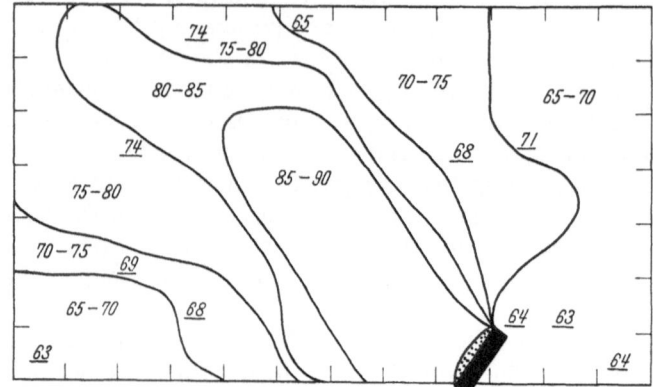

Fig. 1. Sound field in the fish cage during transmission of a 15 kilocycle/second signal, at the 2-foot (.61 m) level (Sound pressure in db re 1 microbar)

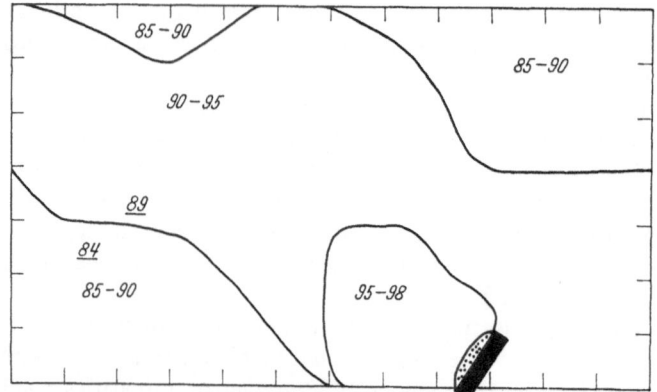

Fig. 2. Sound field in the fish cage during transmission of a 2 kilocycle/second sound signal, at the .61 m level (Sound pressure in db re 1 microbar.)

during presentation of warble tones of 5 to 10, 10 to 15, and 15 to 20 kilocycles/sec. Results with signals above 15 kilocycles were inconsistent even during periods of most active reactions. During one presentation of a warble tone of 12 to 20 kilocycles/sec, the observer noted that strong spurts of activity occurred on the lower side of the warble each time it was presented.

In continuing tests with the same fish at all settings, the violence of the reaction decreased, the fishes tending to behave more and more individually, until after several tests the reaction could not be elicited until a time interval had intervened, and the reaction never returned as strongly as initially in the same group of fishes.

Butterfish formed a rather compact school within the confines of the live car. They tended to head into the tidal current, individuals turning occasionally. In the experiments reported here, a setting of 2 kilocycles/sec was used, since this frequency lies within the hearing range of those species of fishes in which the frequency range of sensitivity has been studied. As a test of sensitivity to the signal, individual butterfishes were conditioned to respond to the signal used by quickened swimming movements between two metal grids in a wooden trough, with a mild electric shock as the unconditioned stimulus — a method similar to one used by GRIFFIN (1950), and adapted from FROLOFF and BULL.

Upon arrival in the cage, butterfish reacted markedly at settings from 1 to 5 kilocycles/sec by a breaking up of the school and the crowding of its members against the screen of the car, in the most marked type of reaction. Reactions to higher settings were inconsistent. Within two hours the fishes had partially adapted to the sound stimuli, and the following procedure was adopted.

In tests at 2 kilocycles/sec, with unconditioned butterfish in the cage, for one minute before initiation of the signal, an otherwise stationary observor made successive drawings in outlines of the cage of the position of the school of butterfish as it appeared from above. These figures were drawn successively for one minute before the signal was transmitted, for two minutes during transmission of 2 kilocycles/sec, and for one minute following termination of the signal. Subsequently, the figures were transferred to graph paper corresponding to the cage grid, and the area of the school before the transducer as seen from above was compared with the area of the school elsewhere in the cage.

Fig. 3 shows a typical series of figures of the behaviour of a group of butterfish in the cage, in this case a situation in which relative areas differed little before the transducer (one half of the car) and elsewhere, but in which the form of the school changed markedly. Whether the areas differed or not, and in this case the average ratio of area before the transducer to the area elsewhere in the live car was 1.0 both before and during the signal, the outline of the butterfish school showed consistent changes correlated with transmission of the signal from the transducer. A summary of the results of several tests is shown in Table 1.

During the change in outline of the butterfish school, individual fishes of the school showed no marked reactions to the sound transmitted. Individual movements were inconsistent, and generally consisted of a slow turning within a narrow radius. Nothing of the frenzied reaction of

Table 1. *A summary of butterfish distribution relative to a 2 kilocycle/second sound field in the fish cage.* Distribution of butterfish *(Poronotus tricanthus)* Before, During, and After 2 k.c. Transmission (School area before/School area behind Transducer)

Transducer directed southwest			Transducer directed northwest		
Before	During	After	Before	During	After
3.7	2.5	1.7	0.9	0.9	1.2
1.4	2.0	1.3	0.4	0.9	0.8
1.2	1.5	1.2	1.2	1.3	1.2
1.1	1.5	1.5	2.0	1.1	1.3
0.6	0.6	0.7	0.4	0.8	0.9
1.7	2.5	1.4	1.0	1.0	0.8
		Averages			
1.6	1.8	1.3	0.9	1.0	1.0

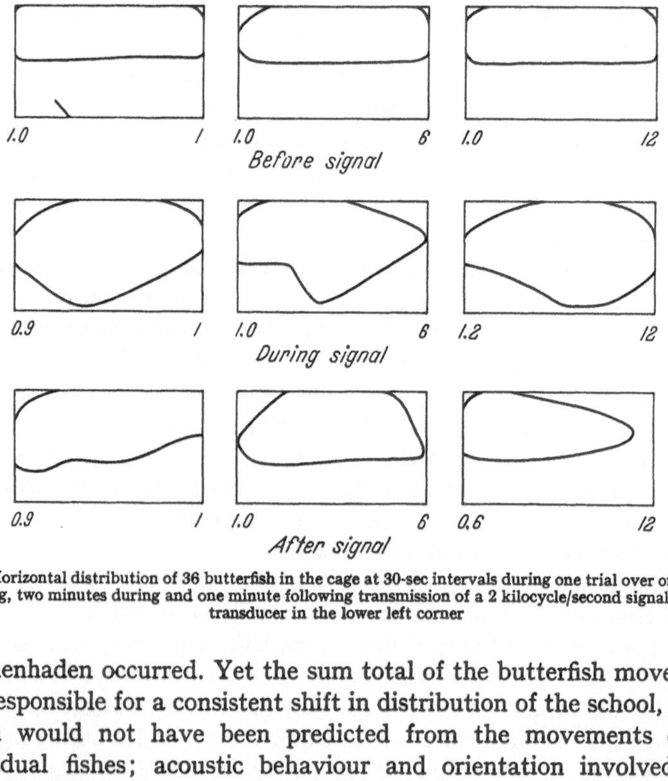

Fig. 3. Horizontal distribution of 36 butterfish in the cage at 30-sec intervals during one trial over one minute preceding, two minutes during and one minute following transmission of a 2 kilocycle/second signal, with the transducer in the lower left corner

the menhaden occurred. Yet the sum total of the butterfish movements was responsible for a consistent shift in distribution of the school, a shift which would not have been predicted from the movements of the individual fishes; acoustic behaviour and orientation involved non-obvious adjustments to the sound field.

The violent nature of the menhaden reaction seems suggestive of a biological significance of sound to the species. According to BIGELOW and SCHROEDER (1953), menhaden are heavily preyed upon by porpoises where the two occur together. That porpoises are extremely noisy predators is well known. When recordings of calls of *Tursiops* made by

WILLIAM SCHEVILL were transmitted into the cage, with a Magnecorder tape recorder substituted for the audio oscillator, the reaction of the menhaden was a violent as during the transmission of signals from the audio oscillator.

The observations reported here add no information as to the hearing sensitivity of menhaden and butterfish, in terms of the physiology of the inner ear. They do illustrate clearly that different species of teleosts, unconditioned, may react very differently to similar sound stimuli. They suggest that where a directional sound exists, reacting teleosts may orient toward its axis as an initial reaction to production of the sound; a possible mechanism for location of sound by fishes has been discussed recently by DIJKGRAAF (1960). The butterfish reactions have indicated that the orientation of at least one schooling marine fish in a sound field may be a function of individual movements which in themselves may not reflect clearly the response of the group as a whole.

Failure of marine fishes and invertebrates to respond to changes in salinity, temperature and food supply in their environment may bring swift retribution; such changes as well as light changes generally stimulate rather abrupt adjustments in position or behaviour — a reorientation on the part of organisms, or may result in extensive mortality if the animals fail to adjust.

The acoustic environment seems less critical to organisms than are temperature, light and chemical changes. As a student of embryology, I wonder whether we may not extend to acoustic biology the concept of "double assurance" mechanisms, first expressed by RHUMBLER (1897) relative to cell division, extended to embryological development by BRAUS (1906) and used by HANS SPEMANN (1927) to indicate those events of development that may be ticked off in more than one way, or by cooperation of relatively unrelated events. This is to suggest that once biological clocks, chemical and light pathways have determined the general distribution of organisms, sound may establish the detail for some species; to extend the idea, having followed appropriate currents to inshore waters, marine fishes may then find their way along a coast familiar through its acoustic characteristics; breeding is accomplished successfully because sound production and/or sensitivity stimulates behaviour patterns of finer and more effective detail than the broad movements congregating the species in a harbour or bay; the remarkable unity of the school or shoal of fish depends on the close interval maintained by members of a species under vibration stimuli, in addition to their aggregation under other influences.

Perhaps we have been seeking too obvious reflections of the importance of the acoustic anatomy and sensitivity of animals.

References

Autrum, H., u. D. Poggendorf: Messung der absoluten Hörschwelle bei Fischen *(Amiurus nebulosus)*. Naturwissenschaften 38, 434—435 (1951).

Bigelow, H. B., and W. C. Schroeder: Fishes of the Gulf of Maine. First revision. U. S. Fish and Wildlife Service. Fishery Bulletin 74 (1953).

Braus, H.: Vordere Extremität und Operculum bei *Bombinator*-Larven. Ein Beitrag zur Kenntnis morphogener Korrelation und Regeneration. Morph. Jahrb. 35, 139—220 (1906).

Brawn, V. M.: Reproductive behaviour of the cod *(Gadus callarias L.)*. Behaviour 18, 177—198 (1961).

— Sound production by the cod *(Gadus callarias L.)*. Behaviour 18, 239—255 (1961).

Bull, H. O., and V. M. Brawn: Reproductive and aggressive behaviour in the cod. Annual Report of the Challenger Society, 1959.

Busnel, R. G.: Étude d'un appeau acoustique pour la pêche, utilisé au Sénégal et au Niger. Bull. de l'I.F.A.N., 21, ser. A (1), 346—360 (1959).

— et A. Dziedzic: Rhytme du bruit de fond de la mer à proximité des côtes et rélations avec l'activité acoustique des populations d'un cirripéde fixé immergé. Cahiers Oceanographiques 14, 293—322 (1962).

Dijkgraaf, S.: Lauterzeugung und Schallwahrnehmung bei der Languste *(Palinurus vulgaris)*. Experientia (Basel) 11, 330—331 (1955).

— Hearing in bony fishes. Proc. roy. Soc. B 152, 51—54 (1960).

Dobrin, M. B.: Measurements of underwater noise produced by marine life. Science 105, 19—23 (1947).

Fish, M. P.: Sonic fishes of the Pacific. Prof. NR. 083-003, Contr. N 6 ori-195, t.o.i. between ONR and Woods Hole Oceanographic Institution, Techn. Report No. 2 (1948).

— The character and significance of sound production among fishes of the western North Atlantic. Bull. Bingham Oceanogr. Coll. 14, Art. 3, 1—109 (1954).

— Bioacoustics project, University of Rhode Island. Proc. First National Coastal and Shallow Water Conf., ed. by D. S. Gorsline. National Science Foundation and Office of Naval Research. Tallahassee, Florida (1962).

— and W. H. Mowbray: The production of underwater sound by *Opsanus* sp., a new toadfish from Bimini, Bahamas. Zoologica 44, 71—76 (1959).

Frisch, K. v.: Über den Gehörsinn der Fische. Biol. Rev. 11, 210—246 (1936).

Gray, G., and H. E. Winn: Reproductive ecology and sound production of the toadfish, *Opsanus tau*. Ecology 42, 274—282 (1961).

Griffin, D. R.: Underwater sounds and the orientation of marine animals, a preliminary survey. Proj. NR 162-429, t.o. 9, between ONR and Cornell University, Techn. Rep. No. 2 (1950).

— Hearing and acoustic orientation in marine animals. Papers Mar. Biol. and Oceanogr., Deep-Sea Research, suppl. to vol. 3, 406—417 (1955).

Hasler, A. D.: Perception of pathways by fishes. Quart. Rev. Biol. 31, 200—209 (1956).

Kleerekoper, H., and E. C. Chagnon: Hearing in fish, with special reference to *Semotilus atromaculatus atromaculatus* (Mitchill). J. Fish. Res. Bd. Canada 11, 130—152 (1954).

Lowenstein, O.: The sense organs: The acoustico-lateralis system. Chap. II, Part 2, in: The physiology of fishes, Vol. II, ed. by Margaret E. Brown. New York: Academic Press, Inc. (1957).

Marshall, N. B.: Aspects of deep sea biology. Philosophical Library. New York. pp. 257—258 (1954).

Moorhouse, V. H. K.: Reactions of fish to noise. Contr. to Can. Biol. and Fish., N. S. 7, 465—475 (1933).

Moulton, J. M.: Influencing the calling of sea robins (*Prionotus* spp.) with sound. Biol. Bull. 111, 393—398 (1956a).
— The movements of menhaden and butterfish in a sound field. Anat. Rec. 125, 592 (1956b).
— Sound production in the spiny lobster, *Panulirus argus* (Latreille). Biol. Bull. 113, 286—295 (1957).
— The acoustical behaviour of some fishes in the Bimini area. Biol. Bull. 114, 357—374 (1958a).
— A summer silence of sea robins. Copeia (1958), 234—235 (1958b).
— Age changes in stridulation and the stridulatory apparatus of the spiny lobster *Panulirus argus* (Latreille). Anat. Rec. 132, 480 (1958c).
— Swimming sounds and the schooling of fishes. Biol. Bull. 119, 210—223 (1960).
— and R. H. Backus: Annotated references concerning the effects of man-made sounds on the movements of fishes. Maine Dept. of Sea and Shore Fisheries, Circ. No. 17 (1955).
Parry, M. L.: The fishing methods of Kelentan and Trengganu. In: Malayan fishing methods, by T. W. Burden and M. L. Parry, J. Malayan Br. Roy. Asiatic Soc., 27, Part 2 (1954).
Poggendorf, D.: Die absoluten Hörschwellen des Zwergwelses (*Amiurus nebulosus*) und Beiträge zur Physik des Weberschen Apparates der Ostariophysen. Z. vergl. Physiol. 34, 222—257 (1952).
Rhumbler, L.: Stemmen die Strahlen der Astrosphäre oder ziehen sie? Arch. Entw. Mech. 4, 659—730 (1897).
Shishkova, E. V.: Concerning the reactions of fish to sounds and the spectrum of trawler noise (translated by J. M. Moulton). Rybnoye Khoziaistvo 34, 33—39 (1958).
Spemann, H.: Organizers in animal development. Croonian Lecture. Proc. roy. Soc. B, 102, 107 (1927).
Tavolga, W. N.: Visual, chemical and sound stimuli as cues in the sex discrimination behaviour of the gobiid fish, *Bathygobius soporator*. Zoologica 41, 49—64 (1956).
— The significance of underwater sounds produced by males of the gobiid fish, *Bathygobius soporator*. Physiol. Zool. 31, 259—271 (1958a).
— Underwater sounds produced by two species of toadfish, *Opsanus tau* and *Opsanus beta*. Bull. Mar. Sci. Gulf and Carribean 8, 278—284 (1958b).
— Underwater sounds produced by males of the blenniid fish, *Chasmodes bosquianus*. Ecology 39, 759—760 (1958c).
Westenberg, J.: Acoustical aspects of some Indonesian fisheries. J. du Conseil pour l'Exploration de la Mer 18, 311—325 (1953).
Winn, H. E., and J. Marshall: Sound production of squirrelfishes. Anat. Rec. 138, 390 (1960).
— J. F. Stout: Sound production by the satinfin shiner, *Notropis analostanus*, and related fishes. Science 132, 222—223 (1960).

The Orientation of Octopus

By M. J. WELLS

Department of Zoology, Cambridge (England)

With 8 Figures

Contents

I. Introduction

Octopus is an animal that learns very rapidly. In the laboratory it can be taught to make a wide variety of visual and tactile discriminations, and systematic studies of the performance of octopuses in training experiments have provided many clues to the organisation of the animals' sensory integrative mechanisms (references see YOUNG, 1961; WELLS, 1962).

The present account is a review of a number of experiments all of which show that octopuses are unable to integrate proprioceptive with other sensory information in learning. These experiments are interesting because they imply a channelling of the sensory input to learning and non-learning parts of the brain according to its nature. This channelling severely limits the performance of the animal in a number of ways and it is worth considering how this state of affairs has come about, and why it has persisted in *Octopus* where it would at first sight appear to be so gravely disadvantageous.

II. Material

All the experiments were made with *Octopus vulgaris* LAMARCK taken, except in the case of SCHILLER's (1949) experiments, from the Bay of Naples. In most cases the octopuses weighed 300—400 grams, which implies at least one year spent in the sea before capture (WELLS and WELLS, 1959). The smaller animals of 10—30 gr. used in the statocyst-removal experiments were possibly hatched early in the year in which they were used — the Naples experiments all being made during July and August. All the animals therefore had had opportunities to learn visual and tactile discriminations in the sea before capture; they were not laboratory animals reared in restricted environments before use in the experiments.

For some of the tests parts of the nervous system were removed or disconnected, the operations being carried out under urethane anaesthesia (WELLS and WELLS, 1957, a, b; WELLS, 1960); there is no apparent trace of postoperational shock in octopuses and apart from those with lesions specifically affecting the control of the mouthparts, all of the animals resumed feeding within an hour or so of recovery from the anaesthetic.

III. Methods

Each animal was kept separately, in a large (100 × 60 × 40 cm) asbestos aquarium, bare except for a heap of three bricks at the far end, forming a shelter into which the animal could retire between trials. After two or three days in such an aquarium an octopus will come out of its 'home' and swim down the tank to attack objects seen moving at the other end. If rewarded with a crab or a piece of fish for attacking one of two figures, and given a small electric shock (6—9 volts A. C. through electrodes attached to a probe) for attacking the other, the animal rapidly (within 20—30 trials) learns to discriminate between the two, remaining in the 'home' when the negative figure is shown.

A similar technique was used for experiments on the tactile sense of *Octopus*, except that in this case the animals were used only after section of the optic nerves, which ensured that they could not learn to recognise the objects visually; test objects were again presented successively, this time by touching them against one of the arms of the octopus which would then twist to grasp the object. The animal was rewarded for passing one of the objects under the interbrachial web to the mouth, and given an electric shock if it did the same with the other.

IV. Experimental results

1. Visual discrimination and statocyst removal

Octopuses can be taught to discriminate between the members of pairs of visual figures, including sometimes those that differ only in orientation. They can, for example, very readily be trained to distinguish between a 2 × 10 cm rectangle cut out of white plastic shown with the long sides horizontal, and a similar rectangle shown with its long axis vertical. They seem, however, unable to learn to distinguish between the same rectangles shown at right angles but obliquely. SUTHERLAND (1957, 1960) has reviewed the results of these and other visual discrimination experiments and concludes that octopuses classify the shapes that they see in terms of differences in the distribution of stimulation along the horizontal and vertical axes, so that what are in effect compared are the projections of the figures shown in Fig. 1.

Any such mechanism implies either that the retina always remains the same way up relative to things seen, or that the animal's brain can take into account the orientation of the retina in assessing the visual input.

Observation of octopuses in aquaria suggests that the retina normally remains constantly oriented with respect to gravity, since the slit pupil of the eye stays horizontal or very nearly so over a wide range of bodily positions (Fig. 2). This constant orientation can be abolished by removal

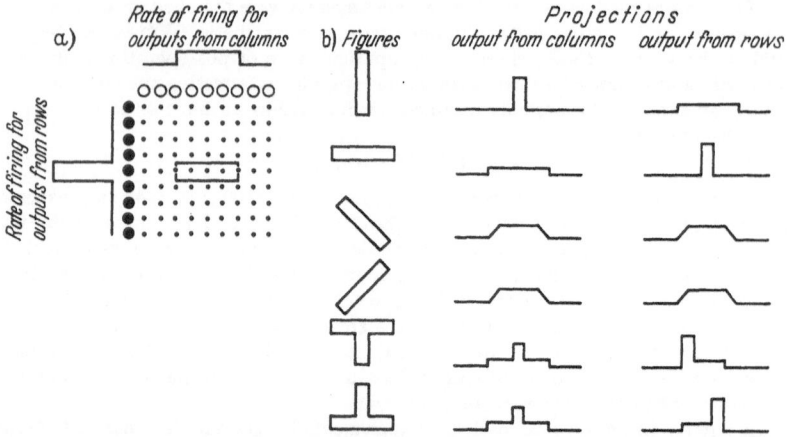

Fig. 1a and b. SUTHERLAND's hypothesis of shape discrimination. a) The small dots represent retinal elements with the image of a horizontal rectangle projected onto them. Open circles at the top represent cells specific to each column and filled circles at the side cells specific to each row. Each element in the retinal array is connected to the output cell for its own row and the output cell for its own column. b) Figures used for investigating discrimination of orientation, with outputs predicted by the theory (redrawn from SUTHERLAND, 1957). Anatomical studies show that the retinal elements are indeed arranged in two planes at right angles, as this theory would require (YOUNG 1960)

of the statocysts, which also stops all compensatory movements of the eyes in response to rotation in the horizontal plane (DIJKGRAAF, 1960); postoperationally the orientation of the eyes depends simply upon how the animals happen to be sitting (Fig. 2).

After statocyst removal trained octopuses cease to discriminate between vertical and horizontal rectangles, but continue to discriminate successfully between black and white discs (Fig. 3). When eye orientation was recorded together with the response at each trial, it was found that the animals' response to the rectangles was determined by the orientation of the figures relative to the retina. A vertical rectangle was mistaken for a horizontal one when the slit pupil was vertical, and so on (WELLS, 1960). It has further been shown that correct responses to two sources of plane polarised light at right angles can be maintained after statocyst removal by adjusting the plane of polarisation to the position of the eye at each trial (Fig. 4). These experiments show that visual discrimination of orientation by *Octopus* depends upon the eyes remaining constantly oriented with respect to gravity. The central nervous machinery that analyses the visual input is constructed on the assumption that the retina remains constantly oriented. If this ceases to be true — as when the statocysts are removed — the animal continues to behave as if the

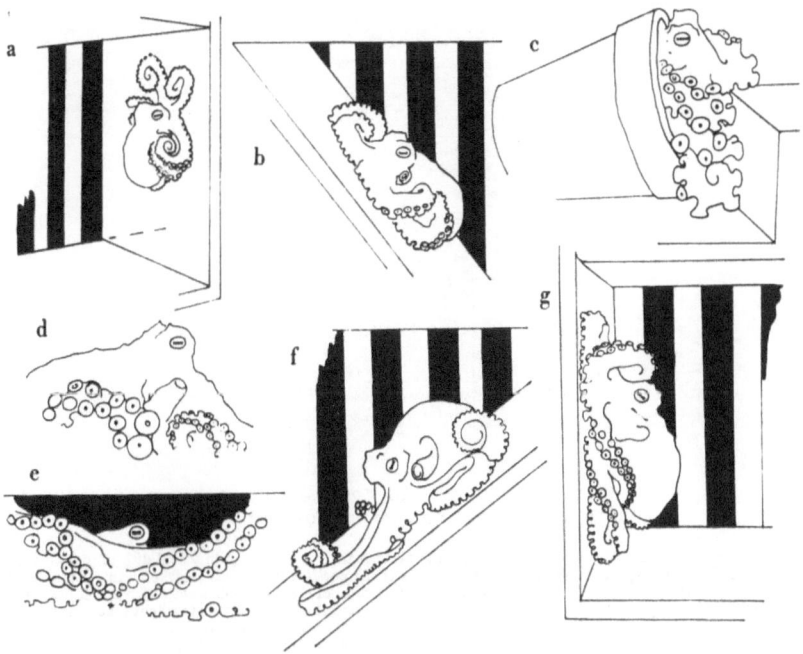

Fig. 2a—g. Orientation of the eyes before and after bilateral statocyst removal. In unoperated animals (a—e) the slit pupil remains horizontal or very nearly so whatever the position of the octopus. After removal of both statocysts this ceases to be true and the orientation of the pupil, and therefore of the retina, thereafter depends on the position in which the animal is sitting (f and g) (from WELLS, 1960)

eyes were still correctly positioned, and discrimination of orientation relative to gravity breaks down.

The breakdown of orientation discrimination is not, of course, the only effect of statocyst removal in *Octopus*. Movement control in general is very severely affected. After the operation the animals are unable to swim effectively, and spiral or loop in the water when they try to do so, apparently lacking the normal refined control of the funnel that directs the jet from the contracting mantle (BOYCOTT, 1960). Walking along the sides or bottom of the tank is scarcely less affected, with considerable postoperational oscillatory movements of the head and a tendency to somersault and fall over the arms when moving rapidly. There seems to be no distinction made between horizontal and vertical surfaces, and the position of the body as the animal moves about is quite erratic, in contrast to the normal octopus where the head is always held uppermost (BOYCOTT, 1960; see also Fig. 2). The animals can still right themselves if they fall on the bottom, but their behaviour on vertical surfaces

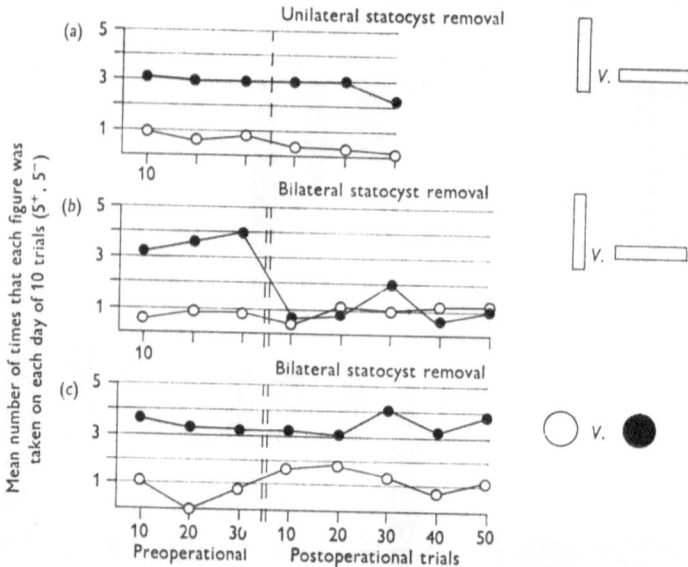

Fig. 3a—c. The effect of statocyst removal on the performance of octopuses in visual discrimination experiments. In each plot ● shows the number of times that the positive figure was attacked and ○ the number of attacks on the negative. (a) A summary of the performance of eight animals from which one statocyst only was removed in the course of training; four of the animals had the left, and four the right statocyst removed; performance in the learned discrimination between horizontal and vertical rectangles was not affected by the operation. (b) A similar plot of the performance of ten octopuses which had both statocysts removed (four of these had the statocysts removed in two successive operations, their performance after removal of the first statocyst being included in plot (a) ● Removal of both statocysts destroys the capacity to discriminate successfully between the rectangles. (c) A similar plot of the performance of five animals trained to discriminate between black and white disks. Removal of both statocysts does not prevent discrimination. These summaries are compounded from training experiments of variable length, and only the thirty trials immediately preceding operation and the first thirty or fifty postoperational trials are plotted (from WELLS, 1960)

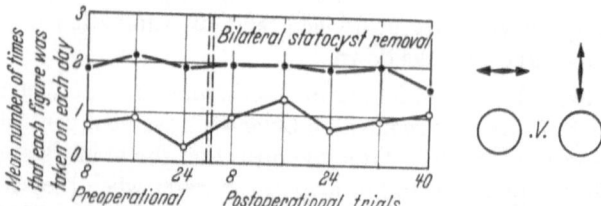

Fig. 4. The effect of statocyst removal on performance in a polarized light discrimination. ● shows the number of attacks in the +ve situation (electric vector normal to the pupil) and ○ the number of attacks in the —ve (vector parallel with the pupil) in each day of 8 trials (4+, 4—). Averages from ten animals, which had 2 days (16 trials) of pre-training to attack in the positive situation before the start of the experiment. On day 11 both statocysts were removed, but correct performance was maintained by appropriate orientation of the electric vector (parallel with, or at right angles to the slit pupil) at each trial. Only the last 24 preoperational and the first 40 postoperational trials are shown. This result incidentally proves that discrimination of the plane of polarisation is intraocular, and not by recognition of reflexion patterns in the animal's surroundings (from ROWELL and WELLS, 1961)

suggests that this is an effect of the ventral distribution of the suckers, which tend to hold on once they have made a contact. There is no return of the capacity to orient in space during at least the first month after statocyst removal; the animals do not relearn to walk or swim effectively.

2. Tactile discrimination

Blinded octopuses can readily be trained to discriminate between the members of pairs of objects that they touch, provided that these differ in texture or in taste (references see WELLS, 1962). But they cannot be taught any discrimination that would require their taking into account

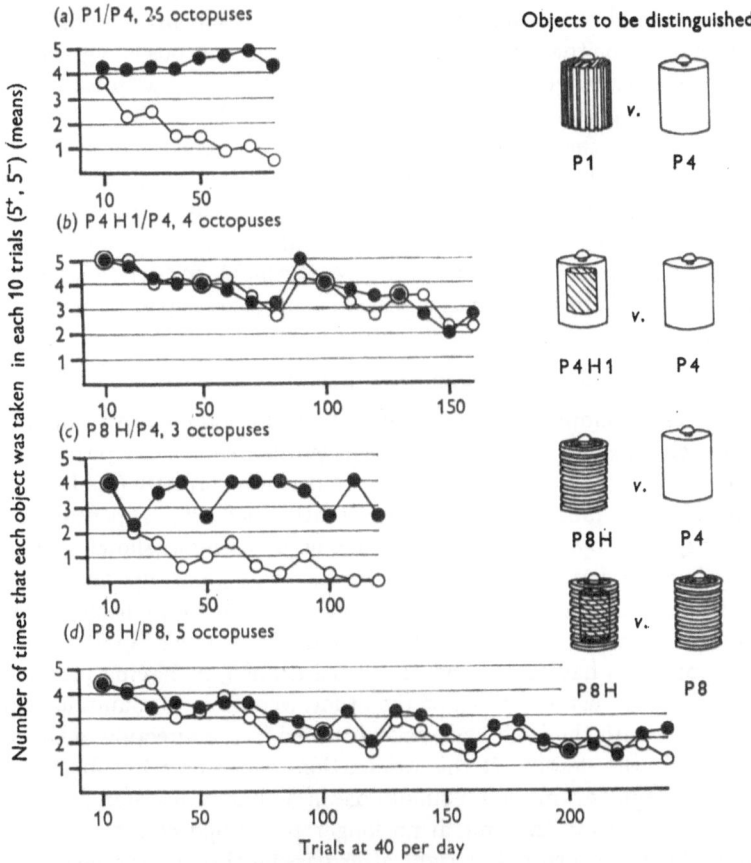

Fig. 5a—d. The results of some weight discrimination experiments. In (a) octopuses were trained to discriminate between the smooth perspex cylinder P4 and the grooved P1, both objects being of the same weight (5 gr.). In (b) an attempt was made to train octopuses to distinguish between P4 and P4H1, a second smooth cylinder nine times as heavy as the first. In (c) animals were trained to discriminate between P8H and P4, which differed in both weight and texture. In (d) an attempt was made to train 5 animals, 2 of them already trained successfully under (c) to discriminate between P8H (weighing 25 gr.) and P8 (5 gr.). In each case ● shows the number of times the positive object and ○ the number of times the negative object was taken (from WELLS, 1961 a)

the position of the arms and suckers making the contact. Thus they do not learn to distinguish between objects differing only in the pattern or orientation of surface irregularities (WELLS and WELLS, 1957a) and they cannot be taught to discriminate between objects differing in shape (WELLS, 1963) or weight (Fig. 5).

In the present context the weight discrimination experiments are particularly interesting because they show clearly that while proprioceptive information plays no part in touch learning, it is certainly utilised for positional adjustments. In these experiments, the blinded octopuses nearly always sat in the angle between the water surface and the side of the tank. Perspex cylinders were presented, one at a time, for the animal to examine and take or reject, each being lowered onto an arm by a line that was let slack as soon as the octopus had grasped the object. Heavy objects (cylinders drilled out and filled with lead) caused considerable passive extension of the arms holding them. Whenever this happened the octopus would contract the relevant arm or arms, increasing muscle tension to support the full weight of the object before taking or rejecting it in the normal way. An observer could easily tell which of the test objects was being handled simply by watching the animal; but the octopus itself was apparently unable to learn to distinguish between objects from the muscle tension needed to support them (WELLS, 1961a).

3. Experiments with octopuses in mazes

Several attempts have been made to train octopuses to run simple mazes with results that have hitherto proved somewhat puzzling, for in nearly every case the animals failed to perform correctly (see BOYCOTT, 1954; WELLS, 1962). The only consistently successful experiments are those reported by SCHILLER (1949). He used the maze shown in Fig. 6, the animal having to learn to go down the central passageway and turn left or right in order to get a crab previously seen through one or other of the windows in the living compartment. SCHILLER's three octopuses are reported as having learned and, once trained, as having practically never made a mistake, *provided* that they were able to maintain bodily orientation with the beak and suckers turned in the direction of the crab throughout the detour. If this orientation were upset by obliging the animal to squeeze through a small hole in a shutter inserted at the end of the passageway, the animal no longer performed correctly and was equally likely to turn left or right after passing through the hole.

These results would be expected from the failure to utilise positional information in other forms of discrimination learning; it should be noted that SCHILLER's octopuses did not learn to run up the passageway and then to turn left or right, they learned to crawl along the wall continuous with that immediately barring them from the crab. This is the only way the correct result could be reliably achieved by them without having to learn to orient themselves correctly in the passageway.

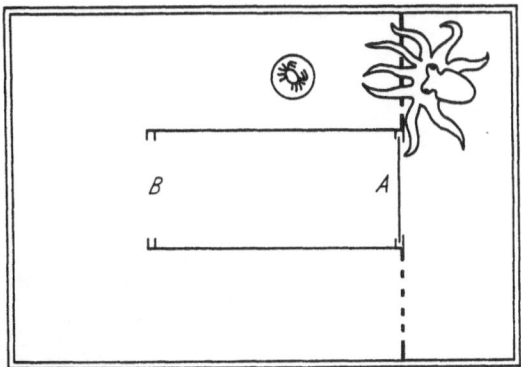

Fig. 6. SCHILLER's detour experiment with *Octopus*. The animal was kept in a compartment at one end of the tank (here seen from above) and through a grille could see, but not touch, a crab in a glass jar. On removal of the shutter at '*A*' the octopus was free to pass down a passageway and turn left or right to collect the crab. A second shutter could be inserted at '*B*' (from WELLS 1962)

4. Brain lesion experiments

Removal of large parts from the supraoesophageal lobes of the brain of *Octopus* does not prevent the animals from learning tactile discriminations. Indeed it can be shown that only the buccal, inferior frontal and subfrontal lobes are essential (Fig. 7) and that considerable parts even of these can be removed before touch learning fails (WELLS, 1959).

Lesions as large as these leave octopuses that are incapable of moving about their tanks. With the higher motor control centres in the basal lobes gone (BOYCOTT and YOUNG, 1950) the animals can no longer swim or move their arms in a coordinated manner. Individual arms can still make the responses needed to take or reject objects touched, (of this, more below) but there is no movement of the animal as a whole.

In contrast to this, lesions limited to the inferior frontal system cause no appreciable changes to the movement or posture of the animals, although they completely destroy the capacity to learn tactile discriminations (WELLS, 1961 b). Such animals appear to be quite normal as they move around their tanks, and it is only when their handling of objects touched is considered that anything unusual is observed. It is then seen that they tend to retain inedible objects that they pick up, passing them in under the interbrachial web to the mouth, and failing to reject them within the few seconds or minutes that an octopus normally requires to decide that an object cannot be eaten or broken up (WELLS, 1961 b). They cannot be trained to recognise and reject specific objects by touch, as normal octopuses can, and if given shocks for failing to reject an object presented and grasped, will flee headlong around the tank without releasing the object, which is passed under the interbrachial web to the mouth as soon as locomotion ceases. Clearly the animal learns something

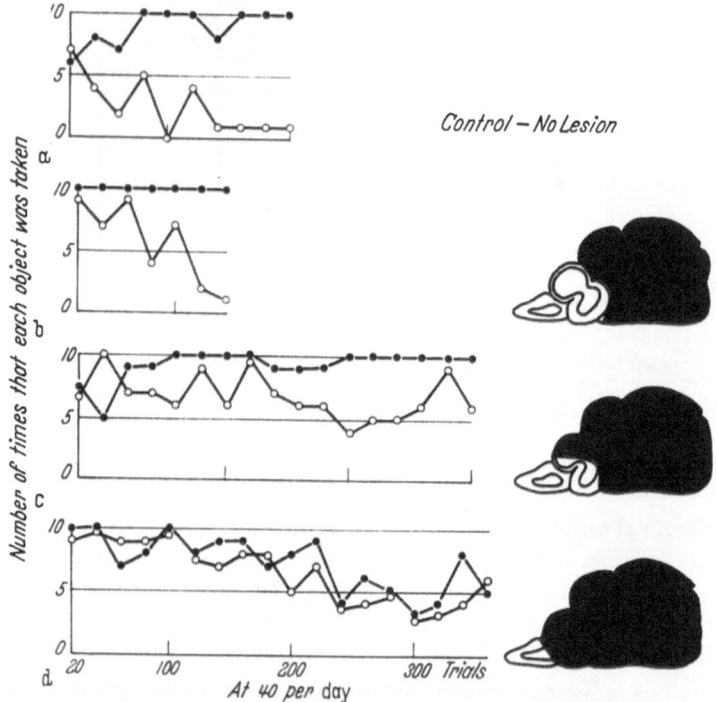

Fig. 7a—d. The effect of lesions on the performance of octopuses in learning to discriminate between P1 and P4 (Fig. 5) by touch: (a) shows the performance of a control animal (all the animals were blinded by optic nerves section but the control had no additional central lesion), (b) the performance of an octopus with the basal lobes removed. (c) and (d) show the effect of partial and total removal of the inferior frontal and sub-frontal lobes. Plotted as Fig. 5; where two points coincide, one is displaced downwards. Parts removed from the brain are shown in black on a diagram representing a median longitudinal section through the supra-oesophageal part of the brain (extract from a longer series in WELLS, 1959)

about the test situation, because it learns to run away when a contact is made. But it does not learn to reject the object.

    This curious behaviour (considered in some detail in WELLS, 1961 b) is perhaps most economically explained in terms of a dual control of arm movements. Thus the arms are known to be controlled at least partly by centres in the basal lobes, without which integrated movement ceases. But we also know that individual and even isolated arms (WELLS, 1959) are capable of making all the movements needed to take an object touched, a chain of motor responses apparently organised within the arm nerve cord, and normally controlled by touch learning mechanisms in the inferior frontal system. With the inferior frontal system damaged or destroyed, there seems to be no means of inhibiting the arm reflexes that cause objects to be taken, other than by keeping the arm musculature fully employed in locomotion. As soon as this ceases, the arm is free to pass an object grasped in towards the mouth.

These results are compatible with what we know about the effects of still larger lesions on the movements of octopuses. Suboesophagael preparations can grasp objects that they touch and regularly pass these in towards the mouth, although with the buccal lobe destroyed they can no longer eat the things they take. They can accurately place an arm to grasp a probe touching the head or back and they show certain reflex responses involving coordinated activity of all the arms at once. Thus when the suckers of an arm are allowed to grasp an object, which is held firm so that the animal cannot pass it in under the web to the mouth, the other arms align themselves along the direction in which the first is pulling or being pulled, and they all contract together.

Lesion experiments thus reveal anyway two levels of motor organisation at which proprioceptive feedback must play a part in regulating movement below the level of the basal lobes. Individual arms grasp and pass objects to the mouth and respond to stretch by contracting, whether or not they are in nervous connexion with the suboesophageal lobes, and at least one more complex response is regulated within the suboesophageal centres. It must be emphasised that these responses appear to be identical with the corresponding movements of intact animals; they are not more exaggerated and they do not appear to be any less accurately organised.

V. Discussion

Visual and tactile discrimination experiments show that octopuses are unable to take the position of parts of their own bodies into account when they learn to recognise things they see or touch. In vision this does not seriously limit the animals performance, since the eyes are normally held in a constant position with respect to gravity by a reflex mechanism working on information received from the statocysts. The visual system (for the anatomy of which see YOUNG, 1960, 1962) appears to be organised in a manner that assumes constancy of retinal orientation, and although this assumption leads the animal to make erroneous responses if the statocysts are destroyed, this is not a serious deficiency so far as the normal behaviour of the animal is concerned; the statocysts are deeply buried in the cartilage surrounding the brain and very unlikely to get damaged.

Octopus, it seems, cannot combine visual and positional information in learning. In maze tests, an octopus cannot learn to turn left or right in order to reach a goal; statocyst information regulates orientation of the body in space and the position of the visual receptors, but the information is not further available for use in learning. Again this is perhaps no very great disadvantage in the normal course of events. The detours that an octopus has to make in order to reach a crab that has fled behind a rock, or to regain its home on the other side of a boulder, can be achieved by the sort of behaviour shown in SCHILLER's experiments, by creeping along the continuous surface that separates the octopus from its goal. Longer distance homing is presumably regulated visually[1] and it is of interest here that the animals can so readily be trained to respond to the plane of polarisation of light. With a fixed orientation retina, this could provide a basis for navigation independent of proprioceptive learning.

Non-use of positional information in learning has much more serious consequences in tactile discrimination and on the animal's capacity to manipulate things that it touches. An octopus cannot learn to recognise objects by their weight, shape or (within limits) by their size; it cannot improve its performance in manipulative tasks nor learn to make movements that are not already a part of its built-in repetoire. With all the necessary muscular machinery and sensory instrumentation to pick up and manipulate the things it touches, *Octopus* can never learn to make a skilled movement because it is unable to put together information from position and surface contact receptors.

Brain lesion experiments suggest that surface tactile and positional information is channelled to quite different parts of the supraoesophageal brain. They show that the proprioceptive input from the arms may be utilised to regulate movements organised at levels (the arm nerve cords and the suboesophageal brain) which play no part in learning. These movements are recognisable components of the normal behaviour of the animal, and it seems reasonable to suppose that the learning systems' control of their performance is limited to switching them on or off.

A picture thus emerges of a hierarchy of control systems regulating the muscles of the animal as it moves about, each stage of which receives a proprioceptive input giving only the minimal degree of detail needed for efficient performance. Movements of the individual suckers, for example, are regulated partly within the subacetabular ganglion (ROSSI and GRAZIADEI, 1958) at the base of the sucker itself, and partly within the arm nerve cord (TEN CATE, 1928). No information about their movements would appear to be relayed to the brain. Arm movements are in

---

[1] Unfortunately we know practically nothing about the activities of these animals in the sea; fishermen in Naples say that they rarely range more than 50 metres or so out from 'home', and it is certainly true that one can find the same octopus in the same hole day after day when skin diving.

turn regulated locally and within the suboesophageal part of the brain. There is no evidence that the basal lobes receive a proprioceptive input from the arms at all — their output could in principle lead to appropriate patterns of movement as a result of information from the statocysts and visual learning system alone, all subsequent adjustments being dealt with at a suboesophageal level. Fig. 8 is an attempt to represent diagrammatically the relationship between the various levels of movement control and between movement control and the parts of the brain concerned in visual and tactile learning.

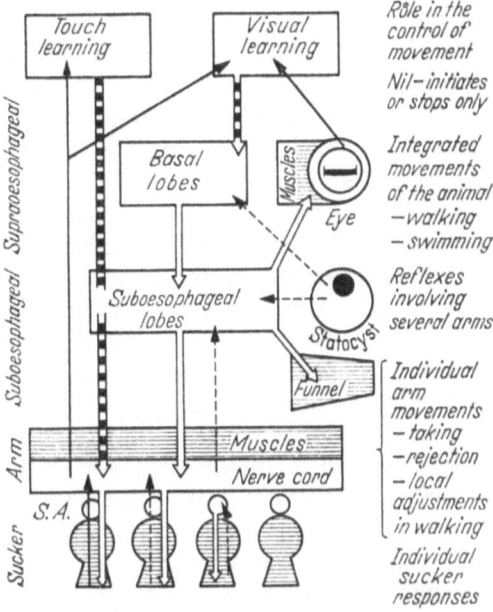

Fig. 8. A summary of the relations between the various levels at which movement is controlled, and learning mechanisms in the octopus. Dotted lines indicate sensory inputs carrying proprioceptive information, solid lines visual or contact information. Motor output channels are indicated by double solid lines, and command outputs from the learning centres by similar pairs of lines with cross-hatching. *SA* = Subacetabular ganglion at the base of each sucker. The diagram shows the minimum pattern of connexions indicated by the results discussed in the text

It is quite possible that this sort of deployment of organisation of movement is not only economic but inevitable in so flexible an animal as *Octopus*. This creature has no skeleton, other than a small cartilagenous box around the brain, in the walls of which the statocysts are buried. The position of parts would have to be defined relative to this and there is no further skeletal restriction of movement; an octopus can extend or contract, twist or bend each arm independently in an almost unlimited number of places at any one time. To define the position in space of one armtip would present a formidable problem for any central integrative mechanism, and the animal has eight arms. The problem, moreover, does not stop here, for the parts actually in contact with an object handled are the suckers, of which there are a hundred or more on each arm. Each sucker is itself able to extend bend, twist or contract. To define the relative positions of the suckers — as would be necessary, for example, for the determination of shape of small objects — may well be impossible simply because of the bulk of central integrative machinery that would be required.

Whether or not a brain capable of utilising a proprioceptive input of such complexity is theoretically possible, the fact remains that the brain of *Octopus* is not built to do so. Those parts of it concerned with altering behaviour as a result of experience do not use this sort of sensory input. It is not difficult to see how this complete separation of movement control and learning may have come about in cephalopods. For the Mesozoic ancestors of the modern cephalopods were pelagic, open water animals living as predators much as the squids do today. Such animals must maintain their bodily orientation in the absence of things to fix on with the eyes and it is logical that bodily orientation be determined by a gravity receptor without reference to the visual input. The use that a squid makes of its arms also previews the condition in *Octopus*, for in such an animal tactile stimulation merely serves to signal the results of visual decisions. What is important is the texture and taste of the object grabbed; it is a little late to start examining the shape of potential prey after it has been taken. Until the time came when the octopuses began to live on the bottom (the earliest known octopus is Cretaceous, and still had fins and a vestigial shell-ROGER, 1944) there was thus little or no advantage in relaying proprioceptive information to the learning parts of the brain, and perhaps by this time in the history of the group it was already too late for any considerable reorganisation of the central nervous system. The brains of decapods very closely resemble the brains of octopods and it seems that all the main features of their organisation were laid down before the two groups became distinct. An octopus is thus structurally a pelagic animal that just happens to live on the bottom; its mechanisms for orienting itself in the world around it are essentially those of an animal living in thé open water and we should not be surprised to find that it is sometimes curiously organised by the standards of a land-living vertebrate.

VI. Summary

1. The eyes of *Octopus* normally remain in a fixed orientation with respect to gravity. Removal of the statocysts abolishes this and visual discrimination of figures differing in orientation breaks down, the animals continuing to behave as if the eyes were still correctly oriented whatever their position.

2. In tactile discrimination experiments octopuses can readily be taught a wide range mechanotactile and chemotactile discriminations, but fail to learn to recognise objects differing only in weight, or shape, or in the orientation or pattern of surface irregularities.

3. In maze tests octopuses only succeed where they can maintain contact with a continuous surface separating them from their goal throughout the detour.

4. The most economical explanation of these results is that *Octopus* is unable to make use of proprioceptive information in learning.

5. Brain lesion experiments show that motor control and learning are dealt with in separate parts of the brain and that there are several levels of motor control, each of which apparently receives proprioceptive inputs giving only the minimal degree of detail necessary to the organisation at that level. Thus, for example, adjustments to movements of the suckers seem to be dealt with wholly at the arm nerve cord level. There is no need to suppose that feedback from the muscles of the individual arms ever penetrates beyond the suboesophageal part of the brain. Deployment of movement control is perhaps inevitable in so flexible an animal.

6. It is argued that the mechanisms of orientation and the relation between movement control and learning found in *Octopus* are more appropriate to an animal living in the open water and a consequence of the pelagic ancestry of the octopoda.

References

BOYCOTT, B. B.: Learning in *Octopus vulgaris* and other cephalopods. Pubbl. Staz. zool. Napoli. **25**, 67—93 (1954).
— The functioning of the statocysts of *Octopus vulgaris*. Proc. roy. Soc. B. **152**, 78—87 (1960).
— and J. Z. YOUNG: The comparative study of learning. Symp. Soc. exp. Biol. **4**, 432—453 (1950).
DIJKGRAAF, S.: The statocyst of *Octopus vulgaris* as a rotation receptor. Pubbl. Staz. zool. Napoli. **32**, 64—87 (1960).
ROGER, J.: Phylogénie des Céphalopodes Octopodes: *Palaeoctopus newboldi* (SOWERBY 1846) Woodward. Bull. Soc. géol. Fr. **5**, 83—98 (1944).
ROSSI, F. et P., GRAZIADEI: Nouvelles contributions à la connaissance du systéme nerveux du tentacule des Céphalopodes: IV. Le Patrimonie nerveux de la ventouse de l'*Octopus vulgaris*. Acta. anat. **34**, suppl. 32, 1—79 (1958).
ROWELL, C. H. F., and M. J. WELLS: Retinal orientation and the discrimination of polarised light by octopuses. J. exp. Biol. **38**, 827—831 (1961).
SCHILLER, P. H.: Delayed detour response in the octopus. J. comp. physiol. Psychol. **42**, 220—225 (1949).
SUTHERLAND, N. S.: Visual discrimination of orientation and shape by the octopus. Nature (Lond.) **179**, 11—13 (1957).
— Theories of shape discrimination in *Octopus*. Nature, Lond. **186**, 848—860 (1960).

TEN CATE, J.: Contribution à l'innervation des ventouses chez *Octopus vulgaris*. Arch. néerl. Physiol. **13**, 407—422 (1928).

WELLS, M. J.: A touch learning centre in *Octopus*. J. exp. Biol. **36**, 590—612 (1959).

— Proprioception and visual discrimination of orientation in *Octopus*. J. exp. Biol. **37**, 489—499 (1960).

— Weight discrimination by *Octopus*. J. exp. Biol. **38**, 127—133 (1961a).

— Centres for tactile and visual learning in *Octopus*. J. exp. Biol. **38**, 811—826 (1961b).

— Brain and behaviour in cephalopods. London: Heinemann (1962).

— Tactile discrimination of shape by *Octopus*. Quart. J. exp. Psychol. (in press) (1963).

— and J. WELLS: Tactile discrimination and the behaviour of blind *Octopus*. Pubbl. Staz. zool. Napoli. **28**, 94—126 (1956).

— — The function of the brain of *Octopus* in tactile discrimination. J. exp. Biol. **34**, 131—142 (1957a).

— — The effect of lesions to the vertical and optic lobes on tactile discrimination in *Octopus*. J. exp. Biol. **34**, 378—393 (1957b).

— — Hormonal control of sexual maturity in *Octopus*. J. exp. Biol. **36**, 1—33 (1959).

YOUNG, J. Z.: Regularities in the retina and optic lobes of *Octopus* in relation to form discrimination. Nature, Lond. **186**, 836—845 (1960).

— Learning and form discrimination in the octopus. Biol. Rev. **36**, 32—96 (1961).

— The optic lobes of *Octopus vulgaris*. Phil. Trans. roy. Soc. B. **245**, 19—58 (1962).

Der Einfluß der Größe bewegter Felder
auf den optokinetischen Augenstielnystagmus
der Winkerkrabbe

Von Peter Kunze*

Department of Zoology, Yale University, New Haven, Conn. (U.S.A.)

Mit 5 Abbildungen

Inhaltsübersicht

I. Einleitung

Bisher liegen nur wenig quantitative Untersuchungen darüber vor, wie die räumliche Lage und die Ausdehnung von Bewegungsreizen die Reaktionen von Arthropoden auf gesehene Bewegung beeinflussen. Crustaceen: Bei *Carcinus* muß etwa die Hälfte des horizontalen Gesichtsfeldes frei sein, damit ein optokinetischer Nystagmus ausgelöst werden kann; die lateralen Sehräume scheinen für das Bewegungssehen bedeutsamer zu sein als der frontale und caudale (v. Buddenbrock und Friedrich, 1933). Ähnliches scheint für *Goniopsis* zu gelten (Waterman, 1961). Insekten: Die Schwelle der optomotorischen Reaktion der Biene wird in bezug auf Sehschärfe und Lichtintensität erhöht, wenn Teile des Facettenauges durch Übermalen geblendet werden (Hecht und Wolf, 1929). Der Käfer *Chlorophanus* verrechnet Bewegungsreize, die verschiedene Augenbereiche treffen, zentralnervös summativ miteinander (Hassenstein, 1957); daraus und aus anderen Versuchen schloß Hassenstein, daß bei konstantem Bewegungsreiz die Stärke der optomotorischen Reaktion proportional der Zahl der beteiligten Ommatidien ist.

---

* Derzeitige Anschrift: Zoologisches Institut der Universität Freiburg, 78 Freiburg, Katharinenstr. 20.

Die vom rechten und linken Auge empfangenen Bewegungsreize wirken sich bei der Biene ebenfalls in summativer Weise auf die Reaktion aus (SCHALLER, 1960).

In der vorliegenden Arbeit soll quantitativ verfolgt werden, wie sich der optokinetische Nystagmus von *Uca* mit der räumlichen Ausdehnung von Bewegungsreizen und ihrer Lage zum Facettenauge ändert.

Herrn Professor T. H. WATERMAN, in dessen Laboratorien an der Yale University die Experimente ausgeführt wurden, danke ich für seine Unterstützung und sein Interesse an der Arbeit. Diese Arbeit wurde ermöglicht von der National Science Foundation der USA (Grant 9690, principal investigator: T. H. WATERMAN, Dept. of Zoology, Yale University, New Haven, Conn.).

II. Material und Methode

Als Versuchstiere dienten Männchen von *Uca pugnax*, die in den Salzmarschen von East Haven, Connecticut, gefangen worden waren. Vor den Versuchen wurde ihnen die große Winkschere abgeschnitten, da sie, wenn sie erhoben wurde, große Teile des Gesichtsfeldes abdecken und durch Bewegung die Messung stören konnte. Die Wunde wurde mit Wachs verschlossen.

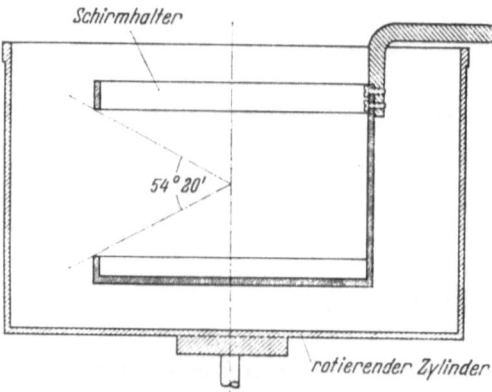

Abb. 1. Schnittbild der Versuchsanordnung, in deren Zentrum das Versuchstier eingebracht wird

Die Tiere befanden sich während der Versuche in einer 24,5 cm weiten Drehtrommel, die über ein Planetengetriebe von einem Synchronmotor angetrieben wurde. Sie waren umgeben von einem runden, nicht rotierenden Schirmhalter von 15,8 cm Durchmesser, der den Boden und den oberen Rand des rotierenden Zylinders für die Tiere verdeckte (Abb. 1). Vom Zentrum dieses Schirmhalters aus gemessen war der rotierende Zylinder in der Vertikalen unter einem Gesamtwinkel von 54°20′ sichtbar. Ein Verbindungsstück zwischen der oberen und der unteren Hälfte des Halters verdeckte in der Horizontalen einen etwa 15° breiten Bereich im Rücken des Tieres. An diesem Halter konnten schwarze Papierschirme befestigt werden, um den Ausblick der Tiere auf den rotierenden Zylinder einzuschränken.

Das Innere des Zylinders wurde durch drei 22 cm zentrisch über ihm angebrachte, mit Gleichstrom gespeiste Mattglasbirnen von je 25 W beleuchtet. Die im Zentrum des oberen Zylinderendes direkt gegen die Birnen gemessene Lichtintensität betrug 120 foot candles.

Die Augenstielreaktionen wurden mit der in einer früheren Arbeit (KUNZE, 1960) beschriebenen Apparatur zur Messung kleiner Drehmomente gemessen. Diese Apparatur erlaubte es, die Drehmomente zu messen, die ein mit ihr verbundener Augenstiel um die durch das Gelenk seines distalen Gliedes führende senkrechte Achse ausübte. Das Versuchstier wurde dorsal am Carapax mit einer Bienenwachs-Colophonium-Mischung an einen justierbaren Bleistab geklebt, der von einem Mikromanipulator gehalten wurde. Mit dem Mikromanipulator wurde das Versuchstier an die Apparatur herangeführt, so daß sein Augenstiel in Normalstellung mit diesem verbunden werden konnte. Dazu war der Augenstiel zuvor median mit einem Wachstropfen versehen worden. Mit einer warmen Nadel ließ sich dieser Tropfen leicht mit einem Wachstropfen verschmelzen, der sich an einem exzentrisch in die das Drehmoment aufnehmende Klammer der Apparatur eingespannten Holzstäbchen befand (Abb. 2). Der Augenstiel brauchte während dieses Vorganges nicht festgehalten zu werden. Der das Tier tragende Mikromanipulator und die Drehmoment-Meßapparatur waren beide am Läufer einer senkrechten Spindel befestigt, mit der das Versuchstier und die Meßapparatur gemeinsam von oben in die Versuchsanordnung hineingeführt werden konnten. Die vom Augenstiel erzeugten Drehmomente wurden mit einem Millivoltschreiber automatisch registriert.

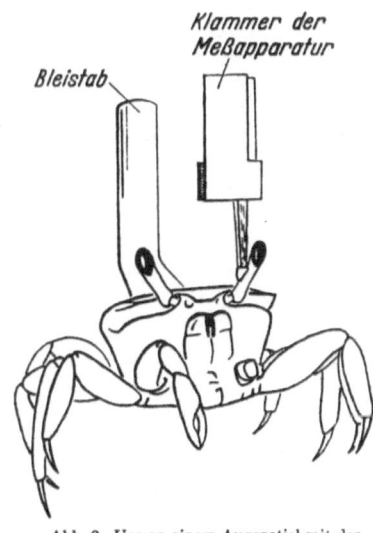

Abb. 2. *Uca* an einem Augenstiel mit der Meßapparatur befestigt

III. Ergebnisse

Zur Erzeugung des Bewegungsreizes wurde ein Muster benutzt, mit dem in Vorversuchen gute Reaktionen hervorgerufen werden konnten. Es bestand aus 0,75° breiten, vertikalen Streifen, die in 21 Graustufen eine sinusförmige Helligkeitsverteilung mit einer Periodenlänge von 30° annäherten. Das dunkelste Grau verhielt sich zum hellsten wie 1:19.

Auf Rotation des Streifenzylinders antwortete das Tier unter Normalbedingungen (größtmögliches bewegtes Feld) mit einem nahezu regelmäßigen Augenstielnystagmus (registriert als ein langsam ansteigendes Drehmoment in der Rotationsrichtung, gefolgt von einer kurzen Drehmomentänderung in der Gegenrichtung). Die Frequenz dieses Nystagmus konnte durch die Zylindergeschwindigkeit geändert werden. Bei dem oben beschriebenen Streifenzylinder reagieren die Tiere bei optimaler Geschwindigkeit (etwa 22°/sec) mit einer Nystagmusfrequenz von etwa 0,8/sec. Bei weiterer Erhöhung der Zylindergeschwindigkeit wird der Nystagmus zunehmend unregelmäßiger, und die mittlere Nystagmusfrequenz nimmt stark ab. Um Sättigungseffekte zu vermeiden, wurde für die folgenden Versuche die unter dem optimalen Wert liegende Zylindergeschwindigkeit von 11,25°/sec gewählt. Sie ruft gegenüber der Optimalgeschwindigkeit etwa die halbe Nystagmusfrequenz hervor. Mit anderen Streifenzylindern lassen sich bei *Uca* Nystagmusfrequenzen von maximal 2—3/sec hervorrufen. Die Drehrichtung des Zylinders hatte keinen Einfluß auf die Nystagmusfrequenz.

1. Änderung der vertikalen Ausdehnung des bewegten Feldes

Die in der beschriebenen Versuchsanordnung maximale vertikale
Erstreckung des bewegten Feldes von 54° 20′ wurde durch Einlegen von
Schirmen verschiedener Höhe in den Schirmhalter (Abb. 1) in bestimmter Weise verkleinert. (Die horizontale Weite des bewegten Feldes
blieb dabei unverändert.) Diese Einengung wurde sowohl in der Weise
vorgenommen, daß sich der Schirm von oben her verschieden tief nach

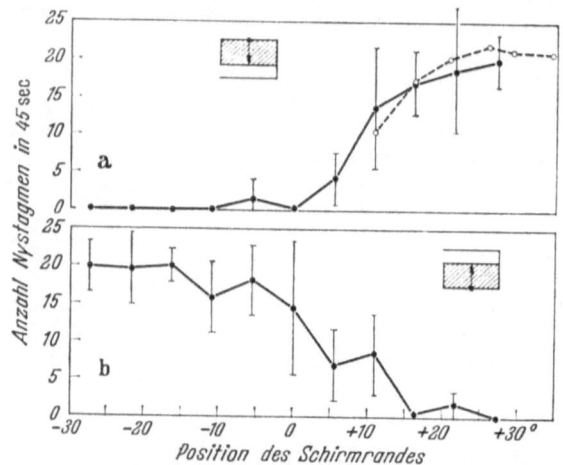

Abb. 3a u. b. Reaktionen von fünf Krabben auf verschiedene vertikale Ausdehnung des bewegten Feldes.
Ordinate: Anzahl Nystagmen während der ersten 45 sec nach Beginn der Zylinderdrehung. Abszisse: Position
des Schirmrandes in Winkelgraden über (+) und unter (—) der horizontalen Ebene, die durch den unteren
Augenrand des Versuchstieres führt, vom Schnittpunkt der Drehachse des Zylinders mit dieser Ebene aus
gemessen. Mittelwerte $\pm s \left(\dfrac{1}{N-1} \Sigma (x - \bar{x})^2 \right)$ von fünf Tieren. a Schirm oben fest, seine Höhe nach unten
variiert. (Gestrichelte Kurve: Mittelwerte von 4 Tieren, Näheres im Text.) b Schirm unten fest, seine Höhe
nach oben variiert

unten in den Zylinder hinein erstreckte (Ergebnisse in Abb. 3a), als auch
von unten her verschieden weit sich nach oben erstreckte (Ergebnisse in
Abb. 3b).

Die Ergebnisse wurden zwar (der Übersichtlichkeit halber) in zwei
Abbildungen dargestellt, im Experiment wurden jedoch bei ein und
demselben Tier die beiden verschiedenen Situationen nacheinander in
unregelmäßigem Wechsel geboten. In der ersten Situation (Schirm von
oben, Abb 3a) traten bei allmählichem Höherrücken des unteren
Schirmrandes höchstens vereinzelt Nystagmen auf, solange der Schirmrand die durch den unteren Augenrand der Krabbe führende horizontale
Ebene nicht erreichte (Abszissenwert 0°). Bei weiterer Vergrößerung des
bewegten Feldes stieg die Reaktion zunächst rasch, später etwas lang-
samer (bis zum Abszissenwert 20° auf 90% des Endwertes) an.

Ein ähnliches Bild, jedoch mit umgekehrtem Vorzeichen des Anstieges,
zeigt Abb. 3b. Mit zunehmender Verringerung der Schirmhöhe von oben

her nimmt die Zahl der Nystagmen zu, bis der obere Schirmrand die Position — 16,2° erreicht hat. Eine weitere Vergrößerung der bewegten Felder nach unten führt zu keiner weiteren Verstärkung der Reaktion (90% des Endwertes werden bei der Position —5° erreicht).

Die Änderung der Reaktionsstärke mit der Position des Schirmrandes besitzt bei den einzelnen Tieren im allgemeinen einen steileren Gradienten, als es in den Mittelwerten der Abb. 3 zum Ausdruck kommt. Die Reaktionen der Einzeltiere beginnen jedoch bei verschiedenen (wenn auch nah beieinanderliegenden) Positionen des Schirmrandes steil abzufallen. Bei einer Mittelung kommt dann ein weniger steiler Kurvenverlauf zustande. Die Unterschiede zwischen verschiedenen Tieren sind wahrscheinlich auf eine individuell verschiedene Haltung der Augenstiele zurückzuführen sowie darauf, daß die verschiedenen Tiere nicht genau die gleiche Position im Zylinder besitzen.

Außer der Änderung der Anzahl von Nystagmen in 45 sec, die besonders auffällig war, traten noch folgende Änderungen der Augenstielbewegungen mit verringerter Ausdehnung des bewegten Feldes auf: Sobald das bewegte Feld soweit verkleinert war, daß weniger als 15 Nystagmen in 45 sec registriert wurden, folgten die Nystagmen einander bei weiterer Verkleinerung der bewegten Felder zunehmend unregelmäßig und häufig zu Beginn des Bewegungsreizes in kürzeren Zeitabständen als im weiteren Verlauf. (Diese Unregelmäßigkeiten treten nicht auf, wenn die Reaktionsstärke durch langsamere Drehgeschwindigkeit verringert wird.) Auch wenn — bei sehr kleinem Gesichtsfeld — keine Nystagmen mehr auftraten, konnten hin und wieder noch schwache Folgereaktionen der Augenstiele gemessen werden.

Es scheint berechtigt, aus den Ergebnissen der Abb. 3 zunächst folgende Schlüsse zu ziehen: 1. Ommatidien, deren Achsen bei normaler Augenstellung mehr als etwa 16° von der Horizontalebene nach unten abweichen, tragen zur gemessenen Reaktion sehr wenig oder nichts bei. Wurde nur ihnen ein bewegtes Feld geboten (Abb. 3a), so war die Reaktion Null; bekamen sie zusätzlich zu anderen Ommatidien einen Bewegungsreiz (Abb. 3b), so erhöhten sie die Reaktion nicht. 2. Die Achsen derjenigen Ommatidien, die hauptsächlich zur Reaktion beitragen, besitzen mit der Horizontalen einen Winkel zwischen etwa 20° nach oben und 5° nach unten.

Es blieb zu prüfen, ob eine weitere Vergrößerung des bewegten Feldes nach oben, als sie im Experiment der Abb. 3a vorgenommen wurde, noch zu einem weiteren Ansteigen der Reaktionen führen würde. Die bisher besprochenen Ergebnisse in Abb. 3a lassen ja noch nicht entscheiden, ob ein Plateau der Reaktionsstärke erreicht war oder nicht. Die Tiere wurden zu diesem Zweck tiefer in den Zylinder hineingebracht; damit war es möglich, eine weitere Ausdehnung des bewegten Feldes nach oben zu erreichen. (Zwar wurde das bewegte Feld nach unten beschnitten; nach dem oben geschilderten Versuchsergebnis sollte dies jedoch ohne Einfluß auf die Reaktion sein.) Die Reaktionen unter diesen Bedingungen (gestrichelte Kurve in Abb. 3a) folgen bei gleicher Lage des Schirmrandes zum Auge der Krabben näherungsweise denjenigen, wie sie unter den oben geschilderten Versuchsbedingungen gemessen wurden. Sie steigen zwar um einen kleinen (nicht signifikanten) Betrag über die zuvor gemessenen Reaktionen an, zeigen jedoch deutlich, daß auch hier ein Endwert erreicht wird.

2. Änderung der horizontalen Ausdehnung des bewegten Feldes

Die Verringerung der horizontalen Weite des bewegten Feldes erfolgte dadurch, daß sich Schirme symmetrisch zur Mediane verschieden weit nach lateral erstreckten. Abb. 4 gibt die dadurch erfolgte Änderung der

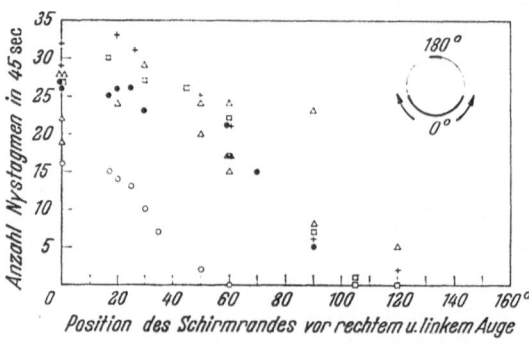

Abb. 4. Reaktionen von fünf Krabben auf verschiedene horizontale Weite des bewegten Feldes

Reaktion von fünf Tieren wieder (jedes Tier durch ein anderes Symbol wiedergegeben). Die Reaktionen des einen Tieres (○) sind zwar durchweg schwächer als die aller anderen, zeigen aber prinzipiell den gleichen Reaktionsverlauf: mit zunehmender Abschirmung des bewegten Feldes zunächst keine oder nur geringe Verminderung der Reaktion, bei einer Position des Schirmrandes zwischen 20 und 40° Fallen der Reaktion und Reaktion Null, wenn die Position des Schirmrandes etwa 120° erreicht (in einem Fall bereits 60°).

Obgleich also das bewegte Feld noch über eine horizontale Gesamtweite von etwa 105° (in einem Fall 225°) der ursprünglich 345° (siehe Methodik) sichtbar war, traten fast keine Nystagmen mehr auf.

Neben der Verringerung der Nystagmuszahl traten die im vorhergehenden Abschnitt besprochenen Unregelmäßigkeiten in eben der gleichen Weise auf, sobald die Anzahl der Nystagmen etwa $^3/_4$ der maximalen Anzahl unterschritt. (Die höhere Maximalreaktion in dieser Versuchsserie gegenüber der im vorhergehenden Abschnitt beschriebenen beruht wahrscheinlich darauf, daß eine andere Gruppe von Tieren benutzt wurde.)

Die Ergebnisse dieses Abschnittes stimmen qualitativ mit den in der Einleitung erwähnten, an anderen Crustaceen gefundenen Ergebnissen überein.

IV. Diskussion

Die Augen von *Uca pugnax* besitzen eine länglich ovale Form mit (bei Normalhaltung des Augenstieles) senkrecht stehender langer Achse (Abb. 5). Das Auge erstreckt sich frontal weiter nach unten als caudal (vgl. Abb. 5a und b). Median ist es geteilt von einer vom oberen Ende des Auges nach unten laufenden blinden Zone. Das Gesichtsfeld beider Augen zusammen beträgt in der Horizontalen 360°. Über die räumliche Verteilung der Ommatidien ist bisher nichts Genaues bekannt. Sie läßt sich unter der Annahme, daß die Ommatidienachsen senkrecht zur Oberfläche verlaufen und die Ommatidien gleich groß sind, abschätzen: Die Winkel zwischen den Ommatidienachsen dürften in vertikaler Richtung kleiner sein als in horizontaler. Mit anderen Worten, die Anzahl von Ommatidien pro Gesichtsfeldwinkel ist in der Vertikalen größer als in der Horizontalen. Dieser Astigmatismus ist auch an der Form der Pseudopupille des *Uca*-Auges deutlich zu erkennen.

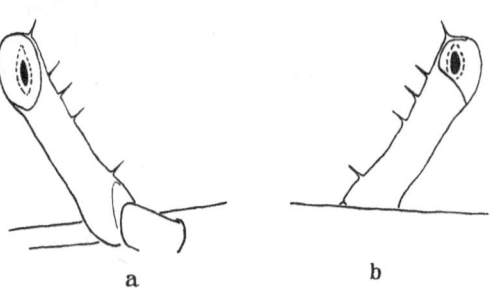

Abb. 5a u. b. Rechter Augenstiel einer *Uca*; a von vorn; b von hinten

Unter der Voraussetzung einer linearen Beziehung zwischen Ommatidienzahl und Reaktion wäre also bei schrittweiser Abschirmung des bewegten Feldes in der Vertikalen ein steilerer Gradient der Reaktionsänderung zu erwarten als in der Horizontalen. Dies ist in den in dieser Arbeit wiedergegebenen Ergebnissen deutlich der Fall.

Daß im wesentlichen nur Ommatidien, deren Achsen bis zu 20° nach oben und 5° nach unten weisen, zum optokinetischen Nystagmus beitragen, wäre zum Teil durch die oben beschriebene äußere Form des Auges zu erklären. Nach unten gerichtete Ommatidien sind hauptsächlich auf den frontalen Augenbereich beschränkt, der Zahl nach also den nach oben gerichteten, die frontal und caudal vorhanden sind, unterlegen.

Trotzdem reicht die Annahme einer proportionalen Beziehung zwischen Nystagmusfrequenz und Anzahl der vom Bewegungsreiz getroffenen Ommatidien nicht aus zur Deutung der Ergebnisse. Dies wird deutlich bei der horizontalen Verkleinerung des bewegten Feldes. Die Nystagmuszahl beginnt hier bereits zu sinken, wenn erst 75° des 360° weiten Gesichtsfeldes abgeschirmt sind, und ist fast Null, selbst wenn nahezu noch ein Drittel des horizontalen Gesichtsfeldes frei ist. Von Bedeutung könnte in diesem Zusammenhang die Beziehung zwischen Bewegungs-richtung und Ausdehnungsrichtung des bewegten Feldes sein. (Experi-mente, in denen der Bewegungsreiz in senkrechter Richtung verläuft, würden hier weiterhelfen.)

Von vergleichendem Interesse sind in diesem Zusammenhang Messungen an dem Flußkrebs *Procambarus* (Kunze, in Vorbereitung). Bei diesem wurde eine annähernd proportionale Beziehung zwischen der Anzahl der vom Bewegungsreiz getroffenen Ommatidien und der Stärke des optokinetischen Nystagmus gefunden, die unabhängig davon war, ob die Ausdehnung des bewegten Feldes in horizontaler oder in verti-kaler Richtung geändert wurde.

V. Zusammenfassung

Der von *Uca pugnax* in der Drehtrommel ausgeführte Augen-stielnystagmus ändert sich mit der Ausdehnung des bewegten Feldes. Zum Augenstielnystagmus tragen vor allem diejenigen Ommatidien bei, deren Achsen mit der Horizontalen einen Winkel zwischen etwa 20° nach oben und 5° nach unten bilden. Die Anzahl von Nystagmen pro Zeit-einheit sinkt bei Verringerung der vertikalen Ausdehnung des bewegten Feldes schneller als bei Verringerung der horizontalen Ausdehnung. Dies wird zum Teil auf den astigmatischen Bau des *Uca*-Auges zurückgeführt.

Summary

The eye stalk nystagmus of *Uca pugnax* in the rotating drum depends on the size of the moving field. Those ommatidia contribute mainly to the eye stalk nystagmus whose axis form an angle of between 20° upward and 5° downward from the horizontal. The number of nystagmi per unit time decreases faster when the size of the moving field is vertically reduced than horizontally. This can be explained partially on the basis of the astigmatism of the eye of *Uca*.

Literatur

Buddenbrock, W. v., u. H. Friedrich: Neue Beobachtungen über die kompen-satorischen Augenbewegungen und den Farbensinn der Taschenkrabben *(Carcinus maenas)*. Z. vergl. Physiol. **19**, 747—761 (1933).

Hassenstein, B.: Über die Wahrnehmung der Bewegung von Figuren und unregel-mäßigen Helligkeitsmustern. Z. vergl. Physiol. **40**, 556—592 (1958).

Hecht, S., and E. Wolf: The visual acuity of the honeybee. J. gen. Physiol. **12**, 727—760 (1929).

Kunze, P.: Untersuchung des Bewegungssehens fixiert fliegender Bienen. Z. vergl. Physiol. **44**, 656—684 (1961).

Schaller, F.: Die optomotorische Komponente bei der Flugsteuerung der Insekten. Zool. Beitr. **5**, 483—496 (1960).

Waterman, T. H.: Light sensitivity and vision. In: The physiology of crustacea. Vol. II. T. H. Waterman editor. New York and London: Academic Press 1961.

Statocysten als Drehsinnesorgane

Von Sven Dijkgraaf

Institut für Vergleichende Physiologie der Universität Utrecht (Holland)

Mit 2 Abbildungen

Den „Otocysten" der Wirbellosen wurde im vorigen Jahrhundert in der Regel eine Hörfunktion zugeschrieben. Seit den bekannten Kreidl-

schen Versuchen betrachtet man sie hingegen allgemein und mit Recht als Organe des Schweresinnes. Daß aber die heute übliche Bezeichnung „Statocyste" dem Aufgabenbereich des Organs wenigstens in manchen Fällen doch nur zum Teil gerecht wird, hat sich erst in neuerer Zeit herausgestellt. So ergab sich in Drehversuchen mit Krabben und Tintenfischen, daß auch geblendete Tiere auf Winkelbeschleunigung mit kompensatorischen Reflexbewegungen ansprechen, die nach operativer Ausschaltung beider Statocysten ausbleiben. Dasselbe gilt für die beim Abstoppen nach längerer Drehung auftretenden Nachreflexe, die stark an

a

b

Abb. 1 a u. b. Geblendete, beinlose *Maja verrucosa*, in Bleiklemme fixiert für Drehung um die Längsachse, von vorne gesehen. Haltung der Augenstiele a in Normalstellung, b nach Rechtsdrehung des Tieres um etwa 45°

ähnliche Erscheinungen bei Reizung der Bogengänge im Labyrinth der Wirbeltiere (Nachnystagmus) erinnern.

Bei der Krabbe *Maja verrucosa* konnte an Hand der kompensatorischen Augenstielreflexe (Abb. 1 und 2) durch selektive Nervendurch-

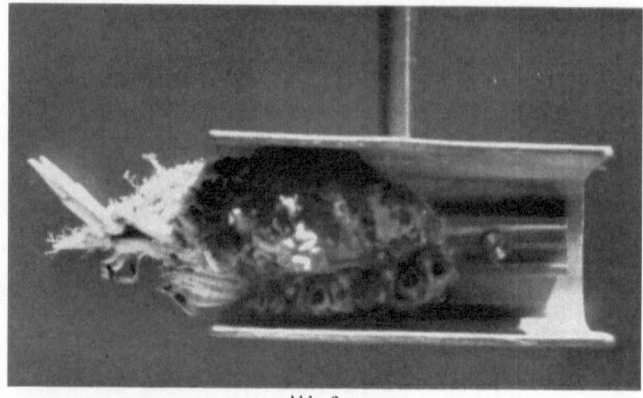

Abb. 2a

schneidung gezeigt werden, daß die langen, frei in die Statolymphe hineinragenden Fadenhaare an der Innenwand der Statocyste bei Drehung des Tieres um alle Hauptachsen ansprechen und als reine Rotationsreceptoren

fungieren, während die im Kontakt mit Statolithen stehenden Hakenhaare Positionsreceptoren darstellen. Eine ähnliche morphologischphysiologische Duplizität der Statocystenreceptoren findet sich beim Tintenfisch *Octopus vulgaris*. Als Drehsinnesorgane dienen in diesem Falle die „Cristae" an der Innenwand der Statocyste.

An Hand von zwei kurzen Filmen wird das Verhalten von *Maja* und *Octopus* bei Drehung vor und nach verschiedenen operativen Eingriffen demonstriert[1]. Nähere

---

[1] Die Filme befinden sich unter den Nummern D 842 (Rotationssinn von *Octopus vulgaris*) und D 843 (Physiologie der Statocyste von *Maja verrucosa*) in der Sammlung des Instituts für den wissenschaftlichen Film in Göttingen.

Abb. 2a u. b. Dasselbe Tier, fixiert für Drehung um die Querachse, von links gesehen. Haltung der Augenstiele a in Normalstellung, b nach Drehung des Tieres kopfaufwärts um etwa 90°

Abb. 2b

Einzelheiten über die hier nur angedeuteten Versuchsergebnisse finden sich in der nachstehend zitierten Literatur.

Literatur

COHEN, M. J.: The function of receptors in the statocyst of the lobster *Homarus americanus*. J. Physiol. (Lond.) **130**, 9—34 (1955).

— The response patterns of single receptors in the crustacean statocyst. Proc. roy. Soc. B. **152**, 30—49 (1960).

— and S. DIJKGRAAF: Mechanoreception. In: The physiology of crustacea, Vol. 2, 65—108. New York: Academic Press Inc. 1961.

DIJKGRAAF, S.: Rotationssinn nach dem Bogengangsprinzip bei Crustaceen. Experientia (Basel) **11**, 407—409 (1955).

— Structure and functions of the statocyst in crabs. Experientia (Basel) **12**, 394—396 (1956).

— The statocyst of *Octopus vulgaris* as a rotation receptor. Pubbl. Staz. Zool. Napoli **32**, 64—87 (1961).

— Nystagmus and related phenomena in *Sepia officinalis*. Experientia (Basel) **19**, 29—30 (1963).

La variabilité des comportements taxiques
Ses principales conditions écologiques et organiques

Par Jean Médioni

Laboratoire de Psycho-physiologie, Faculté des Sciences, Toulouse (France)

Sommaire

I. Introduction

On a trop souvent écrit, à la suite de Loeb, que les réactions d'orientation des Animaux à des agents physiques et chimiques sont des *mouvements forcés*, inconditionellement soumis aux variations quantitatives et qualitatives des stimuli adéquats.

Mon propos n'est pas de tenter de ranimer cette ancienne querelle, mais, en partant du fait positif que représente la variabilité des comportements taxiques, d'analyser les principaux aspects de cette variabilité et d'en examiner les conditions les plus importantes.

Je précise que j'ai emprunté à Richard (1960) le terme de «comportement taxique», en vue de désigner les déplacements des Animaux s'accompagnant de mouvements d'orientation autour de leur axe dorso-ventral, en réponse à un agent physique ou chimique extérieur, indépendamment de toute motivation endogène spécifique.

Cette délimitation exclut évidemment: les tropismes des Animaux fixés, les réflexes d'équilibration et, d'autre part, les mouvements d'approche et d'évitement à caractère instinctif (c'est-à-dire soustendus par une motivation interne). Je laisserai également de côté la ménotaxie, qui constitue le thème du rapport du Professeur LINDAUER.

Toujours en suivant RICHARD, on distingue, dans les comportements taxiques, une «cinèse fondamentale» ou composante locomotrice non-orientée et une «taxie» ou composante d'orientation. Ces deux composantes seront examinées séparément quant à leur variabilité.

II. Les différents aspects de la variabilité des comportements taxiques

1. Caractère transitoire des comportements taxiques. Sous des conditions invariables de stimulation, les réponses taxiques peuvent être évoquées à certains moments et non à d'autres. Tel est l'aspect le plus immédiat de leur variabilité.

URBAN (1932) signale l'occurrence, chez des Abeilles placées dans un faisceau lumineux horizontal, d'une alternance de trajets photopositifs et de trajets non-orientés.

De son côté, PRECHT (1942) observe, chez l'Hémiptère aquatique *Naucoris*, que la phototaxie positive se manifeste quand l'insecte manque d'air et monte respirer à la surface; il est possible de le dérouter en éclairant l'aquarium latéralement ou par en-dessous aux moments où le besoin d'oxygène se fait sentir. Et l'état photopositif cesse dès que la réserve d'air est reconstituée.

Chez les Bourdons et les Abeilles, phototaxie positive et géotaxie ascensionnelle se manifestent au moment de l'envol. Le sens de ces mouvements d'orientation s'inverse lors du retour à la ruche. Entre-temps, les animaux sont indifférents, ou bien ils s'orientent ménotactiquement vis-à-vis de la lumière et de la pesanteur (JACOBS-JESSEN, 1959). Il en va de même pour la phototaxie du Scolyte *Ips curvidens* (HIERHOLZER, 1950).

Bien entendu, la question se pose de savoir si une telle autonomie peut être observée dans l'animalité inférieure. On sait, par exemple, que les Unicellulaires vivent dans une dépendance étroite à l'égard des stimulations externes. Cependant FOLGER rapporte qu'*Amoeba proteus* ne réagit pas à un excitant lumineux pendant qu'elle s'alimente (cité par MAST, 1936). On sait aussi que les Paramécies ne réagissent pas à la lumière dans les conditions normales, bien qu'elles y soient sensibles, comme l'ont montré les expériences de conditionnement de SOEST (1937) et de WAWRZYNCZYK (cité par DEMBOWSKI, 1955). Mais on ne connaît pas d'observations établissant le caractère temporaire des comportements taxiques chez les Protistes.

2. Variations portant sur la cinèse fondamentale. Les variations de la cinèse fondamentale sont particulièrement évidentes dans le cas des champs de stimulation directionnels, sans gradient d'intensité.

C'est ainsi que les Planaires, dans un faisceau horizontal de lumière parallèle, présentent une diminution graduelle de leur vitesse de locomotion, en fonction du temps d'exposition à la lumière (VIAUD, 1950). Le phénomène inverse s'observe chez *Drosophila melanogaster*, sous des conditions des stimulation identiques (MÉDIONI, 1961).

3. Variations portant sur la composante taxique. Alors que les variations de la cinèse sont quantitatives, celles de la composante taxique sont essentiellement qualitatives: elles portent sur le détail, sur les modalités de l'orientation.

Dans une étude portant sur la phototaxie positive de l'Etoile de mer *Asterina gibbosa*, KALMUS observe que les animaux se réorientent vers une deuxième source lumineuse sans pivoter autour de leur axe de symétrie, par simple changement de «bras directeur». Des résultats comparables ont été obtenus par VIAUD (1955) pour la galvanotaxie cathodique d'*Asterias rubens*.

Selon RIMET (1960), *Daphnia pulex*, quand elle est photopositive, peut aller vers la lumière, dans un faisceau horizontal, soit en rétroprogression (réflexe dorsal), soit, plus rarement, la face ventrale du corps tournée vers la source lumineuse.

Dans l'expérience des deux lumières, URBAN (1932) observe que des Abeilles peuvent s'orienter en se tournant vers l'une des deux sources et, à d'autres moments, en se conformant à la règle de la résultante. En outre, les Abeilles éborgnées peuvent effectuer, soit des mouvements de manège vers l'oeil intact, soit des déplacements rectilignes vers la lumière. Il y aurait donc, chez l'Abeille, passage réversible de la tropotaxie à la télotaxie.

4. Les comportements taxiques polyphasiques (ou alternants). Le sens (ou «signe») du déplacement orienté d'un organisme dans un champ de stimulation directionnel peut s'inverser à plusieurs reprises dans un laps de temps relativement court (de l'ordre de 30 mn, par exemple). En pareil cas, on parlera, avec VIAUD (1951), d'un comportement *polyphasique*.

Il semble que la première observation d'un tel comportement ait été faite par GROOM et LOEB (1890), sur la phototaxie des larves de Balanes. Depuis, la phototaxie polyphasique a été signalée chez un grand nombre d'espèces très diverses: l'Hydre (WILSON, cité par MAST, 1936), des Planaires (VIAUD, 1950), *Limax* (FRANDSEN, cité par ROSE,

1929), divers Copépodes (PARKER, 1901), l'Ostracode *Cypridopsis* (TOWLE, 1900), les Daphnies (nombreux auteurs dont: VON FRISCH et KUPELWIESER, 1913; VIAUD, 1938; HARRIS et WOLFE, 1955; RIMET, 1960, etc...), la larve du Trichoptère *Molanna angustata* (GREBECKI, 1955), *Drosophila melanogaster* (MÉDIONI, 1954; DÜRRWÄCHTER, 1957), des Insectes aquatiques comme l'Hydrophile (AMOURIQ, 1961) et *Naucoris* (PRECHT, 1942).

Un aspect particulier de la phototaxie polyphasique n'a été signalé que plus rarement: c'est l'occurrence de mouvements de manège à point de départ visuel *de sens alternant*, chez des animaux éborgnés, qui se rencontre notamment chez l'Amphipode *Orchestia* (BRUNDIN, 1913), chez le Lépisme (MEYER, 1932), chez la Drosophile (MÉDIONI, 1961).

Pour des excitants autres que la lumière, les réponses polyphasiques ont été moins souvent observées; on peut citer, pourtant, la géotaxie des Phasmes (PRECHT, 1942) et celle des Souris (TABOURET-KELLER, 1959); la galvanotaxie des Astéries (SCHEMINSKY, 1931), des têtards de Grenouille et des Carassins (VIAUD et DREYFUS, 1956, 1957), la rhéotaxie alternante de *Planaria lugubris* (SCHWARTZ, 1952).

L'exposé et la discussion méthodiques des explications proposées pour rendre compte des comportements polyphasiques n'entrent pas dans le cadre de ce rapport. Mais j'aurai l'occasion d'en examiner certains aspects particuliers à propos des conditions neurologiques de la variabilité des comportements taxiques.

III. Les principales conditions écologiques et organiques de la variabilité des réactions d'orientation

A. Conditions écologiques

La plupart des facteurs externes susceptibles d'intervenir dans le biotope des Animaux et de modifier, par l'intermédiaire de leur état organique, leurs comportements taxiques ont été récemment inventoriés par PARDI et PAPI (1961) et nous n'examinerons que les plus importants.

1. L'influence de la lumière n'a guère été étudiée de façon systématique. KEENLEYSIDE et HOAR (1954) ont montré que la rhéotaxie anadromique des Saumons du Pacifique peut s'inverser sous l'effet d'un fort éclairement.

D'autre part, on sait que la géotaxie ascensionnelle de la Drosophile est activée, dans sa composante cinétique, par l'éclairement des ocelles (MÉDIONI, 1961).

2. L'influence de la température est mentionnée par beaucoup d'auteurs, soit qu'elle agisse sur la composante cinétique des réactions d'orientation (fait sur lequel il est inutile d'insister), soit qu'elle inverse le sens des déplacements, en particulier dans le cas de la phototaxie.

Ainsi, Müller (1931) signale que l'Abeille et la Fourmi *Camponotus* sont photopositives au-dessus de 16° C., photonégatives en-dessous.

Plus récemment, l'effet de la température sur le signe de la phototaxie a fait l'objet d'une étude approfondie de Perttunen (1958, 1959, 1960), à propos du Scolyte *Blastophagus piniperda*. Les imagos nés au printemps sont photopositifs entre 10 et 35° C., négatifs hors de ces limites. Chez les animaux d'automne, la marge thermique des réactions photopositives est beaucoup plus restreinte : de 20 à 30° C. seulement. Ainsi, une température de 15° C. est «positivante» au printemps et elle pousse les Scolytes à rechercher les endroits ensoleillés propices à l'envol. A l'automne, la même température est «négativante» et maintient les insectes dans leur habitat. La signification biologique de ces variations saisonnières est évidente.

3. L'influence de l'humidité aérienne est une condition écologique importante du comportement taxique des organismes terrestres. Là encore, il convient de citer les travaux récents de Perttunen (1958, 1961) sur les réactions à la lumière de *Tenebrio molitor* et de la Ligie italique.

Ces Arthropodes, normalement très lucifuges, parviennent à supporter un éclairement relativement intense après un séjour prolongé en atmosphère très sèche. Comment cette dessiccation agit-elle ? Il semble que, dans les conditions naturelles au moins, elle puisse déclencher une hygro-orthocinèse impérieuse, qui arrache littéralement les animaux à leur environnement obscur et qui les conduit «n'importe où, mais ailleurs»; en sorte qu'il n'est pas certain que l'on ait affaire, en pareil cas, à une inversion véritable du sens de la phototaxie.

4. La composition chimique du milieu aquatique peut influencer également les comportements taxiques. Il y a bien longtemps que l'on connaît l'effet de sensibilisation à la lumière exercé par le gaz carbonique dissous et ce paraît être une condition biologiquement importante des migrations verticales du zooplankton : le renforcement de la phototaxie positive permettrait aux organismes planktoniques de gagner la «zone euphotique», dans laquelle s'effectue au mieux l'assimilation chlorophyllienne des Algues microscopiques (Harris, 1953).

D'autres substances chimiques en solution dans l'eau exercent également une sensibilisation ou une désensibilisation vis-à-vis de la lumière : ainsi, *Amoeba proteus* est sensibilisée par KCl, $CaCl_2$ et $MgCl_2$, désensibilisée par HCl (Mast, 1936).

La phototaxie positive de divers Entomostracés d'eau douce est activée par des alcools et des acides dilués. Celle des nauplii de Balanes est inactivée par le permanganate de potasse, l'acétone, l'acide picrique (Rose, 1929).

Le problème du mécanisme physiologique des effets de sensibilisation sera envisagé plus loin (voir : Conditions neurologiques).

B. Conditions organiques

1. Conditions diverses. Je me bornerai à énumérer certaines des conditions organiques de la variabilité des comportements taxiques, dans certains cas parce que leur mode d'action va de soi, dans d'autres au contraire parce qu'il demeure encore totalement inexpliqué.

On sait, par exemple, que le comportement phototactique de *Drosophila melanogaster* peut être profondément modifié par les facteurs suivants :

— le stade de développement, les larves étant photonégatives, tandis que les adultes présentent un comportement polyphasique (GROSS, 1913 ; DÜRRWÄCHTER, 1957) ;

— l'âge, le sexe et l'état de nutrition des imagos (McEWEN, 1918) ;

— le rythme nycthéméral de la photo-réactivité (DÜRRWÄCHTER, 1957) ;

— l'état d'adaptation des photorécepteurs (HAMILTON, 1922 ; FINGERMAN et BROWN, 1953).

D'autres conditions encore se sont révélées importantes pour les comportements taxiques d'autres organismes : par exemple, la gravidité, qui rend les Daphnies franchement lucifuges, ou le rythme saisonnier de la photo-réactivité, déjà signalé à propos de *Blastophagus*.

2. Conditions génétiques. L'intérêt manifesté par les généticiens envers les problèmes d'hérédité du comportement est récent et les données concernant le conditionnement génétique des réactions taxiques sont encore relativement peu nombreuses.

Dès 1934, cependant, CROZIER et HOAGLAND mettaient en évidence des différences héréditaires entre trois souches de Rats, quant à certains aspects quantitatifs de leur comportement géoménotactique sur un plan incliné.

De son côté, HERTER (cité par HALL, 1951) a montré que le thermopreferendum d'une lignée de Souris albinos était significativement moins élevé que celui d'une lignée sauvage. Cette différence est conditionnée par une seule paire d'allèles, ayant entre eux des rapports de dominance, et l'effet comportemental de ces gènes paraît dépendre de variations portant sur l'épaisseur de la peau et celle du pelage.

Mais les données les plus nombreuses concernent le conditionnement gènique des réactions à la lumière et à la pesanteur chez *Drosophila melanogaster*. Pour la commodité de l'exposé de ces résultats, j'envisagerai successivement les expériences qui mettent en évidence des différences héréditaires entre deux souches et les expériences de sélection artificielle.

Différences héréditaires inter-souches (Drosophila). Du point de vue de leurs facteurs chromosomiques, il y a lieu de distinguer les variations des comportements taxiques dues à l'effet d'un (ou éventuellement de plusieurs) *gènes majeurs* de celles qui résultent de l'action de *l'ambiance génétique* (ou ensemble de facteurs n'ayant aucune influence immédiatement décelable sur le phénotype morphologique).

(a) Les variations dues à l'action de gènes majeurs ont été généralement démontrées en comparant à une souche sauvage de référence une souche mutante ayant, autant que possible, la même ambiance génétique.

Certains de ces effets dépendent directement de l'effet majeur du gène muté: par exemple, chez les Drosophiles *Bar*, dont les yeux et les ganglions optiques sont fortement réduits, le seuil absolu des réactions à la lumière blanche est cent fois plus élevé que chez les *wild* (MÉDIONI, 1961).

D'autres variations de la phototaxie, par contre, sont liées à un effet *pléiotrope* du gène majeur muté. Ainsi, le gène *yellow* conditionne une réactivité à la lumière plus élevée que son allèle sauvage et cela, dans des conditions où ni la pigmentation corporelle particulière, ni l'ambiance génétique ne peuvent être mises en cause (MÉDIONI, 1961). De même, le gène *tan*, qui commande une très discrète variante de la pigmentation du corps, a pour effet pléiotrope de rendre les mouches absolument indifférentes à l'influence taxique de la lumière (McEWEN, 1918).

(b) Le rôle de l'ambiance génétique devient manifeste, chaque fois qu'il est possible de faire disparaître une différence comportementale par des croisements d'*isogénisation*, donnant lieu à des *crossing-over* répétés entre la souche mutante et une souche sauvage de référence.

SCOTT (1943) a par exemple montré qu'une différence portant sur la photocinèse, entre une souche sauvage et une souche porteuse du gène *brown*, ne résistait pas à une série de 15 croisements d'isogénisation. En pareil cas, le gène muté fait seulement office de marqueur de souche et n'influe nullement sur le comportement.

Il est possible, également, de démontrer l'influence de l'ambiance génétique en comparant entre elles deux ou plusieurs souches, toutes de phénotype sauvage. C'est ainsi que j'ai réussi à établir, à l'aide d'un appareil à choix lumineux à deux compartiments, l'existence de différences héréditaires dans les photo-réactions de vingt souches sauvages de *D. melanogaster*, prélevées dans des populations naturelles d'origines géographiques très diverses (MÉDIONI, 1961). J'ai montré aussi que les variations observées ont un caractère systématique, et non fortuit. En effet, les données obtenues permettent de dissocier un effet de longitude et un effet de latitude: en moyenne, les animaux se montrent de plus en plus lucifuges d'Est en Ouest (depuis le Japon jusqu'à l'Amérique du Nord) et du Nord au Sud (dans l'Hémisphère boréal). La nature exacte de ces variations géographiques et leur importance écologique ne sont pas encore connues.

Expériences de sélection artificielle (Drosophila). Les expériences de sélection artificielle portant sur la phototaxie et sur la géotaxie de *Drosophila melanogaster* sont dues à JERRY HIRSCH et à ses collaborateurs.

Dans un premier travail (HIRSCH et BOUDREAU, 1958), il a été possible de sélectionner, en moins de 30 générations, à partir d'une population polymorphique au point de vue de la phototaxie, réagissant positivement à un signal lumineux d'intensité déterminée une fois sur deux en moyenne, deux lignées dont l'une réagit positivement au même signal 8 fois sur 10, l'autre 1 à 2 fois seulement.

Une autre expérience, portant sur la géotaxie de la Drosophile, est plus remarquable encore, car elle a permis la sélection artificielle de deux lignées, dont l'une est systématiquement géonégative, l'autre systématiquement géopositive (HIRSCH et ERLENMEYER-KIMLING, 1961). Dans certains cas, il semble donc démontré que les variations héréditaires des comportements taxiques puissent affecter des caractères aussi essentiels que le sens (ou «signe») de la taxie.

Enfin, les plus récents travaux de HIRSCH et de ses collaborateurs leur ont permis, en utilisant la méthode du «titrage chromosomique» de MATHER et HARRISON:

(1) de démontrer que les déterminants héréditaires du signe de la géotaxie sont des *polygènes*, c'est-à-dire des «blocs de gènes» physiologiques répartis sur l'ensemble du génome;

(2) d'évaluer la contribution respective des chromosomes I, II et III de *D. melanogaster* à la commande du sens de la géotaxie, tant dans la population fondatrice que dans les deux lignées sélectionnées;

(3) de montrer que les contributions de ces trois paires chromosomiques sont additives (ERLENMEYER-KIMLING et HIRSCH, 1961).

A ma connaissance, il n'existe actuellement aucun autre ensemble d'expériences aussi avancé dans la voie de l'analyse des conditions chromosomiques des comportements taxiques, ni d'aucun autre comportement, d'ailleurs.

On peut regretter, cependant, que les travaux de HIRSCH ne comportent aucune donnée sur les aspects de la géotaxie qui donnent prise à la sélection. Une analyse psycho-physiologique plus serrée serait nécessaire pour compléter ce bel ensemble de travaux.

3. Conditions neurologiques. Parmi les anciens auteurs qui se sont préoccupés des mécanismes physiologiques de l'orientation taxique, principalement dans le cas de la phototaxie, certains ont insisté sur l'importance des *efférences* et sur l'état de contraction tonique des effecteurs (LOEB, GARREY, cités par MAST, 1936).

D'autres (TAGLIAFERRO, 1920; MAST, 1923; LÜDTKE, 1935) ont montré que l'orientation phototactique s'opère par le jeu de réflexes cloniques coordonnés, à point de départ visuel; ils ont émis l'hypothèse d'une correspondance topique entre les aires rétiniennes excitées et les groupes musculaires mis en jeu. Ces auteurs, ainsi que CLARK (1928) et VIAUD (1938) — qui ont insisté sur l'importance de l'état d'adaptation des photorécepteurs pour les réactions des Animaux à la lumière — mettent tous l'accent sur les processus *afférents*.

Mais tous ces auteurs ont méconnu, à des degrés divers, le rôle intégrateur des centres nerveux, qui est d'une importance capitale, au moins dans le cas des Métazoaires Coelomates. Plus récemment, l'importance des conditions nerveuses centrales des comportements taxiques a été particulièrement soulignée par VON HOLST et MITTELSTAEDT (1950).

Inférences tirées de l'analyse des comportements taxiques. (a) On peut évoquer en premier lieu l'opinion (désormais classique) de VON BUDDEN-BROCK (1937), concernant l'orientation intermédiaire des Animaux entre deux sources lumineuses:

«La loi des résultantes, écrit-il, concerne un problème neurophysiologique très important, mais nullement un problème de physiologie sensorielle. Elle nous enseigne que, contrairement aux Animaux supérieurs, beaucoup d'Inférieurs, mis en présence d'une double stimulation, ne sont pas capables de déconnecter („abschalten") l'un des stimuli dans leur cerveau et de se guider exclusivement sur l'autre.»

(b) De même, on peut inférer de la clinotaxie, caractérisée par FRAENKEL et GUNN (1940), l'existence d'une *intégration spatio-temporelle* des stimulations perçues par l'animal, à la faveur des balancements de la partie antérieure de son corps, compte tenu de leur ordre de succession. Il va de soi qu'une telle intégration est de nature centrale. La clinotaxie ne se rencontre, d'ailleurs, que chez les organismes ayant atteint un certain niveau de centralisation de leur système nerveux.

(c) Il faut rappeler, dans le même ordre d'idées, une observation de CLARKE et WOLF (cités par MAST, 1936): chez *Daphnia*, il existe une corrélation étroite entre le sens de la phototaxie et celui de la galvanotaxie. Les Daphnies nagent vers la cathode d'un champ galvanique quand elles sont en disposition photopositive, vers l'anode quand elles sont en disposition photonégative.

Dans le domaine de l'orientation directe, ce fait rappelle les phénomènes de transposition ménotactique signalés par BIRUKOW (1953) et par VOWLES (1954), à propos de l'orientation transverse. Les deux ordres d'observations ne peuvent s'expliquer sans invoquer des processus d'intégration nerveuse centrale, dont la nature exacte est encore absolument inconnue.

Inférences tirées des phénomènes de sensibilisation. (a) *Sensibilisation chimique.* A propos des conditions écologiques de la variabilité des comportements taxiques, j'ai déjà mentionné l'effet de sensibilisation ou de désensibilisation exercé par certaines substances diluées dans le milieu aquatique.

On sait par exemple que le gaz carbonique augmente la tendance à la phototaxie positive chez les Daphnies. Un effet identique peut d'ailleurs être obtenu par l'action de la strychnine (CLARKE et WOLF, cités par MAST, 1936).

De son côté, PROSSER (cité par MAST, 1936) a montré que l'injection de drogues stimulantes à des Lombrics accentue leur photonégativité normale, tandis que des drogues dépressives, qui tendent à mettre hors jeu les centres nerveux supérieurs, rendent les Lombrics photopositifs. Il serait intéressant de savoir si l'influence du CO_2 peut être comparée, au point de vue physiologique, à celle des drogues neurotropes.

(b) *Sensibilisation par des agents physiques.* Il paraît bien établi, au moins pour les Arthropodes, que certains comportements taxiques sont activés par des ébranlements mécaniques. Pour la phototaxie positive des Crustacés, ce fait a été signalé dès 1900 par Miss TOWLE, à propos de l'Ostracode *Cypridopsis* et retrouvé récemment par RIMET (1960) sur *Daphnia pulex* (bien que l'auteur donne une toute autre interprétation du phénomène).

Dans le cas des Insectes, CARPENTER (1905) a observé, le premier, que la phototaxie positive et la géotaxie ascensionnelle de *Drosophila melanogaster* sont facilitées par des secousses et par des ébranlements mécaniques. J'ai précisé les modalités de cet effet, en ce qui concerne la phototaxie : les ébranlements du support augmentent considérablement l'incidence, la vitesse et la rectitude des trajets orientés vers une source lumineuse. Ils tendent à abolir la photo-indifférence et réduisent fortement l'incidence des parcours photonégatifs, ainsi que la précision de leur orientation axiale et générale (MÉDIONI, 1961).

Mais cet effet de *stimulation banale*, tendant à élever le niveau général de la réactivité, à la lumière, peut résulter de l'action d'autres excitants : par exemple, la chaleur, le froid, les courants d'air, comme EHRLICH (1943) l'a montré sur la Blatte américaine — mais également la lumière elle-même. BOZLER (1926) a montré en effet que, chez la Drosophile, les réactions phototactiques guidées par les yeux composés sont fortement déprimées par le vernissage des ocelles frontaux, qui apparaissent comme des *organes de stimulation* au sens de VON BUDDENBROCK (1937).

Chez la Drosophile, il a été possible d'évaluer l'effet de facilitation exercé par les afférences ocellaires, tant sur la phototaxie positive que sur la géotaxie ascensionnelle (MÉDIONI, 1961). Il existe, entre les excitations lumineuses ocellaires et les excitations du sens vibratoire, des rapports de vicariance et de sommation hétérogène : une Drosophile dont les ocelles sont aveuglés par vernissage est peu apte à se déplacer et à s'orienter vers une source lumineuse ou à grimper sur une paroi verticale. Le défaut des stimulations ocellaires peut être compensé, dans une très large mesure, par des stimulations vibratoires. Dans le comportement taxique des animaux aux ocelles intacts, les deux ordres d'excitants ont des effets qui s'ajoutent.

En résumé, les phénomènes de sensibilisation des comportements taxiques par des agents physiques ou chimiques semblent pouvoir être expliqués, non par un abaissement des seuils sensoriels (dont le mécanisme physiologique serait d'ailleurs assez problématique), mais par la mise en branle d'un *dispositif d'activation centrale non-spécifique*, qui élèverait le niveau général de réactivité aux stimuli les plus divers. Il vaut mieux, semble-t-il, abandonner le terme classique de «sensibilisation» et parler d'une facilitation (ou d'une activation) des comportements taxiques. — Sur le plan neurophysiologique, se pose le problème de la nature exacte du dispositif d'activation centrale, qui apparaît comme un *analogue fonctionnel* du système réticulaire ascendant du tronc cérébral des Vertébrés (MAGOUN, 1960).

Résultats des interventions portant sur le système nerveux central. J'examinerai à présent certaines expériences de destruction des centres nerveux qui ont permis de dissocier différents aspects des comportements taxiques.

(a) Les Lombrics sont normalement photonégatifs, sauf aux très faibles éclairement; or, après décérébration, ils réagissent positivement à la lumière du jour (W. N. HESS, cité par MAST, 1936).

(b) Chez *Planaria*, les photo-réactions, dans un gradient lumineux non-directionnel, sont très affaiblies par la décérébration (THIEMANN, 1957). D'autre part, en lumière horizontale dirigée, des Planaires decapitées présentent une phototaxie positive permanente. Le comportement polyphasique normal se rétablit au bout d'une vingtaine de jours, quand les yeux, ainsi que les voies et les centres optiques, sont régénérés (VIAUD et MÉDIONI, 1949).

(c) Chez les Etoiles de mer, la section de l'anneau nerveux abolit la galvanotaxie biphasique normale et ne laisse subsister que la phase cathodique initiale (SCHEMINSKY, 1931).

(d) Le Phasme présente des mouvements géotactiques alternativement ascendants et descendants. Après décérébration, seule la géotaxie ascensionnelle demeure (PRECHT, 1942).

Ces faits permettent de penser, avec VIAUD (1951), que les orientations taxiques de sens opposé ne sont pas, comme on l'a souvent écrit, des réactions identiques «au signe près». La phototaxie négative, la galvanotaxie anodique, la géotaxie descendante dépendent du fonctionnement normal des centres nerveux supérieurs. Par contre, la phototaxie positive, la galvanotaxie cathodique, la géotaxie ascensionnelle seraient des réponses plus fondamentalement inscrites dans l'organisme («tropismes vrais» au sens de VIAUD). Les expériences de VIAUD (1959) ont montré que la galvanotaxie dépend d'une polarisation permanente et invariable des tissus, condition du «signe primaire cathodique» de l'orientation. Mais nous manquons encore de données positives concernant les autres comportements taxiques.

Un autre effet des décérébrations doit être cité ici: HERTER (cité par VIAUD, 1955) a montré que, chez le Lézard, l'orientation thermotactique consiste dans la recherche active d'un optimum, au cours de laquelle l'animal se repère avec précision sur le gradient thermique. La décérébration abolit cette faculté d'orientation directe et ne laisse subsister que des réponses phobotactiques grossières, déclenchées seulement par des températures extrêmes.

Toutes ces observations montrent que les centres nerveux supérieurs exercent une action *régulatrice* sur les comportements taxiques.

Ils assurent en effet une distribution adéquate de ces comportements parmi les autres activités des Animaux, inhibent certaines réactions d'orientation dans des conditions écologiques défavorables, contrôlent l'inversion des réponses taxiques primaires et conditionnent l'exercice des modalités les plus perfectionnées de l'orientation taxique (par exemple, la télotaxie).

VON BUDDENBROCK écrivait en 1937: «Il se peut que, chez les Animaux les plus inférieurs, avant tout les Protistes, la phototaxie ne soit autre chose qu'un réflexe. Mais pour les Arthropodes, qui ont fait l'objet de la majorité des travaux dans ce domaine, on peut dire en toute sécurité que la phototaxie n'est pas un réflexe, mais une action d'ordre supérieur (eine höhere Handlung) (...) C'est bien l'animal qui vole, rampe ou nage vers la lumière, ce ne sont pas ses extrémités qui le transportent vers la lumière, en raison de la répartition du stimulus dans ses yeux et des mouvements de ses membres qui résultent de cette répartition.»

En vérité, on ne saurait mieux exprimer l'importance de l'activité intégratrice du système nerveux central pour les comportements taxiques. Tout ce que l'on peut ajouter, c'est que ces considérations s'appliquent, non seulement à la phototaxie et aux Arthropodes, mais également à d'autres réactions taxiques et — à des degrés divers — à tous les organismes Coelomates.

IV. Conclusions

Il est évident que l'examen des conditions neurologiques de la variabilité des comportements taxiques n'épuise pas la série des conditions, étrangères au stimulus adéquat, qui régissent cette variabilité.

En particulier, il y aurait lieu d'envisager certaines conditions psychologiques (ou comportementales) d'une importance extrême, dont l'examen nous entraînerait trop loin.

Je citerai seulement, pour mémoire, des facteurs tels que les relations *d'interférence et d'inhibition réciproque* entre les taxies et d'autres comportements (par exemple l'exploration, de nombreux actes instinctifs); les *effets de groupe* et les *facteurs proprement sociaux*; l'influence de la *répétition* du stimulus adéquat; le rôle de *l'expérience individuelle*, etc. . .

Pourtant, en dépit de mainte lacune, le but de cet exposé aura été atteint, s'il s'en dégage l'idée que les comportements taxiques ne sont pas des «mouvements forcés», ni des «actes simples» ou «élémentaires»; qu'au demeurant, il n'existe pas d'actes élémentaires, puisque, de tous, les taxies sont encore les moins complexes; enfin et surtout, que l'étude expérimentale de l'orientation fondamentale des Animaux n'est pas un thème de recherche périmé, mais qu'elle pose, et qu'elle posera encore longtemps des problèmes importants, difficiles et, par là-même, stimulants.

V. Summary

1. The aim of the present report is to analyse some of the main aspects of the variability of taxic behaviour patterns.

2. The concept of taxic behaviour pattern is defined as follows: locomotory movements of animals, accompanied by oriented movements around a dorso-ventral axis of the body, in response to some external agent, independently of every endogenous specific motivation. This delimitation excludes: tropisms of sessiles animals; equilibration reflexes; instinctive approach-avoidance movements.

3. The following modalities may be distinguished in the variation of taxic behaviour patterns:

— transitory nature of certain taxic responses;

— quantitative variations of the kinetic component;

— qualitative variations of the taxic component;

— alternating or polyphasic taxic responses, whose "sign" is periodically reversed.

4. If one excepts the variations of the adequate stimuli, the principal external (ecological) factors responsible for the variations of the taxic responses are:

— the light, which exerts an essentially kinetic influence;

— the heat, which plays a similar role and may furthermore reverse the sign of the taxis;

— the air humidity;

— the chemical composition of aquatic medium, which has some properties of "sensitization", especially to light-stimuli.

5. Various internal (organic) conditions are known to modify taxic responses in animals: for instance, age, sex, nutritional state, diurnal and seasonal organic rhythms ... These conditions are not discussed in

detail in this paper, but more attention is devoted: (a) to genetic; (b) to neurologic conditions.

6. In the field of the genetic conditions which influence the variability of taxic behaviour, most available data concern phototaxis and geotaxis in *Drosophila melanogaster*.

Comparison between strains enables one to distinguish between:

— firstly, the influence of *major genes*, which results either from their chief structural manifestation or from a pleiotropic effect;

— secondly, the influence of the *genetic background*, which can be possibly demonstrated by a treatment of isogenisation between a mutant strain and a wild one, or by means of comparisons between many wild strains differing largely in their geographical origin.

Artificial selection was recently used and enabled one to establish that the sign of geotaxis is largely dependent upon chromosomal factors, presumably polygenic ones.

7. Neurological conditions are reviewed, especially in order to emphasize the importance of the integrative action of the central nervous system for a satisfactory coordination of taxic behaviour patterns.

Three classes of facts are taken into consideration:

— The inferences which can be extracted from an experimental analysis of taxes, with special reference to the rule of resultant and to clinotaxis.

— The inferences which can be extracted from the phenomena of "sensitization" of taxes by chemical and physical agents. From this point of view the particular importance of the "stimulatory organs" (VON BUDDENBROCK) is stressed. In every animal which possesses a certain degree of nervous centralization, it is possible to distinguish between two types of sensory inputs, which play quite different roles in determining the variations of taxic behaviour: the specific sense organs are important for the perception of the adequate stimuli, whereas other, non-specific sense organs increase by their afferences the general reactivity of the living system.

— Decerebration experiments seem to show, generally speaking, that in polyphasic taxes only one phase (one direction) of the oriented response is suppressed by decerebration, whereas the other is not. Therefore it seems unlikely that positive and negative orientations in response to the same agent (light, gravity, electric current) are identical in their nature.

8. It is finally stressed that taxis behaviour patterns are by no means "elementary reactions", nor do they consist in "forced movements", because of the large number of external and internal conditions which may make them vary.

Bibliographie

AMOURIQ, L.: Action d'un faisceau divergent de lumière blanche sur des Hydrophiles mâles. Bull. Soc. Ent. Fr. **66**, 9 (1961).

BIRUKOW, G.: Photogeomenotaktische Transpositionen bei *Geotrupes silvaticus*. Rev. Suisse Zool. **60**, 535 (1953).

BOZLER, E.: Experimentelle Untersuchungen über die Funktion des Stirnauges der Insekten. Z. vergl. Physiol. **3**, 145 (1926).

BRUNDIN, T. M.: Light reactions of terrestrial Amphipods. J. anim. Behavior **3**, 334 (1913).

BUDDENBROCK, W. VON: Grundriß der vergleichenden Physiologie. Tome 1, 2 ème éd., 1937.

CARPENTER, F. W.: The reactions of the pomace-fly *(Drosophila ampelophila* LOEW*)* to light, gravity and mechanical stimulation. Amer. Natur. **39**, 157 (1905).

CLARK, L. B.: Adaptation versus experience as an explanation of modification in certain types of behavior (circus movements in *Notonecta*). J. exp. Zool.**51**,37 (1928).

CROZIER, W. J., and H. HOAGLAND: The study of living organisms. In: Handbook of General Experimental Psychology, ed. by CARL MURCHISON. Worcester 1934.

DEMBOWSKI, J.: Tierpsychologie. Trad. allemande. Berlin 1955.

DÜRRWÄCHTER, G.: Untersuchungen über Phototaxis und Geotaxis einiger *Droso phila*-Mutanten nach Aufzucht in verschiedenen Licht-Bedingungen. Z. Tierpsychol. **14**, 1 (1957).

EHRLICH, H.: Verhaltensstudien an der Schabe *Periplaneta americana*. Z. Tierpsychol. **5**, 497 (1943).

ERLENMEYER-KIMLING, L., and J. HIRSCH: Measurement of the relations between chromosomes and behavior. Science **134**, 1068 (1961).

FINGERMAN, M., and F. A. BROWN JR.: Color discrimination and physiological duplicity of *Drosophila*-vision. Physiol. Zool. **26**, 56 (1953).

FRAENKEL, G. S., and D. L. GUNN: The orientation of animals. Kineses, taxes and compass reactions. Oxford 1940.

FRISCH, K. VON, u. H. KUPELWIESER: Über den Einfluß der Lichtfarbe auf die phototaktischen Reaktionen niederer Krebse. Biol. Zbl. **33**, 517 (1913).

GREBECKI, A.: Sur la réponse de la larve d'un Trichoptère *Molanna angustata* CURTIS à l'action de la lumière. (En polonais, résumé français). Folia biol. **3**, 95 (1955).

GROOM, T. T., u. J. LOEB: Der Heliotropismus der Nauplien von *Balanus perforatus* und die periodischen Tiefenwanderungen pelagischer Tiere. Biol. Zbl. **10**, 160 (1890).

GROSS, A. O.: Reactions of Arthropods to monochromatic lights of equal intensities. J. exp. Zool. **14**, 467 (1913).

HALL, C. S.: The genetics of behavior. In: Handbook of Experimental Psychology, ed. by S. S. STEVENS. New York 1951.

HAMILTON, F. W.: A direct method of testing color vision in lower animals. Proc. nat. Acad. Sci. (Wash.) **8**, 350 (1922).

HARRIS, J. E.: Factors involved in the vertical migration of plankton. Quart. J. microsc. Sci. **94**, 537 (1953).

— and U. WOLFE: A laboratory study of vertical migration. Proc. roy. Soc. B. **144**, 329 (1955).

HIERHOLZER, O.: Ein Beitrag zur Orientierung von *Ips curvidens* GERM. Z. vergl. Physiol. **7**, 588 (1950).

HIRSCH, J., and J. C. BOUDREAU: Studies in experimental behavior genetics: I. The heritability of phototaxis in a population of *Drosophila melanogaster*. J. comp. physiol. Psychol. **51**, 647 (1958).

HIRSCH, J., and L. ERLENMEYER-KIMLING: Sign of taxis as a property of the genotype. Science 134, 835 (1961).

HOLST, E. VON, u. H. MITTELSTAEDT: Das Reafferenzprinzip. Naturwissenschaften 37, 464 (1950).

JACOBS-JESSEN, U. F.: Zur Orientierung der Hummeln und einiger anderer Hymenopteren. Z. vergl. Physiol. 41, 597 (1959).

KEENLEYSIDE, M. H. A., and W. S. HOAR: Effects of temperature on the responses of young salmons to water-currents. Behaviour 7, 77 (1954).

LÜDTKE, H.: Die Funktion waagerecht liegender Augenteile des Rückenschwimmers und ihr ganzheitliches Verhalten nach Teillackierung. Z. vergl. Physiol. 22, 67 (1935).

MAGOUN, H. W.: Le cerveau éveillé. Trad. française. Paris 1960.

MAST, S. O.: Photic orientation in insects, with special reference to the drone-fly Eristalis tenax and the robber-fly Erax rufibarbis. J. exp. Zool. 38, 109 (1923).

— Motor responses to light in the invertebrate animals. In: Biological effects of radiation, p. 573. New York 1936.

MC EWEN, R. S.: The reactions to light and to gravity in Drosophila and its mutants. J. exp. Zool. 25, 49 (1918).

MÉDIONI, J.: Analyse expérimentale du phototropisme en lumière blanche de Drosophila melanogaster MEIG. (race sauvage). C. R. Soc. Biol. (Paris) 148, 2071 (1954).

— Contribution à l'étude psycho-physiologique et génétique du phototropisme d'un Insecte: Drosophila melanogaster MEIGEN. Thèse Fac. Sci. Strasbourg 1961.

MEYER, A. E.: Über Helligkeitsreaktionen von Lepisma saccharina L. Z. wiss. Biol. A, 142, 254 (1932).

MÜLLER, E.: Experimentelle Untersuchungen an Bienen und Ameisen, über die Funktionsweise der Stirnocellen. Z. vergl. Physiol. 14, 348 (1931).

PARDI, L., and F. PAPI: Kinetic and tactic responses. In: The Physiology of Crustacea, ed. by T. H. WATERMAN, t. 2, p. 365 (1961).

PARKER, G. H.: The reactions of Copepods to various stimuli and the bearing of these on daily depth-migrations. Bull. U. S. Fish. Comm. 1901, 103 (1901).

PERTTUNEN, V.: The reversal of positive phototaxis by low temperatures in Blastophagus piniperda L. (Col., Scolytidae). Ann. Ent. Fennici 24, 12 (1958).

— Reversal of negative phototaxis by dessiccation in Tenebrio molitor L. (Col., Tenebrionidae). Ann. Ent. Fennici 24, 69 (1958).

— Effect of temperature on the light reactions of Blastophagus piniperda L. (Col., Scolytidae). Ann. Ent. Fennici 25, 65 (1959).

— Seasonal variations in the light reactions of Blastophagus piniperda L. (Col., Scolytidae) at different temperatures. Ann. Ent. Fennici 26, 86 (1960).

— Réactions de Ligia italica F. à la lumière et à l'humidité de l'air. Vie et milieu 12, 219 (1961).

PRECHT, H.: Das Taxis-Problem in der Zoologie. Z. wiss. Zool. 156, 1 (1942).

RICHARD, G.: Les comportements élémentaires: tropisme et réflexes. In: L'Encyclopédie Française 4, 483 (1960).

RIMET, M.: Contribution à l'étude du phototropisme de Daphnia pulex DE GEER. Arch. Zool. Expér. et Gén. 99, 44 (1960).

— La phototaxie polyphasique de Daphnia pulex DE GEER. J. de Physiol. 52, 769 (1960).

ROSE, M.: La question des tropismes. Paris 1929.

SCHEMINSKY, F.: Zur Analyse der zweiphasischen Galvanotaxis der Echinodermen. Pflügers Arch. ges. Physiol. 226, 366 (1931).

SCHWARTZ, E.: Le rhéotropisme de Planaria lugubris O. SCH. Etude quantitative. C. R. Soc. Biol. (Paris) 146, 768 (1952).

SCOTT, J. P.: Effects of single genes on the behavior of *Drosophila*. Amer. Nat. **77**, 184 (1943).

SOEST, H.: Dressurversuche mit Ciliaten und rhabdocölen Turbellarien. Z. vergl. Physiol. **24**, 720—748 (1937).

TABOURET-KELLER, A.: Analyse expérimentale du géotropisme de la Souris. Publ. de l'U. I. S. B., Section de psychol. exp. et comport. animal, **1961**, 33.

TAGLIAFERRO, H.: Reactions to light in *Planaria maculata*, with special reference to the function and structure of the eyes. J. exp. Zool. **31**, 59 (1920).

THIEMANN, W.: Die Phototaxis als Präferendumeinstellung. Versuche an *Planaria gonocephala* DUGÈS. Zool. Jb., Physiol. **67**, 177 (1957).

TOWLE, E.: A study in the heliotropism of *Cypridopsis*. Amer. J. Physiol. **3**, 345 (1900).

URBAN, F.: Der Lauf der entflügelten Honigbiene *(Apis mellifica)* zum Licht und der Einfluß von Eingriffen an Receptoren, Centralnervensystem und Effectoren. Z. wiss. Biol. **140**, 291 (1932).

VIAUD, G.: Recherches expérimentales sur le phototropisme des Daphnies. Thèse Fac. Lettres. Strasbourg 1938.

— Recherches expérimentales sur le phototropisme des Planaires. Behaviour **2**, 163 (1950).

— Les tropismes. Paris 1951.

— Le thermopreferendum et les thermotropismes. Arch. Sci. Physiol. **9**, p. C-35 (1955).

— Le galvanotropisme animal sous son nouvel aspect. Strasbourg-Médical, **1959**, n° 10, 605.

— et E. DREYFUS: Le galvanotropisme polyphasique des têtards de *Rana temporaria*. C. R. Soc. Biol. (Paris) **150**, 1255 (1956).

— — Le galvanotropisme du Poisson *Carassius vulgaris* NILSSON. C. R. Soc. Biol. (Paris) **151**, 1590 (1957).

— et J. MÉDIONI: Phototropisme et régénération chez *Planaria lugubris* O. SCHM. C. R. Soc. Biol. (Paris) **143**, 1221 (1949).

VOWLES, D. M.: The orientation of ants: I. The substitution of stimuli. J. exp. Biol. **31**, 341 (1954).

Gibt es eine echte skototaktische Orientierung?

(Zusammenfassung)

Von Štefan Sušec-Michieli

Inštitut za biologijo univerze v Ljubljani (Jugoslawia)

Mit dem Wort „Skototaxis" bezeichnet man nach ALVERDES (1930) und DIETRICH (1931) eine telotaktische Orientierung gegen die scharf umgrenzten dunklen Flächen, die sich innerhalb eines beleuchteten Feldes befinden. In einer diffus beleuchteten weißen Arena mit schwarzen Schirmen an den Wänden kriechen viele Tiere den Schirmen zu. Das betrachteten die erwähnten Autoren als Beweis für das Bestehen einer besonderen Dunkelheitsreaktion. Dieses Reagieren wurde ursprünglich nur einigen photonegativen Crustaceen — Isopoda (DIETRICH, 1931), Decapoda (ALVERDES, 1930) —, Myriapoden (GÖRNER, 1959; KLEIN, 1934) und Insekten (KLEIN, 1934; MEYER, 1932) zugeschrieben, scheint aber im Tierreich sehr weit verbreitet zu sein. Skototaxis hätte zwar mit der negativen Phototaxis eine allgemeine (natürlich negative) Photodisposition gemeinsam, sollte sich aber von ihr dadurch unterscheiden, daß sie eine positive Reaktion sei. Diese, jedenfalls nicht besonders beweiskräftige Trennung der beiden Taxien hat ALVERDES beim Begründen der neuen Orientierung in folgende Worte gekleidet: „Bei negativer Phototaxis ist ein Schwarz zugunsten der Lichtwirkung, bei positiver Skototaxis die Lichtwirkung zugunsten eines Schwarz intrazentral ausgeschaltet." Einfacher ausgedrückt: Skototaxis sollte ein aktives Suchen der Dunkelheit bedeuten. Wie man sieht, liegt in der Alverdesschen Definition eine sehr starke Betonung auf dem inneren, psychologischen Motiv des Reagierens, während der Reiz allein eine untergeordnete Rolle spielt. Vielleicht müßte man eben hier den Hauptgrund suchen, daß einige Taxienforscher Skototaxis als eine eigene Taxienart bereits formell abgelehnt haben. Es soll nur erwähnt sein, daß sich auch O. KOEHLER (1950) sehr energisch gegen ein solches Aufstellen der Taxien ausgesprochen hat. Der Ausdruck Skototaxis hat sich trotz der erwähnten Einwände in der Fachliteratur eingebürgert und ist immer noch in den meisten zoophysiologischen Lehrbüchern zu finden (VON BUDDENBROCK, 1952; ROEDER, 1953; WIGGLESWORTH, 1955).

6*

Zu einer systematischen Analyse der skototaktischen Orientierung haben uns einige zufällige Beobachtungen angeregt (Michieli, 1957). Abgesehen von den bereits erwähnten Bedenken gegen diese Orientierungsart, die einen mehr prinzipiellen Charakter haben, bemerkten wir zahlreiche Tatsachen, die man mit dem klassischen Konzept über das skototaktische Reagieren nicht in Einklang bringen konnte. Um die Schirmorientierung zu studieren, haben wir uns, ähnlich wie die meisten anderen Autoren, hauptsächlich der Arenaversuche bedient, es wurde aber auch die Lebensweise der Versuchstiere berücksichtigt.

Die Tiere befanden sich während der Versuche in der Mitte einer diffus beleuchteten Trommel und strebten den Wänden zu. Die innere Seite und den Boden der Trommel bedeckten wir mit dem normierten Papier verschiedener Helligkeit; an den Wänden der so vorbereiteten Arena wurden verschiedene Schirme befestigt, deren Lage dann im Laufe der Versuche regelmäßig gewechselt wurde. Die Bewegungen der Versuchstiere wurden beobachtet und in eine Arbeitsskizze eingetragen. Auf das endgültige Ergebnis des Versuches konnte man natürlich erst nach einer Reihe der Teilversuche mit verschiedenen Individuen gleicher Art schließen. Gleichzeitig befand sich in der Arena nur je ein einziges Tier, so daß das gegenseitige Beunruhigen der Tiere im Versuche vermieden wurde. Bewegungen (Kriechspuren), die die Tiere am Anfang der Versuchsserie und im erregten Zustand machten, haben wir nicht berücksichtigt. Fast bei allen Tierarten konnte man im Verhalten einzelner Individuen große Unterschiede beobachten, und das Markieren der Tiere hat sich oft als notwendig erwiesen. Bei den empfindlicheren Arten (z. B. vielen Bodentieren), wenn eine Austrocknung vermieden werden mußte, setzen wir auf die Lauffläche eine befeuchtete Papierschicht und milderten das Erwärmen von oben mit Hilfe des Wärmefilters unter den Glühbirnen.

Noch vor Beginn der Arenaversuche wurden die Tiere bezüglich ihrer Photodisposition im Lichtfelde einer Mikroskopierlampe und in einer zur Hälfte beleuchteten Glaskammer geprüft. Dann starteten die Tiere in einer weißen Arena mit vier schwarzen, quadratischen und kreuzweise befestigten Schirmen. Damit wollte man erkennen, ob sie überhaupt eine Schwarzschirmreaktion zeigen. Im Falle des positiven Reagierens folgte der Versuch in einer zweiten, gleichgroßen Trommel, deren Wände schwarz und deren Schirme weiß waren. Bei den meisten Tieren konnte man im vollen Widerspruch mit der Grundthese der skototaktischen Orientierung auch jetzt die charakteristische, gegen die weißen Schirme gerichtete Bewegung beobachten. Es lag also auf der Hand, daß bei der „skototaktischen" Orientierung die Dunkelheit selbst nur einen untergeordneten Faktor darstellt und daß das Wesen des Reagierens auf der kontrastierenden Wirkung beruht. Um eine möglichst starke Schirmorientierung auszulösen, durften die Schirme nicht zu klein, aber auch nicht zu groß sein. Die besten Resultate ergaben uns die Flächen mit dem Zentralwinkel 13°, obwohl auf einige Arten (*Chrysopa vulgaris* L., *Cassida vibex* L.) auch viel kleinere Schirme wirkten (4,5°). In dem nun folgenden Versuch starteten die Tiere aus der Mitte einer halb weißen und halb schwarzen Arena. Die beiden kontrastierenden Flächen waren jetzt so groß, daß sie nicht mehr als Ganzes fixiert sein konnten. Unter diesen Umständen beobachtete man bei den photonegativen Tieren zwar oft eine Bevorzugung der schwarzen Arenahälfte, man sah aber auch, daß die schwarz-weiße Grenze sehr anlockend wirkt. Diese sehr charakteristische Erscheinung wurde auch von anderen Autoren bemerkt (HUNDERTMARK, 1936, 1937a, 1937b; KALMUS, 1937; TISCHLER, 1936); wir nennen sie „Kontureffekt". Endlich war noch zu prüfen, wie sich die Versuchstiere in einer optisch nicht homogenen Umgebung verhalten, wenn keine stärker kontrastierenden Flächen vorhanden sind. Es wurde deshalb eine Versuchstrommel verfertigt, an deren Wände verschieden helle, graue Streifen dicht nebeneinander geklebt wurden. Die Graustufen, mit verschieden verdünnten Tuschlösungen gemacht, wurden nach der Methode des Maxwellschen Kreisels geeicht. In der Arena wurden die Streifen so verteilt, daß die Helligkeitsdifferenz zwischen je zwei benachbarten Streifen immer gleich war. Auch in diesem Versuch orientierten sich photopositive und photonegative Tiere verschieden. Wie zu erwarten war, wurden bei den einzelnen Arten entweder dunklere oder hellere Felder der Trommelwand bevorzugt. Unter diesen Versuchen konnte man in einigen Fällen auch keine ausgesprochenere Bevorzugung der dunklen bzw. hellen Teile der Wand sehen, und die Kriechspuren verliefen mehr oder weniger gleichmäßig zu allen Graustufen. Schließlich führte man diesen Versuch auch so aus, daß man die Trommelwände mit dem weißen oder schwarzen Papier teilweise bedeckte und so die viel stärkere, jetzt durch den Kontrastfaktor unterstützte Wirkung der einzelnen Graustufen auf die Tiere mit den vorigen Ergebnissen vergleichen konnte. Es folgten einige Versuche, die wir aber nur mit sehr sehtüchtigen und gut reagierenden Arten ausführten. So wurden statt der kontrastierenden schwarzen oder weißen Schirme an den Wänden der Arena nur deren Umrisse mit einer dickeren Linie aufgezeichnet, man konnte auch die Versuchsergebnisse bei den gleichflächigen, aber verschieden gegliederten Schirmen (Kreis, Dreieck, Stern usw.) miteinander vergleichen usw. Schließlich wurde mit den Schirmreaktionen auch das Farbunterscheidungsvermögen einzelner Arten überprüft. Zu diesem Zweck mußte man die Kriechbewegungen zu den verschieden hellen, farbigen Schirmen auf einem konstanten grauen und dann kontrastfarbigen Grunde protokollieren. Die notwendigen Farbflächen wurden ähnlich wie die Graustufen mit verschieden verdünnten Farblösungen gemacht und dann spektrophotometrisch geeicht (Spectronic Colour Analyser). Zuvor wurde auch das Reagieren der Tiere auf verschiedene Helligkeitsstufen der Grauskala bestimmt. Als helligkeitsgleich betrachteten wir im Farbunterscheidungsversuch diejenigen Farbstufen, in denen die Schirmreaktionen am schwächsten waren bzw. gänzlich ausfielen. So gelang es uns, bei den farbtüchtigen Tieren auch ohne bedeutendere Helligkeitsdifferenzen zwischen den Schirmen und der Unterlage ein positives Reagieren auszulösen.

Im ganzen haben wir mehr als 50 verschiedene Arthropoden in den Arenaversuchen auf das skototaktische Reagieren untersucht; darunter befanden sich auch alle Arten, die bei der Begründung der Skototaxien untersucht wurden: *Oniscus asellus* L., *Carcinus maenas* (L.), *Iulus terrestris* L., *Lithobius forficatus* L., *Lepisma saccharina* L., *Forficula auricularia* L. Man konnte sicher nicht erwarten, daß bei einem solchen Reichtum der Arten mit sehr verschiedener Lebensweise das Verhalten in den Schirmtesten ähnlich sei. Und doch waren die Unterschiede verhältnismäßig gering. Wie ich bereits erwähnt habe, orientierten sich die meisten der untersuchten Arten zu den schwarzen Schirmen in einer weißen Arena wie auch zu den weißen Schirmen in der schwarzen Arena. Nur auf schwarze Schirme reagierten lediglich einige Myriapoden: *Iulus terrestris* L., *Glomeris conspersa* KOCH, die Raupen von *Dendrolimus pini* L. und die entflügelten Stechmücken; jedoch konnte man auch hier das Fixieren der hellen Schirme an der schwarzen Wand beobachten. Diese Tiere, die jedenfalls sehr stark negativ phototaktisch sind, kriechen nämlich oft zuerst den weißen Flächen zu; während des Näherkommens verschwindet aber die fixierte Kontur, die abstoßende Wirkung der hellen Fläche wächst, und das Tier biegt endlich aus seiner ursprünglichen Richtung ab gegen die dunkle Wand. Die Orientierung gegen die schwarz-weißen Grenzen kommt besonders häufig bei den dekapoden Krebsen und Insekten vor. Eine Schirmorientierung zu den farbigen Flächen, die mit der Unterlage helligkeitsgleich sind, gelang bei der Spinne *Xysticus cristatus* (CLERK), bei den Wanzen *Palomena prasina* (L.) und *Mesocerus marginatus* (L.), den Käfern *Coccinella septempunctata* L. und *Leptinotarsa decemlineata* L., bei den Raupen von *Vanessa io* L. und *Araschnia levana* L. und bei der Hausfliege *Musca domestica* L. Für die Arenaversuche haben wir die Tiere aus möglichst verschiedenen Biotopen ausgewählt, und es hat sich gezeigt, daß die Schirmorientierung nie nur auf der Dunkelheit der Schirme beruht; vielmehr hat die Helligkeit der fixierten Fläche allein beim Reagieren der Tiere nur eine untergeordnete Rolle. Als entscheidenden Faktor in der sog. skototaktischen Orientierung muß man nach wie vor statt der Dunkelheit die kontrastierende Silhouette betrachten. So kommen wir wieder zurück zu der Frage, ob es zulässig ist, die Schirmorientierung oder „*Skototaxis*" als eine eigene Taxienart anzusehen. Man kann nicht mehr zweifeln, daß der Name *Skototaxis* als Folge der irrigen Vorstellungen über das Wesen der Schwarzschirmorientierung entstanden ist. Das Bestehen eines Reagierens im Sinne von ALVERDES und DIETRICH — also ein aktives Suchen der Dunkelheit — kann man im Tierreich zwar nicht a priori verneinen; es ist aber mehr als fraglich, ob man immer noch zwischen der negativen *Phototaxis* und der positiven *Skototaxis* unterscheiden sollte, da diese sich eigentlich nicht exakt nachweisen läßt. Der Name *Skototaxis* könnte ja beim weiteren Gebrauch zu vollständig falschen Vorstellungen über die Schirmorientierung führen. In einer früheren Arbeit (MICHIELI, 1959) haben wir als Bezeichnung der Schirmorientierung den Ausdruck *Perigrammotaxis* vorgeschlagen. Darunter sollte man alle taktischen Fixierbewegungen verstehen, die auf der Wahrnehmung der Konturen beruhen. Zu einer solchen Taxisart könnten vielleicht auch die bisher bekannten und mit besonderen Namen bezeichneten Spezialfälle der Konturorientierung zugezählt werden. Es ist ja schon lange bekannt, daß viele Insekten beim Gang den kontrastierenden Linien auf dem Papier folgen (TISCHLER, 1936); KALMUS (1937) hat eine ähnliche Erscheinung bei den frisch geschlüpften Larven von *Dixippus morosus* BR. als *Photohorotaxis* beschrieben. Daß eine taktische Konturorientierung sogar im Fluge vorkommen kann, beobachtete SCHNEIDER (1952) beim Maikäfer *Melolontha vulgaris* F. Die Tiere orientierten sich gegen dunkle Linien am Horizont, und diese Orientierung benannte SCHNEIDER *Hypsotaxis*. Es scheint also, daß die Konturorientierung bei den Arthropoden eine sehr verbreitete Erscheinung ist.

Schließlich noch einiges über die biologische Bedeutung der analysierten Schirmreaktionen (Konturorientierung). Bei der alten Auffassung der skototaktischen Orientierung konnte man die Bewegungen zu den schwarzen Flächen verhältnismäßig einfach erklären: die photonegativen Tiere suchten in der Dunkelheit einen Zufluchtsort. Mit einer solchen Erklärung kommt man aber nicht mehr durch, wenn man auch die verbreitete Schirmorientierung der photopositiven Tiere in Betracht zieht. Naturgemäß müssen die gesehenen Konturen für sie eine sehr verschiedene Bedeutung haben. So könnten wohl die phytophagen Insektenarten Umrisse der Schirme als Ränder der Nahrungspflanzen betrachten. Eine Übereinstimmung zwischen der Schirmorientierung und der Lebensweise beobachteten wir bei den Schmetterlingsraupen. Die Raupen von *Syntomis phegea* L. sind Bodentiere, und sie zeigen praktisch keine Konturorientierung. Raupen von *Vanessa io* L. leben auf Brennesseln, kriechen oft vom Boden an den Pflanzen empor und orientieren sich auch in den Arenaversuchen gut. Am besten fielen jedoch die Schirmversuche bei *Dendrolimus pini* L. aus. Er lebt auf den Kiefern, die Raupe überwintert in der Erde und klettert im Frühling wieder in die Baumkronen. Bei *Dendrolimus pini*-Raupen beobachteten wir eine sehr ausgeprägte Schirmorientierung und sogar eine intrazentrale Interferenz der einzelnen Konturreize. Die Beziehungen zwischen der Biologie der Versuchstiere und der beobachteten Konturorientierung waren aber nicht immer so klar wie in diesem Falle. Man kann sich dem Eindruck nicht entziehen,

daß die Schirmreaktionen oft nur durch die Versuchsbedingungen ausgelöst worden sind. In einer optisch homogenen, diffus beleuchteten
Arena, wo außer den kontrastierenden Schirmen alle anderen Reize
möglichst eliminiert sind, konnte man das Fixieren der Schirme und das
Kriechen zu ihnen schon erwarten. Es ist nicht immer möglich, für das
Verhalten der Tiere in einer vollkommen unnatürlichen Umgebung auch
eine sinnvolle biologische Erklärung zu finden, und eben die vermutete
skototaktische Orientierung könnte ein gutes Beispiel sein, wie leicht
man sich hier täuschen läßt.

Literatur

ALVERDES, F.: Tierpsychologische Analyse der intrazentralen Vorgänge, welche bei
  decapoden Krebsen die lokomotorischen Reaktionen auf Helligkeit und Dunkel-
  heit bestimmen. Z. wiss. Zool. 137, 403—475 (1930).
BUDDENBROCK, W. VON: Vergleichende Physiologie. Bd. I. Basel 1952.
DIETRICH, W.: Die lokomotorischen Reaktionen der Landasseln auf Licht und
  Dunkelheit. Z. wiss. Zool. 138, 187—232 (1931).
GÖRNER, P.: Optische Orientierungsreaktionen bei Chilopoden. Z. vergl. Physiol. 42,
  1—5 (1959).
GUNN, D. F., J. S. KENNEDY and D. P. PIELOU: Classification of taxes and kineses.
  Nature (Lond.) 140, 1064 (1937).
HIERHOLZER, O.: Ein Beitrag zur Frage der Orientierung von *Ips curvidens* GERM.
  Z. Tierpsychol. 7, 589—620 (1950).
HUNDERTMARK, A.: Helligkeits- und Farbenunterscheidungsvermögen der Ei-
  raupen der Nonne *(Lymantria monacha)*. Z. vergl. Physiol. 24, 42—57 (1936).
— Das Helligkeitsunterscheidungsvermögen bei der Stabheuschrecke *(Dixippus
  morosus)*. Biol. Zbl. 57, 228—233 (1937a).
— Das Formenunterscheidungsvermögen der Eiraupe der Nonne *(Lymantria
  monacha)*. Z. vergl. Physiol. 24, 563—582 (1937b).
JACOBS-JESSEN, U. F.: Zur Orientierung der Hummeln und einiger anderer Hymen-
  opteren. Z. vergl. Physiol. 41, 597—639 (1959).
KALMUS, H.: Photohorotaxis, eine neue Reaktionsart, gefunden an den Eilarven von
  Dixippus morosus. Z. vergl. Physiol. 24, 644—655 (1937).
KENNEDY, J. S.: The visual responses of flying Mosquitoes. Proc. Zool. Soc. London
  109, 221—242 (1939).
KLEIN, K.: Über die Helligkeitsreaktionen einiger Arthropoden. Z. wiss. Zool. 145,
  1—38 (1934).
KOEHLER, O.: Die Analyse der Taxisanteile instinktartigen Verhaltens. Symp. Soc.
  exp. Biol. 4, 269—302 (1950).
MEYER, A. A.: Über Helligkeitsreaktionen von *Lepisma saccharina*. Z. wiss. Zool.
  142, 254—312 (1932).
MICHIELI, Š.: Beobachtungen bei skototaktischen Versuchen mit Landarthropoden.
  Bull. sci. 3, 70 (1957).
— Analiza skototaktičnih (perigramotaktičnih) reakcij pri artropodih. Razprave
  SAZU, Cl. IV., 5, 237—286 (1959).
RAO, T. R.: Visual responses of Mosquitoes artifically rendered flightless. J. exp.
  Biol. 24, 64—78 (1947).
ROEDER, K. D.: Insect Physiology. New York-London 1953.

SCHNEIDER, F.: Untersuchungen über die optische Orientierung der Maikäfer *(Melolontha vulgaris* F. und *M. hippocastani* F.*)*. Mitt. Schweiz. Ent. Ges. 25, 269—340 (1952).

TISCHLER, W.: Ein Beitrag zum Formensehen der Insekten. Zool. Jb., Abt. Allg. 57, 157—202 (1936).

WALLACE, C. G.: Some experiments on form perception in the nymphs of the desert locust *Schistocerca gregaria* FORSKAL. J. exp. Biol. 35, 765—775 (1958).

WEYRAUCH, W. K.: Untersuchungen und Gedanken zur Lichtorientierung von Arthropoden I. Zool. Jb., Abt. Allg. 47, 291—328 (1930).

— Untersuchungen und Gedanken zur Orientierung von Arthropoden. Zool. Anz. 113, 115—125 (1936).

WIGGLESWORTH, V. B.: The sensory physiology of the human louse *Pediculus humanus corporis* DE GEER. Parasitology 33, 67—109 (1941).

— Physiologie der Insekten. Basel-Stuttgart 1955.

Effect of Desiccation on the Light Reactions of some Terrestrial Arthropods

By Vilho Perttunen

Zoological Department, University of Helsinki (Finland)

With 6 Figures

Relatively little is yet known about the effect of desiccation on the intensity and direction of the light reactions in terrestrial arthropods. BREITENBRECHER (1918) reported that specimens of *Leptinotarsa decemlineata* taken from wet soil reacted positively to light, whereas specimens from dry soil reacted negatively. In certain terrestrial isopods, *Armadillidium cinereum* (HENKE, 1930) and *Oniscus asellus* (WALOFF, 1941; CLOUDSLEY-THOMPSON, 1952), the negative reaction to light is affected by water loss. It is interesting to note that the moisture conditions of the substrate also affect the astronomical orientation of a terrestrial amphipod, *Talitrus saltator* (PARDI and PAPI, 1953), and of a tenebrionid beetle, *Phaleria provincialis* (PARDI, 1956): In their attempts to escape, the animals accumulate in the direction of the sea if the experimental vessel is dry and in the direction of the land if some water is present on the bottom of the vessel.

The experiments reported here were carried out with a terrestrial isopod, *Ligia italica* (PERTTUNEN, 1961), the fruit fly *Drosophila*, Berlin Wild strain (PERTTUNEN and ROPONEN, unpublished work), the mealworm beetle *Tenebrio molitor* and its larva (PERTTUNEN and LAHERMAA, 1958), and four other beetles, *Calandra granaria, Calandra oryzae, Rhizopertha dominica*, and *Acanthoscelides obsoletus* (PERTTUNEN, unpublished work)[1].

The dark-light alternative chamber used in the experiments has been described by PERTTUNEN (1958). The animals, usually ten at a time, were inserted into the chamber through a hole in the lid, on the mid-line between the dark and illuminated sides, and the positions of the animals recorded at half-minute or one-minute intervals for a period of 30 minutes (*Calandra granaria, C. oryzae, Rhizopertha, Acanthoscelides*) or 15 minutes (*Ligia, Tenebrio, Drosophila*) in each experiment. Each experiment was then repeated five or ten times with different individuals. The tem-

[1] The experiments with the last four species were carried out in the "Laboratoire d'Évolution des Êtres Organisés" in Paris. As a "Boursier du Gouvernement Français", the author is deeply indebted to the Head of this laboratory, Prof. P.-P. Grassé, for all his kindness and for excellent working facilities.

peratures, air humidities, and light intensities on the illuminated side, all kept constant during the experiments, as well as the desiccation times and temperatures, can be seen from the legends to figures 1—6.

The experiments were carried out in the darkroom, the light source being about 50 cm. above the choice chamber, and neutral filters were used for smaller light intensities when needed. In some experiments a small temperature difference of around 0.5° C. was noticed between the dark and light sides of the chamber but was shown in control experiments not to affect the light reaction. All the animals, even those stored in the darkroom, were always allowed to adapt themselves to the artificial light for at least 15 minutes before the experiments were made. Desiccation was always carried so far that a considerable number of the specimens were dead, and the last experiments in each series were carried out with those still surviving. The maximum desiccation time was naturally very different in different species.

The results are seen from figs. 1—6. The intensity of the light reaction is expressed as the excess percentage on the illuminated side of the chamber, $\frac{100\,(L-D)}{N}$, where L represents the number of animals (position records) on the light-exposed side, D the number of records on the dark side, and N the total number of position records, including the small number of specimens observed on the narrow mid-line. The broken line at 0 is the zero line of no reaction, the percentages above this (+) indicating a positive reaction to light, and the percentages below it (—) a preference for the dark side of the choice chamber. In figs. 1, 3, 4 and 6 each separate open or solid circle represents a mean of all the reaction intensities obtained during the 15-minute or 30-minute experimental series after each period of desiccation. Fig. 5 shows the time course of the light reaction during the 15-minute period, in both undesiccated and desiccated specimens.

It is to be noted that in this investigation the reaction intensities obtained are always the mean reactions of the groups of specimens, and the reactions of the individuals have not been studied as such. Therefore a possible polyphasic light reaction, such as MÉDIONI (1961) has reported for the individuals of *Drosophila melanogaster*, has not been recorded in any of the species studied.

As to the effects of desiccation, the light reactions of the animals studied can be divided into four categories:

1. A photopositive mean reaction is reversed to a photonegative one. This effect is shown in *Drosophila melanogaster* (Fig. 1), whose mean reaction in undesiccated flies is strongly photopositive at the experimental temperature of 25° C, the intensity of the reaction being over 90% in both males and females. As desiccation proceeds, the intensity of the photopositive reaction gradually diminishes; after 6 hours' desiccation the sign of the reaction is still positive, but after 7 to 10 hours' desiccation, when a considerable number of individuals have already died, the reaction is reversed and the surviving specimens show a definite photo-

negative mean response. Fig. 2 shows the percentage loss of weight at progressive hours of desiccation in individuals of the same population from which the flies were taken for the light experiments. The curves in Figs. 1 and 2 nicely show the correlation between the water loss and the

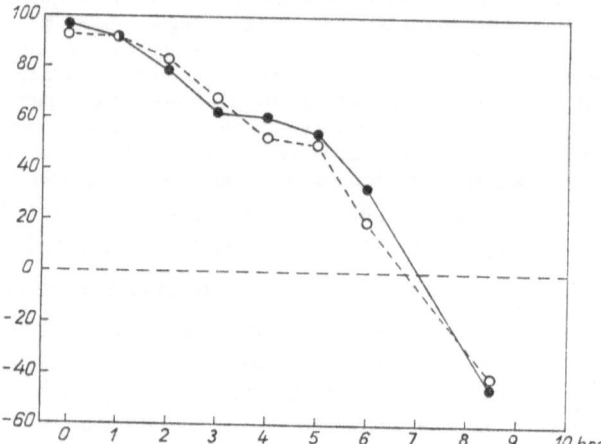

Fig. 1. Effect of desiccation on the light reaction of *Drosophila melanogaster* in the dark-light alternative chamber. Ordinate: intensity of reaction expressed as excess percentage on the illuminated side $\frac{100\,(L-D)}{N}$. Abscissa: desiccation time (at 25° C over silica gel) in hours. Open circles: males. Solid circles: females. Experimental temperature 25° C, humidity 77% R.H. Light intensity on the illuminated side: 1000 lux. Age of the flies: 7 days. Altogether, the curves are based on 1300 individuals and 39000 position records

intensity of the mean light reaction when the temperature, the light intensities, the humidity, and the age of the flies are kept constant: for

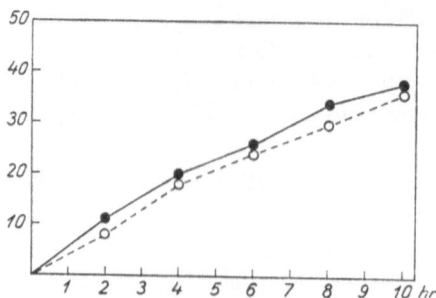

Fig. 2. Loss of weight of *Drosophila melanogaster* on desiccation at 25° C over silica gel. Ordinate: loss of weight as percentage of the initial weight. Abscissa: dessication time in hours. Age of the flies: 7 days. Open circles: males; solid circles: females

a certain percentage of loss of weight there is a certain corresponding intensity of the mean light reaction.

2. A photonegative mean reaction is reversed to a photopositive one. This effect is found in *Ligia italica* (Fig. 3), *Calandra granaria* and *C. oryzae* (Fig. 4), and the adult *Tenebrio molitor* (Fig. 5). In *Ligia*, which is very sensitive to desiccation (25% of the individuals are already dead after desiccation for 4 hours at 28° C; see PERTTUNEN, 1961), the effect of a dry atmosphere on the light reaction is already seen in a 15-minute experiment without previous desiccation: the reaction is less photonegative at 0% R.H. than at 77% or 100% R.H.

(Fig. 3). After desiccation for 3 hours the photoreaction is about indifferent, but after 4 hours the reversal has taken place and the sign of the reaction is definitely positive. The same effect is seen in the two *Calandra* species (Fig. 4) which, however, stand desiccation much better than *Ligia*, *Calandra granaria* being more resistant than *C. oryzae*. The early stages of the desiccation effect have not been studied, but after 24 hours the negative sign is reversed to a positive one in both species, and the photopositive tendency increases as desiccation proceeds. When after desiccation for 48 hours, water, but no food, was given to these beetles, the sign of the

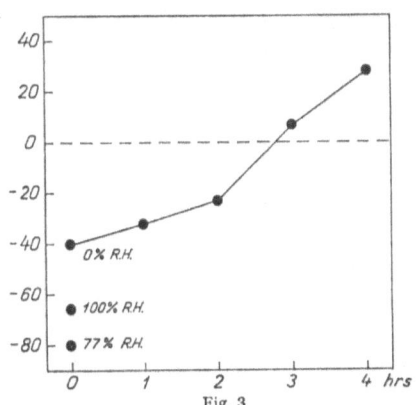

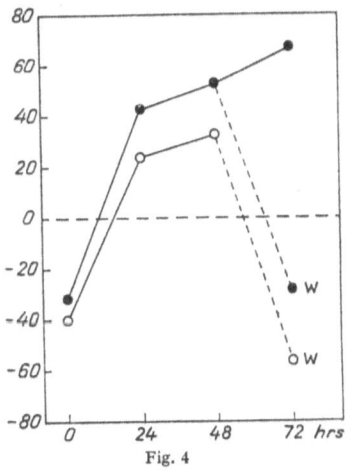

Fig. 3. Effect of desiccation on the light reaction of *Ligia italica*. Ordinate: intensity of reaction. Abscissa: desiccation time (at 28° C) in hours. Temperature 28—30° C, humidity 0% R.H., light intensity on the illuminated side 10 lux. 270 specimens, 18900 position records

Fig. 4. Effect of desiccation and subsequent drinking of water on the light reaction in *Calandra granaria* (solid circles) and *C. oryzae* (open circles). Ordinate: intensity of reaction. Abscissa: desiccation time (at 30° C over silica gel) in hours. Temperature 22—23° C, humidity 0% R.H., light intensity on the illuminated side 100 lux. *W* means experiments carried out with specimens kept for 24 hours before the tests on wet blotting paper after 48 hours' desiccation. *C. granaria*: 351 specimens, 10530 position records. *C. oryzae*: 320 specimens, 9600 position records

light reaction was again reversed and the animals became photonegative once more (Fig. 4). Since it is difficult to distinguish between the effect of desiccation and the effect of starvation, which has been reported to modify the light reactions in certain lepidopterous and hymenopterous larvae and in the beetle *Pissodes strobi* (WELLINGTON, 1948; SULLIVAN and WELLINGTON, 1953; GREEN, 1954; WELLINGTON, SULLIVAN and HENSON, 1954; SULLIVAN, 1959), these experiments with *Calandra* show clearly that water alone, given to the animals after desiccation, is enough to establish the original reaction to the light which the animals had before desiccation began. Therefore it can be concluded that in the reversal from a photonegative to a photopositive reaction in the two *Calandra* species after desiccation, water loss is probably the most important factor, if not the only effective one.

And finally, in *Tenebrio molitor* it was actually shown (Fig. 5) that mere starvation for 6 days without desiccation did not reverse the photonegative mean reaction, whereas in beetles desiccated for the same length of time reversal of the photonegative sign to a positive one took place. Further experiments are needed to show whether very prolonged starvation, even without desiccation, would cause a reversal of the reaction; in *Tenebrio*, at least, starvation for 6 days already weakens the photonegative tendency to some extent (Fig. 5).

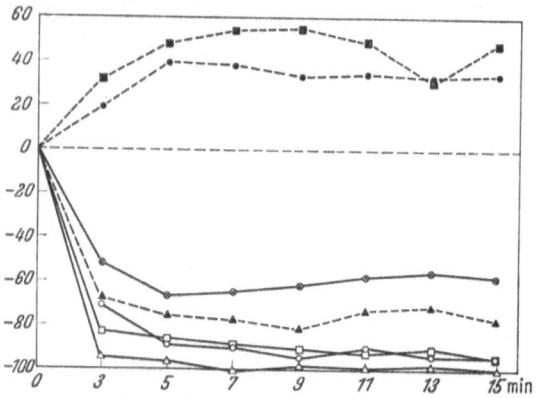

Fig. 5. Light reactions of adults and larvae of *Tenebrio molitor* in the dark-light alternative chamber. Ordinate: intensity of reaction. Abscissa: time in minutes. Open circles: normal undesiccated males. Open squares: normal undesiccated females. Black circles: males desiccated for 6 days at 25° C. Black squares: females desiccated for 6 days. Open circles with cross: males and females starved for 6 days at 25° C, but not desiccated. Open triangles: normal undesiccated larvae. Black triangles: larvae desiccated for 21 days at 25° C. Altogether, the curves are based on 700 specimens and 10500 position records. Experimental temperature: ca. 20° C, humidity: ca. 40% R.H. (from PERTTUNEN and LAHERMAA, 1958)

3. No reversal of sign takes place in the photopositive mean reaction. When *Acanthoscelides obsoletus* is desiccated (Fig. 6), the intensity of the photopositive reaction remains about the same as in the undesiccated specimens even after desiccation for 3 or 4 days, by which time a large proportion of the population is dead. And if water is given to the surviving specimens after desiccation for 3 days, the intensity of the reaction remains at about the same level.

4. No reversal takes place in the photonegative mean reaction. This is seen in *Rhizopertha dominica* (Fig. 6) in which, in spite of some change in the intensity of the reaction, the sign always remains photonegative. Giving water to the beetles after desiccation for 3 days perhaps somewhat intensifies the weakened photonegative reaction. That the intensity of the reaction, although not the sign, can be affected to some extent by desiccation in an animal belonging to this category, can also be seen from the reactions of the larva of *Tenebrio molitor*, which is extremely resistant to desiccation. In larvae desiccated for 21 days the

photonegativity was somewhat weakened compared with the intensity of the reaction in the undesiccated larvae (Fig. 5).

As regards the species of categories (1) and (2), in which actual reversal takes place, further experiments are needed to find out whether the photopositive or photonegative tendency in each individual gradually weakens and finally results in reversal of the reaction when desiccation has proceeded far enough, or whether the change in the reaction in an individual is more or less sudden, in which case the increase in the pro-

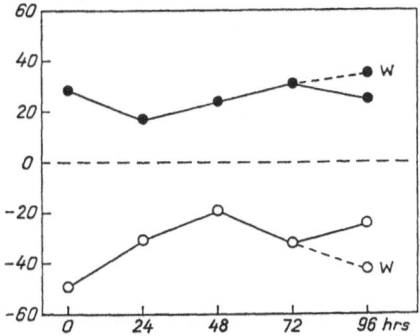

Fig. 6. Effect of desiccation and subsequent drinking of water on the light reaction of *Acanthoscelides obsoletus* (solid circles) and *Rhizopertha dominica* (open circles). Ordinate: intensity of reaction. Abscissa: desiccation time (at 30° C over silica gel) in hours. Temperature 22—23° C, humidity 0% R.H., light intensity on the illuminated side 100 lux. W = specimens given water to drink for 24 hours after 72 hours' desiccation. *Acanthoscelides:* 349 specimens, 10470 position records. *Rhizopertha:* 424 specimens, 12720 position records

portion of such individuals in an experimental group would be the decisive factor in the gradual change of the mean light reaction as desiccation proceeds.

Whatever the internal mechanism of the desiccation effect on these reactions, the reversals of the direction of the response to light must have an important ecological significance in the natural habitats of many animals. To mention one example, the terrestrial isopod *Ligia italica* normally avoids light and seeks a moist atmosphere (PERTTUNEN, 1961), and is usually found during the day-time only under the stones and in the holes and crevices of the rocks in the immediate neighbourhood of the sea. However, during the long periods of drought in the Mediterranean climate, many holes and crevices in the rocks dry up and the Ligias hiding there would die of desiccation in a few hours if they were continually trapped there by their photonegative behaviour. The reversal of the photonegative reaction to a photopositive one by desiccation must be one factor which makes it possible for these animals to leave the dried up crevices, come to the surface of the rock, and search, in part at least by means of their hygropositive reactions, for new hiding places where the moisture conditions are more favourable. In fact, specimens of *Ligia* can sometimes be seen running on the rock surface in blazing sunshine, but soon disappearing into some crevices again.

The reversal of the photonegative reaction to a photopositive one in *Tenebrio molitor* after desiccation has also been suggested to be of definite survival value to this species (PERTTUNEN and LAHERMAA, 1958). By becoming photopositive when the degree of desiccation is about to reach fatal proportions, *Tenebrio* would perhaps be guided by the light to another environment which might offer better chances for regaining the lost water balance. The reversal of the humidity reaction similarly caused by desiccation would also aid the beetle in this search. The same would apply to the two *Calandra* species.

It would be even easier to understand the significance of the reversal of the light reaction in the normally photopositive *Drosophila melanogaster*. In nature, the darker places often also mean a more humid atmosphere, and the photonegative reaction in desiccated *Drosophila* would perhaps be able to guide the flies to moister places even quicker than the hygropositive reaction itself.

It has been pointed out (FRAENKEL and GUNN, 1940) that light as such, in contrast to humidity and temperature, is in nature seldom a factor dangerous to the life of an animal. It can rather be regarded as a "token" stimulus which guides the animals to places where the other more important factors are favourable. And, as has been emphasized in the case of *Ligia italica* (PERTTUNEN, 1961), it is very important that the direction of the light reaction is not too rigid; in normal environmental conditions the photonegative reaction (or in some other species photopositive behaviour) guides the animals to places where the other factors are favourable, but in certain situations, in which the normal photoreaction would, on the contrary, guide them to or keep them in an environment which would even be fatal to them if the animals were to remain there for long, it is important that the reaction to light should be readily modifiable. One of these modifying factors causing reversals and changes in the intensity of the reaction is temperature (see, for instance, PERTTUNEN, 1958—1961), and another important factor, as has been pointed out in this report, is desiccation. Fine adjustments must exist between the nature of the light reaction, on the one hand, and the environmental temperature, humidity, light intensity, state of water balance, state of nutrition, and perhaps some other factors, on the other. The physiological mechanisms involved are for the present almost completely unknown, however.

References

BREITENBRECHER, J. K.: The relation of water to the behavior of the potato beetle in a desert. Carnegie Inst. Washington Publ. No. 263, 341—384 (1918).

CLOUDSLEY-THOMPSON, J. L.: Studies in diurnal rhythms. II. Changes in the physiological responses in the woodlouse *Oniscus asellus* to environmental stimuli. J. exp. Biol. **29**, 295—303 (1952).

FRAENKEL, G. S., and D. L. GUNN: The orientation of animals. Oxford: Clarendon Press 1940.

GREEN, G. W.: Some laboratory experiments on the light reactions of the larvae of *Neodiprion americanus banksianae* ROH. and *N. lecontei* (FITCH) (Hymenoptera: Diprionidae). Canad. Entomologist **86**, 207—222 (1954).

HENKE, K.: Die Lichtorientierung und die Bedingungen der Lichtstimmung bei der Rollassel *Armadillidium cinereum* ZENKER. Z. vergl. Physiol. **13**, 534—625 (1930).

MÉDIONI, J.: Contribution à l'étude psycho-physiologique et génétique du phototropisme d'un Insecte: *Drosophila melanogaster* MEIGEN. Thèses présentées à la Faculté des Sciences de l'Université de Strasbourg No. 203, Série E, 1—157 (1961).

PARDI, L.: Orientamento solare in un Tenebrionide alofilo: *Phaleria provincialis* FAUV. (Coleopt.). Boll. Ist. Mus. Zool. Univ. Torino **5**, 1—39 (1956).

— e F. PAPI: Ricerche sull'orientamento di *Talitrus saltator* (MONTAGU) (Crustacea-Amphipoda). I. L'orientamento durante il giorno in una popolazione del litorale tirrenico. Z. vergl. Physiol. **35**, 459—489 (1953).

PERTTUNEN, V.: The reversal of positive phototaxis by low temperatures in *Blastophagus piniperda* L. (Coleoptera, Scolytidae). Ann. Ent. Fenn. **24**, 12—18 (1958).

— Effect of temperature on the light reactions of *Blastophagus piniperda* L. (Col., Scolytidae). Ann. Ent. Fenn. **25**, 65—71 (1959).

— Seasonal variation in the light reactions of *Blastophagus piniperda* L. (Col., Scolytidae) at different temperatures. Ann. Ent. Fenn. **26**, 86—92 (1960).

— Réactions de *Ligia italica* F. à la lumière et à l'humidité de l'air. Vie et Milieu **12**, 219—259 (1961).

— and M. LAHERMAA: Reversal of negative phototaxis by desiccation in *Tenebrio molitor* L. (Col., Tenebrionidae). Ann. Ent. Fenn. **24**, 69—73 (1958).

SULLIVAN, C. R.: The effect of light and temperature on the behaviour of adults of the white pine weevil, *Pissodes strobi* PECK. Canad. Entomologist **91**, 213—232 (1959).

— and W. G. WELLINGTON: The light reactions of larvae of the tent caterpillars, *Malacosoma disstria* HBN., *M. americanum* (FAB.), and *M. pluviale* (DYAR). (Lepidoptera; Lasiocampidae). Canad. Entomologist **85**, 297—310 (1953).

WALOFF, N.: The mechanisms of humidity reactions of terrestrial Isopods. J. exp. Biol. **18**, 115—135 (1941).

WELLINGTON, W. G.: The light reactions of the spruce budworm, *Choristoneura fumiferana* CLEMENS (Lepidoptera: Tortricidae). Canad. Entomologist **80**, 56—82 (1948).

— C. R. SULLIVAN and W. R. HENSON: The light reactions of larvae of the spotless fall webworm, *Hyphantria textor* HARR. (Lepidoptera: Arctiidae). Canad. Entomologist **86**, 529—542 (1954).

The Analysis of Spatial Orientation

By Talbot H. Waterman

Department of Biology, Yale University, New Haven, Conn. (U.S.A.)

With 8 Figures

Contents

I. Introduction

The orientation of an animal in space consists of its active main-tenance of a particular body position relative to some direction-indicating feature of the environment. Hence orientation data are measurements of the angular relations between the organism's own axes of symmetry and the reference co-ordinates of the external directional stimulus. Such measurements are secured by recording simultaneously or repeatedly the spatial positions of the individuals in a population or at successive instants of time the positions of a single animal. As a result, one obtains a frequency distribution in angular intervals whose size is determined by the fineness of the organism's discrimination and the accuracy of its response to the orienting stimulus as well as the limits of the observer's method of measurement.

While a free swimming or flying bilaterally symmetrical animal has in fact three degrees of rotational freedom (Waterman, 1950) the following discussion will be restricted to one of these, rotation about the dorsoventral axis which determines azimuthal orientation. The same principles apply in the three dimensional case, and future analyses should be extended to this more general situation.

When a frequency distribution for an animal's azimuth directions has been obtained, the first question to answer is whether orientation is present. If so, its direction (or directions) and intensity or accuracy must be determined. When comparative data are available, it is important to know whether two or more sets of data differ significantly in the direction or intensity of the orientation present. Clearly, objective answers to these 5 points must be found by the application of appropriate statistical tests which also permit the estimation of the answer's reliability.

Unfortunately, powerful and well developed methods for accomplishing this are not yet available to biologists. But some partially successful and promising approaches exist and should be more widely applied and developed. The present paper reviews briefly some of these and reports on the preliminary computer application of a Fourier curve-fitting method which may prove helpful in the further development of such analysis.

Acknowledgements. Heartfelt thanks are due to Dr. C. I. BLISS and Professor M. S. BARTLETT as well as to Dr. H. L. SEAL, Dr. HENRY QUASTLER, Dr. A. T. JAMES and Mr. CHRISTOPHER BINGHAM for their stimulating discussions and help on various aspects of this analysis. In addition the essential contributions of two programmers for the computer project should be gratefully acknowledged, Mr. DAVID BLEVINS who undertook some of the preliminary work and Mrs. LOIS FRAMPTON who developed the final computer program and supervised its execution by the IBM 709 at the Yale Computer Center. Financial aid for this aspect of the work was provided in part by the Higgins Fund of Yale University.

The author's research on animal orientation has been aided by grants G-9690 and G-24055 from the U.S. National Science Foundation and his work on underlying visual mechanisms by grant B-3076 from the U.S. Public Health Service.

II. Analytical methods

A. General. A generally appropriate but not very sensitive method of testing the points raised above is to use the χ^2-test on various properties of the data and possible hypotheses (general discussion: PINCUS, 1956; some specific applications: BAINBRIDGE and WATERMAN, 1957; JANDER and WATERMAN, 1960; WATERMAN, 1960). Thus if enough counts are available one can test the null hypothesis of no-orientation against the actual distribution or the reality of the difference between two distributions with peaks in slightly different locations, etc. An application of this technique in testing goodness-of-fit is made in Section IV below.

7*

Another approach which is interesting, but not particularly powerful in this connection, is to use information theory to measure the degree of orientation by estimating how much of the maximal directional information theoretically available is being utilized by the animal (QUASTLER and BLANK, 1954).

If the number of angular groups discriminated is symbolized by n, information theory defines the condition of maximum uncertainty (i.e., no orientation) in bits as

$$H_{max} = - \log_2 n .$$

The observed uncertainty is

$$H_{obs} = - \sum_1^n \frac{O}{N} \log_2 \frac{N}{O}$$

where N is the total number of observations and O are those in one angular group. The certainty or effectiveness of the orientation defined in this way can be calculated in bits from:

$$H_{max} - H_{obs} .$$

Obviously this method of estimation does not take into account the distribution of the angular groups comprising the data but may, nevertheless provide a useful figure for comparisons.

As an illustration, an analysis of the degree of orientation by *Daphnia* swimming in a vertical beam of lineary polarized light may be given for three different percentages of polarization (WATERMAN and JANDER, unpublished). For each condition 100 measurements of azimuth orientation were made successively on 10 individual animals and the angular distribution totaled in 5° intervals over 180° (3000 observations in all). In the three distributions the mean orientation and the largest count for an angular group were at 90° to the *e*-vector or plane of polarization. A minor peak at 0° appears in the curves, which in addition are much more sharply peaked than a normal distribution (Table 1). The information measures of their degree of orientation are shown in Table 2. For comparison the corresponding standard deviations and the reciprocals of the variances (KUIPER, 1960) are shown. These latter provide two means of estimating the degree of orientation using statistical techniques for a linear independent variable. Such methods, of course, have the advantage that obtaining the mean, standard deviation and their standard errors are simple well-known procedures.

B. Linear and circular distributions. Note, however, that for orientation data the estimation of the degree of scatter about the mean as well as the location of the orientation peak by methods valid for the linearly distributed normal case raises questions of statistical propriety, whatever its convenience or even apparent empirical appropriateness may be. The orientation data are, of course, circularly distributed. Their treatment as linear normal may be reasonable if the overall distribution angle concerned is not too large (and hence the departure of the independent variable from linearity and the problem of where to break the circle therefore not too great) and if their differences from a Gaussian distribution are not too marked. However, in the present example and in most interesting cases of animal orientation neither of these restrictions is met. Certainly in the 360° case such treatment seems quite inappropriate.

A possible way of avoiding such objections is to utilize the von Mises circular normal distribution for determining the peak and scatter of the data (GUMBEL, GREENWOOD and DURAND, 1953; GUMBEL, 1954; WATSON and WILLIAMS, 1956; KUIPER, 1960; WATSON, 1961, 1962). In this case the distribution is represented by an exponential cosine function of the angle. It has the general form

$$y = C\,e^{k\cos\theta}$$

where k expresses the degree of concentration around the peak. For uniform circular distribution $k = 0$; to demonstrate the occurrence of orientation k must be greater than some minimal value. Unfortunately the mathematics involved is not easily handled so that the best method

Table 1. *Influence of % polarization on Daphnia orientation. Frequency of swimming directions observed in 5° groups* (Data from WATERMAN and JANDER, unpublished)

Angular Group	Degree of Polarization		
	27%	52%	68%
0°	45	29	32
5	20	28	18
10	10	4	9
15	7	8	8
20	7	7	10
25	5	4	14
30	8	12	7
35	8	8	7
40	10	12	12
45°	12	14	25
50	24	9	19
55	27	14	14
60	34	10	15
65	24	24	14
70	33	25	29
75	34	31	33
80	47	68	74
85	58	147	138
90°	107	179	155
95	101	80	74
100	43	58	48
105	38	35	39
110	47	32	29
115	26	16	16
120	20	9	13
125	23	12	18
130	27	13	10
135°	23	13	15
140	19	13	12
145	16	12	15
150	14	15	14
155	10	7	20
160	15	12	14
165	14	5	6
170	10	16	5
175	34	17	19

Table 2. *Estimates of degree of orientation* (calculated from data in Table 1)

Statistic	27% polarized	52% polarized	68% polarized
H_{max}.	5.170	5.170	5.170
H_{obs}	4.774	4.374	4.500
$H_{max} - H_{obs}$	0.396	0.796	0.670
Standard deviation(s) .	41.58°	36.72°	37.74°
$1/s^2$.	0.000578	0.000741	0.000704

of estimating significance is not obvious. Several solutions have been proposed, however, including some consideration of bird orientation analysis by these means (KUIPER, 1960; SCHMIDT-KOENIG, 1961; WATSON, 1962). This general approach has also been used by geologists in dealing with problems like sedimentary particle orientation (CURRAY, 1956; PINCUS, 1956) where its proper evaluation apparently is also difficult (DURAND and GREENWOOD, 1958).

C. Multimodal distributions. The statistical analysis of orientation data becomes even more troublesome when, as is always the case with basic polarized light orientation[1], more than one peak is present in 360° (e.g. JACOBS-JESSEN, 1959; JANDER and WATERMAN, 1960). Since the stimulus condition is symmetrical about the e-vector, the animal's preferred orientation directions need to be given only within 180° although there will be a corresponding set of peaks in the other half circle. Thus under various conditions with vertical beams of linearly polarized light there may be 1, 2 or 4 peaks in a semicircle. Four patterns are known: 1) 90° to the e-vector, 2) 0° and 90°, 3) 45° and 135°, and 4) 0°, 45°, 90°, 135° (JANDER and WATERMAN, 1960). In other words the animals may orient parallel or perpendicular to the polarization plane or at an oblique angle of 45° between these other two directions.

The occurrence of the intermediate orientation is reminiscent of the perpendicular phototaxis observed in some animal species half-way between positive and negative phototaxis (JANDER, 1957). Some similarity of underlying mechanism may also be seen in polarized light orientation and in the angular transposition of gravity-determined directions to phototactically determined ones in various insects where it may be an unambiguous 1:1 transposition or it may be 1:2 or 1:4 (JANDER, 1960). The physiological origins of such multimodal responses are a matter of considerable interest. In some instances a peripheral sensory origin seems to be inherent in retinal structure (STOCKHAMMER, 1959; JANDER, DAUMER and WATERMAN, 1962; MAYRAT, 1962; JANDER, 1963 unpublished) but in other cases the central nervous system or even the motor output system may be crucially involved (JANDER, DAUMER and WATERMAN, 1962). So far, however, most of the relevant evidence is indirect.

---

[1] Basic and menotactic orientation can be distinguished as follows (JANDER, 1957, 1960, 1963 unpublished; JANDER and WATERMAN, 1961). Menotaxis is defined as compass orientation in which any azimuth direction may be steered with reference to a directional cue in the environment such as the sun's position. The angle between the course and the reference may have any value up to 360° (or ± 180°) and is presumably determined by the interaction of higher coordinating centers with the basic orientation mechanism. The latter gives rise to orientation like positive or negative phototaxis, the dorsal light reflex or typical laboratory responses in a vertical beam of linearly polarized light, in which only one, or two or four symmetrical specific orientation directions are steered from a given directional reference.

A statistical method of dealing with multiple peaks has been described (BROADBENT, 1955, 1956) but has the restrictions that the independent variable must be linear, as opposed to angular or circular, and that either the location of the peaks be known or that the relation of each data point to a particular peak be certain. Although periodic modes in circular distributions have been discussed to some extent (e.g. DURAND and GREENWOOD, 1958) no specific methods for analyzing them have been developed.

At the present time, therefore, there seem to be no highly satisfactory statistical techniques for handling the large body of comparative orientation data which is becoming available. Obviously as indicated above, there are several promising possibilities, but these require considerable further development by the statisticians or in collaboration with statisticians before they can be practical for most biologists. Meanwhile one may well ask whether or not an alternative method may provide some way of approaching this problem.

III. Curve fitting

A. General. One possibility that is worth examining is curve fitting. In this well-known technique, the experimentally determined points, or a simple transform[1] thereof, are fitted with the desired degree of precision to a curve of known mathematical form. At its worst this procedure results in a mere description of the data in which no biological significance can be assigned to the components of the function fitted even though the result may provide simple bases for comparison or suggest further fruitful experiments. At its best, however, this technique yields a mathematical model [a "white box", to use WIENER's (1961) term] in which the variables and constants of the equation fitted correspond to significant parameters in the system under study (the "black box").

To solve the curve fitting problem for a complex non-linear case like the present one there are two alternatives: 1) transform the observed function to a straight line by taking logarithms, probits or making other appropriate conversions, then do a linear regression; 2) fit the non-linear function directly with an appropriate polynomial equation. In the first case the fitted function will have the form

$$y = a + bx$$

while in the second, higher order terms must be added up to the minimum number necessary. Hence this equation would have the form of some

---

[1] The important possibility of using transformations to increase the suitability of enumeration data for analysis of variance and to improve the biological relevance of the fitted curves (BLISS, 1958) is considered below (p. 115).

series such as

$$y = a + bx + cx^2 + dx^3 + \cdots + mx^n \,.$$

Naturally for the effective modelling of orientation phenomena a particular polynomial equation must be chosen not only because it fits the data curve but also because it cogently represents some important properties of the biological system. In the early stages of an analysis a completely rational choice may be difficult because the underlying mechanism itself is poorly known and inadequately conceptualized. Consequently an intuitive, exploratory beginning must be made in choosing among the many possible functions. For complex examples this testing of models requires such laborious calculations that only a few important cases are practical by hand but with high speed electronic computers a much wider range of analysis becomes accessible.

B. Fourier analysis. In the present case the circular nature of the data suggests the use of a trigonometric polynomial, and the regular angular spacing of preferred orientation directions in polarized light responses further indicates that harmonic analysis by the method of FOURIER would be a good way to begin. In addition the repeatedly demonstrated sinusoidal nature of the steering force in directional orientation suggests that sine and cosine functions are not inappropriate models in this field (VON HOLST, 1950; SCHÖNE, 1952, 1962; JANDER, 1957, 1963 unpublished; MITTELSTAEDT, 1963). Some testing of the usefulness of periodic regression (BLISS, 1958) had already been done in this connection with Dr. C. I. BLISS (cited in WATERMAN, 1960), but these hand calculations were not carried far enough at that time to provide any practical results. More recently, stimulated by discussions with Professor M. S. BARTLETT, the author has been able with the help of several collaborators to carry out periodic regression analyses on over 200 sets of orientation data both published and unpublished from our laboratory and from the published work of others.

The derivation of the present computer program was as follows. The underlying periodic regression method was originally described in detail for dealing with periodic data in climatology and biology (BLISS, 1958). This procedure had subsequently been directly programmed for computer execution by the Division of Biostatistics of the UCLA School of Medicine as BIMD Program 21. In adapting this program for present use extensive changes were made to allow for more harmonics to be tested and to make an appropriate graphic printout of the results. Most of the analysis of variance had to be dropped from the original method partly so the computer could accommodate the additional harmonics but mainly because most of our orientation experiments do not have replications in the usual sense since successive measurements are merely accumulated until enough counts are present to define the distribution. Hence our data did not have the constant number of observations per angular group assumed by BIMD 21. However, the future analysis (and additional measurements) of cases where replication will permit further subdivision of the variability and some assessment of the error in the Fourier terms would be a desirable continuation of the present work.

IV. Computer program

A. Method. The basic operation of the computer program can be outlined as follows. By Fourier analysis any single valued continuous curve can be fitted by an equation of the form

$$y = a_0 + a_1 \cos\theta + b_1 \sin\theta + a_2 \cos 2\theta +$$
$$b_2 \sin 2\theta + \cdots + a_n \cos n\theta + b_n \sin n\theta$$

where a_0 is the mean of all the observed values of y, the succeeding pairs of terms are the cosine and sine components of the fundamental period (360°) and the higher harmonics (the 2θ term represents the second harmonic with a 180° cycle and the $n\theta$ term the nth harmonic with a $360°/n$ cycle) and the other a's and b's are the corresponding regression coefficients. In other words any curve, such as our orientation distributions, may be fitted by the sum of an harmonic series of sinusoidal waves of appropriate amplitudes and phases.

The computer program therefore begins by fitting and then calculating the variability accounted for by the fundamental 360° period of orientation (first harmonic). Then it repeats the process for the second harmonic, third harmonic, and so on until the number of components tested reaches $n/2 - 1$ if n is even, or $(n - 1)/2$ if n is odd, where n is the number of angular groups. Thus for the commonest case in our recent data, where 5° angular groups are measured over 180°, $n = 36$ and 17 harmonic components were tested. Note that in a 180° distribution only the even multiplies of θ will enter the analysis; thus the components tested will be the 2θ, 4θ, 6θ etc. harmonics. In each case the mean square due to the first (and higher) Fourier terms is divided by the residual mean square (total variability about the mean in the case of the fundamental) to yield a value of F, the significance of which is checked by the computer from an F-table for the 5% probability level.

In addition, the percentage of the total variation accounted for by each harmonic component is also tabulated. The final column of the computer printout at this stage lists the cumulative percent of the estimated variance corresponding to those components found significant in the F test. The last entry in this column obviously indicates the degree to which the total variability can be accounted for by harmonic analysis.

Next a full scale graph is printed showing both the observed distribution and the fitted curve resulting from summation of all components significant at the level selected (Figs. 1—3, 6—8). This is followed by a series of separate plots for each of the elements in the fitted curve (Figs. 4, 5). The program is then completed with two further tabulations. The first of these gives the phase angle, range, semiamplitude, and the

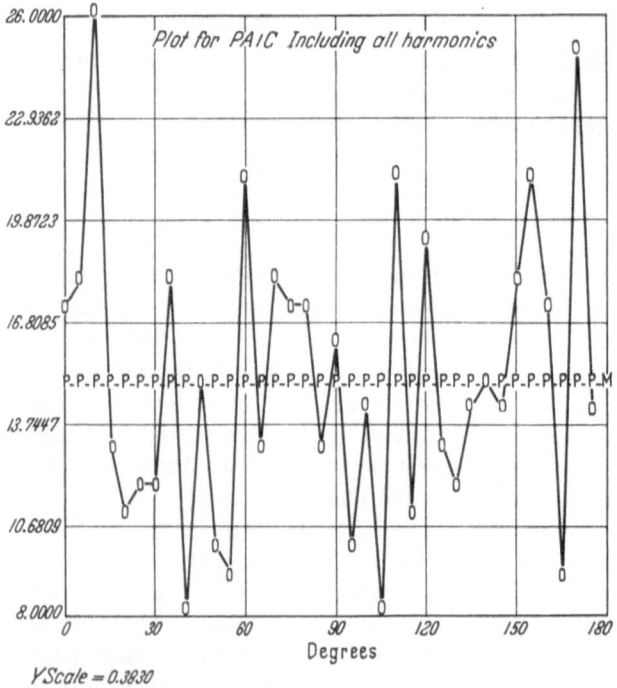

Fig. 1. Azimuth distribution of the transverse body axis of the crab *Podophthalmus* in a vertical beam of unpolarized light. The observed distribution is plotted as O's for each of the thirty-six 5° angular groups; the distribution predicted by the periodic regression method is indicated by the points P. These fall on a horizontal straight line coincident with the mean (M) since none of the first 17 harmonics tested by the computer program account for a significant fraction of the overall variability. On this basis the crab was not orienting, a conclusion supported by a χ^2 goodness-of-fit test which is satisfactory at the $P = 0.05$ level. Except for the lines connecting the points observed the figure has been traced from the graph printed out directly by the computer. As a result accuracy of point position is limited by the quantal jumps between print-out lines and columns. The ordinate represents the numbers of observations in each angular group plotted in Y-scale quanta indicated at the lower left. The awkward decimals result from programming the computer to make full use of the available Y-axis regardless of the number of observations in a particular case. The abscissa records the position of the body axis relative to a reference 0°. Total $n = 537$. Data from Waterman, 1961

position of first maximum and minimum for each harmonic component and the mean for the whole distribution. The second lists the original observations and the corresponding calculated values along with the residuals and the squared residuals in case further hand calculations are needed.

The IBM 709 requires approximately 5 minutes to set up the overall program but takes only about 30 seconds to accomplish the above analysis on each orientation distribution (for a case with 36 angular groups and 3—4 significant harmonic components) presented thereafter.

Obviously, the method of curve fitting by Fourier analysis reduces any set of orientation data to a few parameters which can be directly compared from one experimental condition to another and from one organism to another. Furthermore the resulting parameters will be seen to be those required for an effective analysis of orientation (p. 99 above). 1) The presence of orientation is indicated by the F-tests for the regression on the successive harmonic components. If one or more significant periodic functions are fitted, the answer is affirmative. 2) The number of orientation peaks is shown by the harmonic components in the composite curve and their phase relations to one another. 3) The peak locations are indicated by the phase angle and period of the main components and 4) the degree of orientation depends on the ratio of the half amplitudes $(A_i = \sqrt{a_i^2 + b_i^2})$ of the main in-phase sinusoidal components to the mean number of counts in all the angular groups (a_0).

The relative success of the curve fitting program can be roughly gauged from the total percentage of the estimated variance accounted for by significant harmonic components (this ranges from about 45% to more than 99% in various cases) and from the degree to which components which are required for reasons of curve shape rather than its peak positions, are absent. In other words the closeness of fit is a major criterion, but freedom of the theoretical curve from components which have no obvious operational relation to the orientation is an important additional test. An independent statistical method of testing the goodness-of-fit provided by the periodic regression technique is to compute χ^2 for the hypothesis that the calculated distribution and the observed distributions are not significantly different. In applying this technique the actual numbers of observations, as opposed to percentages, have of course to be used; also angular groups containing less than five observations should be lumped so as to supply this minimum number and the degrees of freedom adjusted accordingly (PINCUS, 1956).

B. Applications. As a first example, data for the Hawaiian crab *Podophthalmus vigil* may be cited for which a comparison is to be made of its angular distribution in a vertical beam of polarized light and in a similar beam unpolarized (WATERMAN, 1961). In the latter case there was considerable variance about the mean and one might wonder whether the distribution had any systematic periodicity (Fig. 1). In fact, the calculations for the first 17 harmonic components show that none of these will provide significant fit at the 5% level so that the best fitting curve predicted with this technique is the mean as shown in the figure.

In the case of the polarized beam, on the other hand, the variance about the mean is considerably larger and has one strong periodic component, the $\cos 8\theta$ element with a wave length of 45°. This accounts for 45.6% of the variance and is the only significant harmonic component in this instance (Fig. 2). Here then is a case of a 4-mode curve where a considerable fraction of the variability is accounted for by the corresponding sinusoidal curve. The phase angle for the fitted curve is 44.9° which means that the peaks are effectively at 0, 45, 90, 135° relative to the e-vector at 0°. The ratio of the semiamplitude to the mean is 0.47.

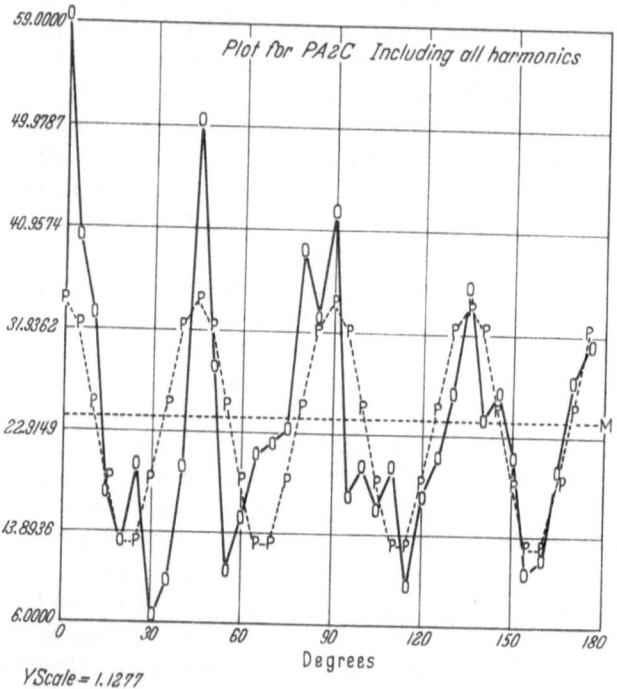

Fig. 2. Comparison of the observed (O's connected with continuous lines) and predicted (P's connected with broken lines) azimuth orientation of the transverse axis of the crab *Podophthalmus* in a vertical beam of linearly polarized light (e-vector at 0°). The predicted curve, comprising only the 45° harmonic with a phase angle of 44.9°, indicates that significant basic orientation was occurring with 4 peaks in 180°. However, the deviations in amplitude are so large that a χ^2 goodness-of-fit test shows the predicted curve in this case not to be a satisfactory fit for the data even though it accounts for 45.6% of the total estimated variance. Details of plotting are similar to those in Fig. 1. Total $n = 848$. Data from WATERMAN, 1961

Thus the computer analysis shows that the crab's distribution is random in unpolarized light but in polarized light is systematically oriented around 4 peaks spaced at 45° including one in a direction parallel to the polarization plane. However, a χ^2 goodness-of-fit test indicates that with polarized light the fourth (8 θ) harmonic does not provide a satisfactory fit since the hypothesis that the two distributions were the same must be rejected. The corresponding hypothesis nevertheless is not rejected in the unpolarized case or in the other examples to follow.

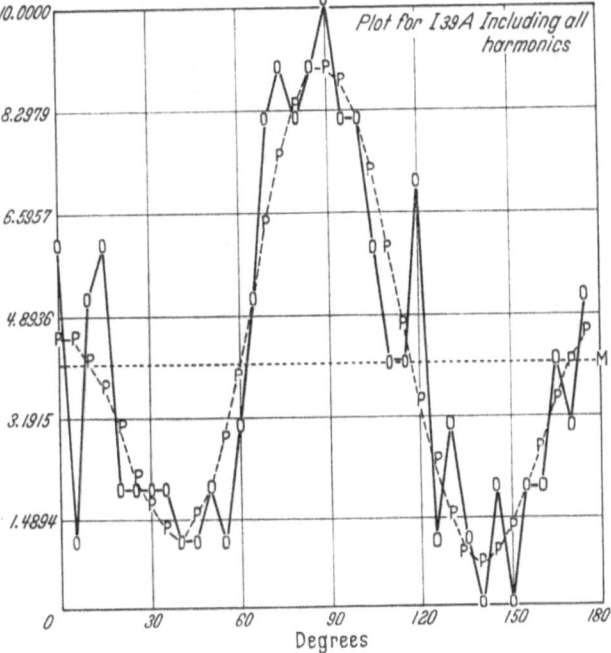

Fig. 3. Comparison of the observed and predicted orientation of the longitudinal axis of the bumble bee *Bombus* in a vertical beam of linearly polarized light (e-vector at 0°). The predicted curve which has two peaks in 180° accounts for 79.1% of the total estimated variance and comprises two components with 180°, 90° periods respectively and both with phase angles not more than 3° from 90°. These are graphed separately in Figs. 4 and 5. Details of plotting are similar to those in Figs. 1 and 2. The original measurements through 360° have been folded into 180°. Total $n=143$. Data from Jacobs-Jessen, 1959

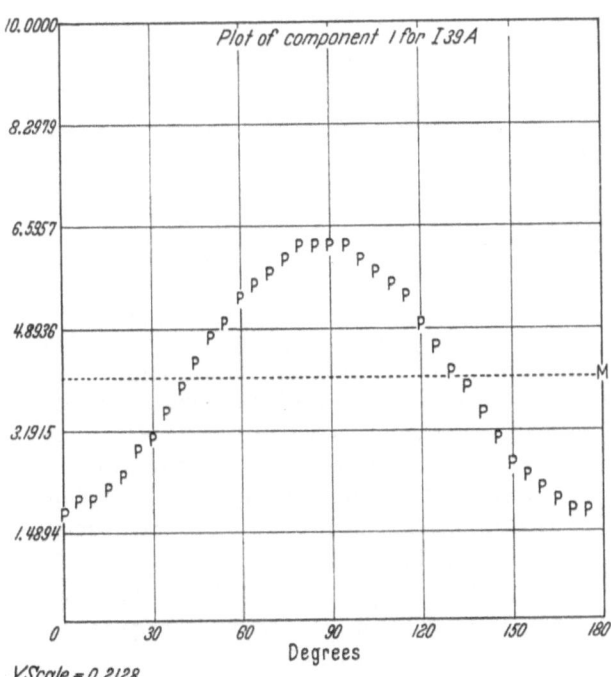

Fig. 4. First of two harmonic components of the predicted curve for *Bombus* polarized light orientation (Fig. 3). This has a period of 180°, a phase angle of 87° and a semiamplitude-to-mean ratio of 0.56. Summing with the other component (Fig. 5) their closely in-phase peaks at 90° will add together while the minimum at 0° in the 180° component will strongly diminish the effect of the 90° component peak in that direction. Details of plotting are similar to those in Figs. 1 and 2

In a second example where the azimuth distribution of the bumble bee *Bombus* (JACOBS-JESSEN, 1959) in a vertical beam of linearly polarized light is analyzed, its distribution curve (Fig. 3) shows two significant harmonic components: the major one (Fig. 5) accounting for 48.6% of the estimated variance is the 4θ (90° period) element, while the less

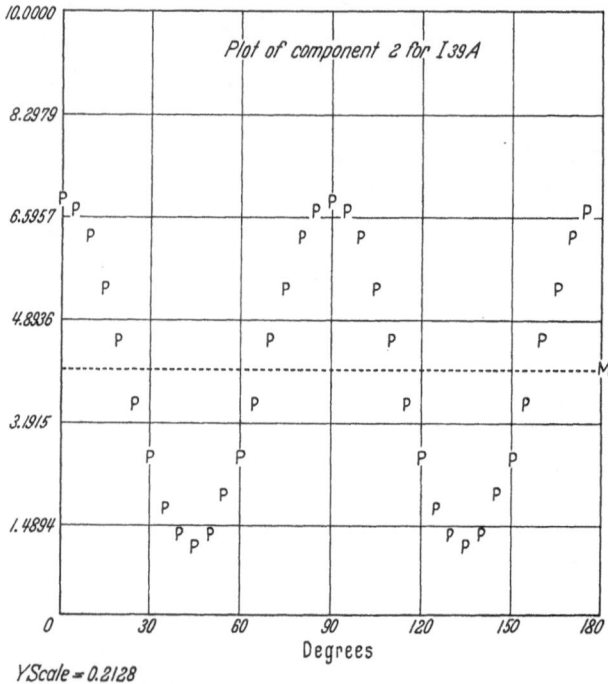

Fig. 5. Second of two harmonic components of the predicted curve for *Bombus* polarized light orientation (Fig. 3). This has a period of 90°, a phase angle of 89.9° and a semiamplitude-to-mean ratio of 0.71. Summing with the other component (Fig. 4) the phase relationships are such that the 0° peak in the 90° component will be strongly reduced and its 90° peak will be enhanced. Details of plotting are similar to those in Figs. 1 and 2

prominent component (Fig. 4), 2θ (180° period) accounts for 30.5%. These are within 3° of being exactly in-phase with a common mode at 90° to the *e*-vector. Hence the significant orientation present comprises a two-peak distribution with a major peak at 90° and a somewhat less one at 0°. The two fitted components together account for 79.1% of the total estimated variance and the ratio of semiamplitude-to-mean (A/a_0) for the major component is 0.71; this ratio summed for the two in-phase components is 1.27.

As a third example a case will be cited for *Daphnia* in which a change in stimulus conditions produced a change in the number of significant peaks induced by polarized light (JANDER and WATERMAN, 1960). With a black screen surrounding the experimental vessel the observed distribution is unimodal and sharply peaked at 90° to the polarization plane (Fig. 6). The computer analysis shows that the first 4 harmonic components are all significant with the first of these (2 θ, 90°) accounting for 64.8% of the total mean square while adding to this the effect of the other 3 elements brings the overall harmonic fit to 88.6%. The phase

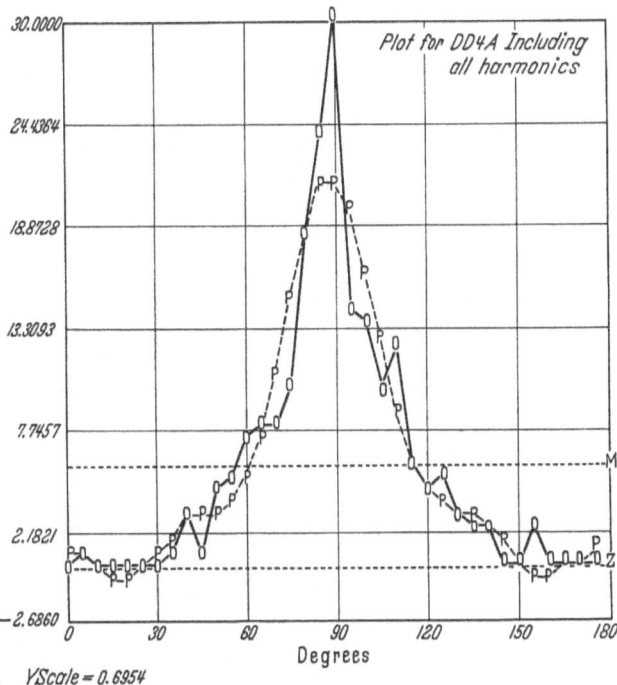

Fig. 6. Comparison of the observed and predicted directions of swimming for *Daphnia* surrounded by a black screen in a vertical beam of linearly polarized light (*e*-vector at 0°). The predicted curve is sharply peaked near 90° and comprises 4 significant harmonic components which together account for 88.6% of the total estimated variance. All 4 have a common mode close to 88° so that they build up strength in this direction and mainly cancel out in other directions. The high degree of orientation is indicated by the sum of the semi-amplitude-to-mean ratios of the in-phase components which is 2.97. Details of plotting are similar to those in Figs. 1 and 2 except that a zero line (*Z*) has been added since this is crossed by the predicted curve. Total *n* = 193. Data from JANDER and WATERMAN, 1960

relations of these 4 sinusoidal components are such that all have a common mode almost exactly at 88° which builds up the central peak and nearly cancels out the various side peaks except for slight effects at 45° and 135° and at 0°. Note that the ratio of semiamplitude-to-mean, 1.50 in this instance, indicates a stronger orientation than in any of the preceding cases. If the relative amplitudes of all four in-phase components are summed, A/a_0 is of course even larger, reaching 2.97.

With the white screen, in comparison, *Daphnia* showed a polarized light orientation (Fig. 7) like that of *Podophthalmus*. Only the 8 θ, 45°

harmonic component was significant and this accounted for 45.8% of the total variance around the mean. The phase angle was 0° so that peaks again occurred at 0, 45, 90, 135° relative to the *e*-vector while A/a_0 was 0.48. Unlike the *Podophthalmus* case, this 45° harmonic does provide for *Daphnia* a satisfactory fit on the χ^2 criterion cited.

As a final example, data on the directions of escape reactions of *Talorchestia deshayesei* under the natural sky are analyzed (PARDI, 1960).

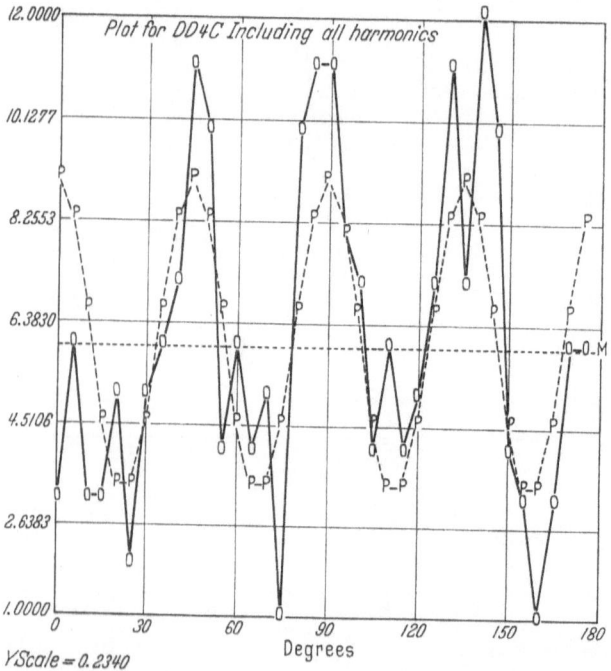

Fig. 7. Comparison of the observed and predicted swimming directions for *Daphnia* surrounded by a white screen in a vertical beam of linearly polarized light (*e*-vector at 0°). The predicted curve has 4 peaks at about 0, 45, 90, 135° and comprises only the 45° harmonic with a phase angle of 0.1° and a semiamplitude to mean ratio of 0.48. This harmonic accounts for 45.8% of the total estimated variance and provides a satisfactory fit on a χ^2 criterion. Details of plotting are similar to those in Figs. 1 and 2. Total $n = 217$. Data from JANDER and WATERMAN, 1960

Note that this is the only case cited here where 360° data are included (Fig. 8). Observe also that with 22.5° angular groups there were only 16 such groups altogether and hence only 7 harmonics could be tested. For this set of data the fitted curve, comprising the sum of the first 4 harmonic elements, accounts for 99.9% of the total estimated variance; the first component (1 θ, 360°) alone absorbed 79.9% of this total. In contrast to all of the other examples, the animal here was orienting menotactically, and the phase angle involved was 135°. The orientation was quite powerful since A/a_0 for the first component was 1.35. In addition the 2 θ, 180° harmonic also had a phase angle near 135° and thus made a significant additional contribution to the orientation intensity about this peak so that the corresponding ratio for the summed in-phase contributions was 1.98.

C. Discussion. Clearly the harmonic curve-fitting technique described above provides a useful basis for analyzing multimodal orientation data

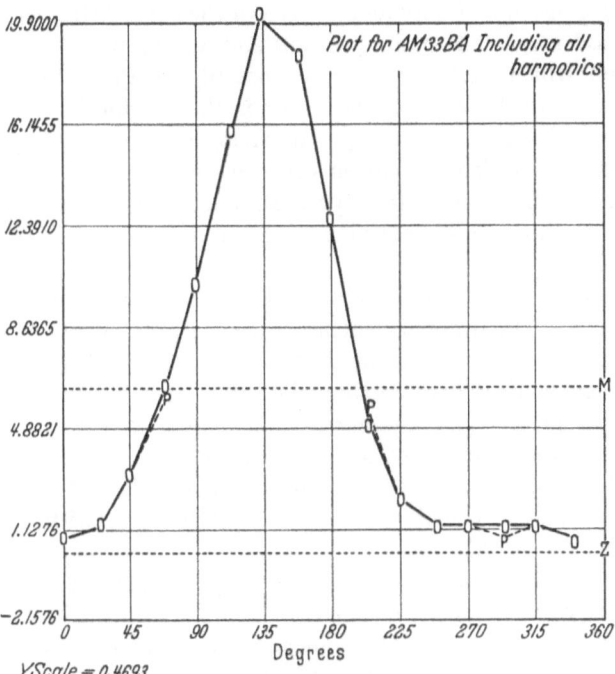

Fig. 8. Comparison of the observed and predicted orientation of fleeing responses of *Talorchestia* under the blue sky (North = 0°). The predicted curve fits the observations very closely, accounting for 99.9% of the total variability. Harmonics 1—4 contribute significantly to the fit although only 1 and 2 have in-phase maxima at 135° which is the position of the peak. The type of plot is closely similar to that of Figs. 1 and 2 but there are several differences. Unlike Figs. 1—7, the angular groups here are fewer, larger and specified over a full 360°. In addition the ordinate is plotted as percentage of the total *n* which was 945. Also a zero line is present since the first harmonic in the Fourier series extends to negative values. The predicted curve fits the observations so closely that only three of the predicted points appear. Data from PARDI, 1960

of the sort tested. It can, as shown, distinguish between random and oriented distributions, between one-, two- and four-peaked curves and can effectively give the position of the modes as well as the degree of orientation about any particular harmonic component. Both basic orientation and menotactic orientation can be equally well described. Hence the use of this technique for the further comparative study of the large amount of data already run through the program would seem thoroughly worthwhile. In addition, the design of new experiments to test certain implications of the Fourier analysis should provide one promising way of advancing our understanding of orientation.

However, there are obviously problems both of interpretation and of statistical method which still need further attention. For example, although the Fourier series equation has several properties which recommend it for the present application, the shape of observed animal distribution functions, particularly when orientation is strong, are not in general sinusoidal[1]. Consequently part of the curve fitting provided by this method is related to the position and amplitude of the modes of orientation, which are biologically meaningful. Yet the rest of the fitting will be related to the shape of the modes in the distribution. While these shapes are of interest, for example in attempting to recognize the same or different underlying mechanisms for the various peaks, one must be able to differentiate shape from position and number factors in a good analytical method. This is not always objectively possible in the present technique.

For example, it is anomalous that in those examples where orientation significance of the harmonic elements is clearest, the overall fit provided is minimal [the polarized light *Podophthalmus* data (Fig. 2) and the white screen polarized light *Daphnia* data (Fig. 7)]. Conversely where the fit is almost perfect [the *Talorchestia* menotactic orientation (Fig. 8)], two or three of the four components in the harmonic analysis are present to force the resultant curve's shape towards that of the distribution peak. No obvious biological significance can be attached a priori to these peak form influences as such.

Naturally if an equation could be discovered which fitted orientation modes well, whether they were weak or strong, and for whose variables and constants some biological significance were readily identifiable, the appropriate circular periodic manipulation of this function would be an improvement over the present method. Such a technique, however, is not presently available and may take some considerable statistical development before it can be used in biological applications.

A fundamental statistical difficulty with the present application of the periodic regression method is that it applies analysis of variance techniques appropriate for measurements to enumeration data. Hence the validity of employing the F-test to select significant harmonics may be questioned. The use of χ^2 to test goodness-of-fit in the fitted curves resulting from the Fourier analysis is free of this objection, however, and demonstrated in five out of the six examples presented that the predicted

---

[1] Note, however, that as mentioned above (p. 104), the turning tendency which regulates orientation to a directional stimulus frequently is a sinusoidal function. Clearly the relationship between such factors regulating basic orientation and the ultimate directional distribution is an important matter for future study.

distribution was a satisfactory fit at the $P = 0.05$ level. Also the least squares method may not be adequate if the distribution of the residuals is far from normal.

In further work the suitability of the present type of data for the analysis desired might be improved by transforming the counts to square roots, logarithms or some other transformation which seems appropriate (BLISS, 1958). In this way it might be possible to avoid some limitations of dealing with enumerations and at the same time to reduce the number of biologically uninteresting harmonics in the fitted curve. Consequently study of the differences between the Fourier analysis of the raw data and that of logarithms of the data should be fruitful.

Obviously the present report is not a final one either in terms of the method used or of the data which have been analyzed so far. Yet it seems desirable to stimulate a more serious attempt than is usually made to provide some quantitative comparative analyses of animal orientation. This area of biology like a number of others in the life sciences has been too hesitant in developing more objective and effectively conceptual ways of dealing with its data (WATERMAN, 1962). With the large number of interesting observations and experiments currently being made in the field of orientation, the time is ripe for their more quantitative and theoretical evaluation (MITTELSTAEDT, 1963; JANDER, 1963, in preparation).

V. Summary

1. Relatively little previous attention has been given to the problem of quantitative comparative analysis of animal orientation data despite widespread interest in the general subject.

2. Among those techniques which have been utilized to some extent, χ^2 and information theory tests are of general but not highly effective use; ordinary linear statistics are valid under restricted conditions; circular normal methods are more appropriate to such angular distributions as orientation but discrimination procedures with their use have not been standardized.

3. Analytical problems are more serious with multimodal orientation such as always occurs in basic polarized light orientation considered over 360° and may occur in other cases such as menotactic transposition.

4. The application of a computer Fourier analysis to such multimodal orientation distributions is proposed and a general description of the periodic regression technique involved is presented.

5. Examples of the kind of results obtainable with this method are given for several arthropods; calculation of the appropriate χ^2 demonstrates that in five of the six examples cited periodic regression provides satisfactory goodness-of-fit.

6. Although the procedure is not definitive, the preliminary analyses indicate that it will permit the effective description and comparison of a considerable amount of available data and should stimulate the design of new experiments to test hypotheses of underlying mechanisms as well as the search for more rigorous and powerful analytical techniques.

References

BAINBRIDGE, R., and T. H. WATERMAN: Polarized light and the orientation of two marine Crustacea. J. exp. Biol. 34, 342—364 (1957).

BLISS, C. I.: Periodic regression in biology and climatology. Conn. Agric. Exp. Station Bull. 615, 3—55 (1958).

BROADBENT, S. R.: Quantum hypotheses. Biometrika 42, 45—57 (1955).

— Examination of a quantum hypothesis based on a single set of data. Biometrika 43, 32—44 (1956).

CURRAY, J. R.: The analysis of two-dimensional orientation data. J. Geol. 64, 117—131 (1956).

DURAND, D., and J. A. GREENWOOD: Modifications of the Rayleigh test for uniformity in analysis of two-dimensional orientation data. J. Geol. 66, 229—238 (1958).

GUMBEL, E. J.: Applications of the circular normal distribution. J. Amer. Statist. Assoc. 49, 267—297 (1954).

— J. A. GREENWOOD and D. DURAND: The circular normal distribution: theory and tables. J. Amer. Statist. Assoc. 48, 131—152 (1953).

HOLST, E. VON: Die Arbeitsweise des Statolithenapparates bei Fischen. Z. vergl. Physiol. 32, 60—120 (1950).

JACOBS-JESSEN, U. F.: Zur Orientierung der Hummeln und einiger anderer Hymenopteren. Z. vergl. Physiol. 41, 597—641 (1959).

JANDER, R.: Die optische Richtungsorientierung der Roten Waldameise *(Formica rufa L.)*. Z. vergl. Physiol. 40, 162—238 (1957).

— Menotaxis und Winkeltransponieren bei Köcherfliegen (Trichoptera). Z. vergl. Physiol. 43, 680—686 (1960).

— and T. H. WATERMAN: Sensory discrimination between polarized light and light intensity patterns by arthropods. J. cell. comp. Physiol. 56, 137—160 (1960).

KUIPER, N. H.: Tests concerning random points on a circle. Konink. Ned. Akad. Wetenschap. Proc. Series A. 63, 38—47 (1960).

MAYRAT, A.: Premiers résultats d'une étude au microscope électronique des yeux des Crustacés. C. R. Acad. Sci. (Paris) 255, 766—768 (1962).

MITTELSTAEDT, H.: Bikomponenten-Theorie der Orientierung. Ergebn. Biol. 26, 253—258 (1963).

PARDI, L.: Innate components in the solar orientation of littoral amphipods. Cold Spr. Harb. Symp. quant. Biol. 25, 395—401 (1960).

PINCUS, H. J.: Some vector and arithmetic operations on two-dimensional orientation variates, with applications to geological data. J. Geol. 64, 533—557 (1956).

QUASTLER, H., and A. A. BLANK: Notes on the estimation of information measures. Control Systems Laboratory, University of Illinois, Urbana, Ill. Rep. No. R.-56, 1—36 (1954).

SCHMIDT-KOENIG, K.: Die Sonne als Kompaß im Heim-Orientierungssystem der Brieftauben. Z. Tierpsych. 18, 221—244 (1961).

SCHÖNE, H.: Zur optischen Lageorientierung („Lichtrückenorientierung") von Dekapoden. Naturwissenschaften 39, 552—553 (1952).

— Optisch gesteuerte Lageänderungen (Versuche an Dysticidenlarven zur Vertikalorientierung). Z. vergl. Physiol. 45, 590—604 (1962).

STOCKHAMMER, K.: Die Orientierung nach der Schwingungsrichtung linear polarisierten Lichtes und ihre sinnesphysiologischen Grundlagen. Ergebn. Biol. 21, 23—56 (1959).

WATERMAN, T. H.: Flight instruments in insects. Amer. Scient. 38, 222—238 (1950).

— Interaction of polarized light and turbidity in the orientation of *Daphnia* and *Mysidium*. Z. vergl. Physiol. 43, 149—172 (1960).

— Polarized light orientation by aquatic arthropods. In "Progress in Photobiology". Proc. 3rd. Internat. Cong. Photobiology. (B. C. CHRISTENSEN and B. BUCHMANN, eds.) pp. 214—216. Amsterdam: Elsevier 1961.

— Revolution for biology. Amer. Scient. 50, 548—569 (1962).

WATSON, G. S.: Goodness-of-fit tests on a circle. Biometrika 48, 109—114 (1961).

— Goodness-of-fit tests on a circle. II. Biometrika 49, 57—63 (1962).

— and E. J. WILLIAMS: On the construction of significance tests on the circle and the sphere. Biometrika 43, 344—352 (1956).

WIENER, N.: Cybernetics; or, Control and Communication in the Animal and the Machine. 2nd ed. 212 pp. Cambridge, Mass.: M. I. T. Press 1961.

The Role of Tidal Streams in the Navigation of Migrating Elvers (*Anguilla vulgaris* Turt.)

Netherlands Institute for Sea Research, Den Helder (Netherlands)

With 4 Figures

There are many species of fish which, in different periods of their lifetime, live in areas where environmental conditions are drastically different. Fishes known as anadromous and katadromous undertake migratory movements from open sea or ocean to small bodies of fresh water such as ponds, lakes and streamlets and movements in the reverse direction at some other stage of development. Besides endocrinological and osmoregulatory problems these migrations provide a rich field of study for ecologists, sense physiologists, ethologists, in a word for everyone who is interested in orientation problems.

We were concerned in Den Helder especially with the migration of young eels (elvers) towards inshore waters. Eels are known as the most classical example of katadromous fishes, which means that the (first) inward migration is undertaken by the offspring which has never met fresh water surroundings before. We may assume that, within the area of occurrence, the dispersal of elvers over the different freshwater outlets is more or less at random.

In the case of some anadromous fishes such as salmon the problem is more complicated by the fact that movements towards inland waters coincide with the homing migration. This implies that from any spot in the ocean or open sea the fishes must be capable of returning to the estuary of their home river (a problem which is studied by Hasler, Braemer and Schwassmann). Once they are back in their own river system they must be able to find their parent stream. Hasler and Wisby (1951, 1954) obtained good evidence for a conditioned odour response.

In the more simple case of the inshore migration of elvers, however, there are still gaps in our knowledge, and in order to shed some light on this problem field observations were made in the Dutch Waddensea and the North Sea near the Dutch coast during the months of February,

March and April of the years 1955, 1956, 1957, 1958 and 1959. Horizontal hauls were made with a ring trawl at the surface as well as at a depth of about 8 metres mainly in the Texelstroom, which is a large tidal gully between the more flat parts of the Waddensea. In the Texelstroom the ingoing flood stream and the outgoing ebb reach velocities of 1.50 m per second.

During these annual surveys we were always confronted with the phenomenon that the flood catches largely exceeded the ebb catches at the surface as well as at a depth of 8 metres (CREUTZBERG, 1958, 1961). During ebb it seemed that the elvers disappear from the water column and probably dwell near the bottom. The significance of this phenomenon was so evident that we believed it to be a special mechanism of inward migration. They are carried in the direction of the inland waters by the flood stream and go to the bottom during the ebb, with the result that they are not washed back towards the sea. This of course implies that elvers must be able to discriminate between certain characteristics of the ebb and flood streams.

The idea that the increasing amount of fresh water passing out during ebb would induce the elvers to keep near the bottom got some confirmation from observations made in the North Sea near Scheveningen. This area is characterised by its situation at the northern extremity of the estuarine area of the rivers Rhine and Meuse. On the flood tide brackish water is carried from that estuarine area in a north-easterly direction along the coast. Consequently in the coastal waters off Scheveningen the flood stream resembles in this respect the ebb tide in the Waddensea. On the other hand, the ebb stream is comparable to the flood stream in the Waddensea. Hauls made at the surface during two nights in that area yielded more elvers during the ebb than during the flood tide, contrary to our experience in the Waddensea, but in accordance with the expectation that a current which supplies fresh water induces the elvers to go to the bottom.

To support this view and to analyse the phenomena an experimental approach was considered. With this aim a circular stream apparatus was constructed in which it was possible to generate water currents of different velocities by means of rotating paddle boards (Fig. 1). In the first place it was assumed that decreasing salinity during ebb and increasing salinity during flood tide might stimulate the elvers to show a different behaviour. Salinity changes in the circular stream apparatus could easily be produced by changing the salinity of the water supply in the central part without interfering in the current velocity. Mixtures of sea water and tap water were used.

No results were obtained with this device. During the artificial tides, imitated by salinity changes, the elvers failed to show a behaviour suggestive of migration towards fresh water outlets.

Although some authors (van Heusden, 1943; Fontaine and Callamand, 1941) had established a distinct preference of elvers for water of
lower salinities, our negative results with the circular stream apparatus
called for an attempt to reproduce their findings. To this end experiments
were performed with a selection testing apparatus, in which elvers can
choose between four streamlets of sea water, fresh water and water of

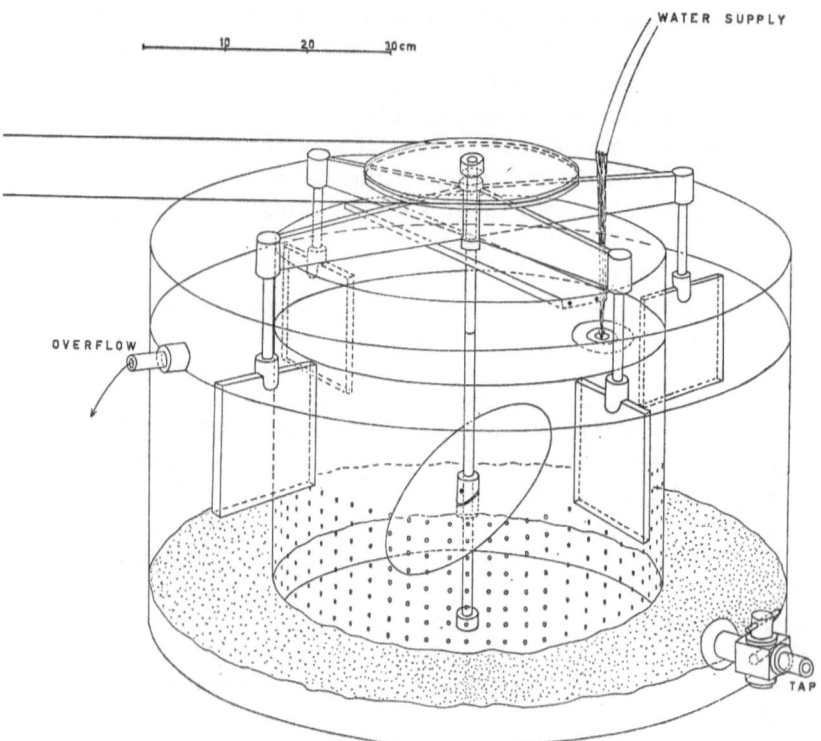

Fig. 1. Apparatus for circular streams. Gradual changes in the composition of the water can be effected by
changing the composition of the water supply in the central part. From here the gradually changing water
penetrates the circular channel, equally distributed through the perforations in the wall of the inner cylinder

intermediate salinities respectively. The results were rather surprising
because the elvers did not show any preferences, whichever degree of
salinity they were offered. Most probably this fact will explain the failure
of the experiments with the tidal stream apparatus, in which the tides
were imitated by salinity changes only.

In further experiments, however, it was found that, contrary to
their indifferent behaviour with respect to salinity, the elvers showed an
obvious preference for natural fresh water originating from inland
sources as the Yssel Lake and the Noord Hollands Kanaal near Den

Helder. Even when natural fresh water was mixed with sea water it was preferred by the elvers to pure tap water. On the other hand, when filtered over a charcoal filter the natural fresh water loses its attractiveness. Good evidence was found that no repellent factor in tap water or in water filtered over charcoal was responsible for these results; there must, then, be a stimulating factor, which is removable by means of a charcoal filter.

Considering all these features, there can be hardly be any doubt that the preference for inland water is based on olfactory stimuli. This view might be supported by the findings of TEICHMANN (1957) who demonstrated the remarkably high sensitivity of the olfactory sense in the eel. Most probably elvers show an innate response to an odour, specific to inland water in general, wherever it occurs.

With this new result another attempt was made to reproduce a migratory behaviour of elvers in the circular tidal apparatus, but now the artificial ebb was imitated by a supply of untreated natural inland water from the Yssel Lake, while flood was reproduced by gradual increase of the amount of sea water filtered over charcoal. With this new device it was indeed found possible to induce a migratory behaviour in elvers. When the circular stream apparatus contained pure sea water only (freed from odour by a charcoal filter) the greater part of the elvers swam with the current, were transported passively or were hidden in the sandy bottom. As soon, however, as untreated inland water was introduced into the apparatus, the number of elvers swimming counter current increased significantly. In some cases this even happened before the salinity recorder could register the salinity changes; the negative rheotaxis reversed to a positive rheotaxis and the elvers hidden in the bottom slipped out of the sand and swam actively against the current. When after about two hours the flood tide was introduced by the supply of filtered sea water a gradual return to negative rheotaxis and passive transport was observed.

In some respects the mechanism is quite clear. As long as the elvers stay near the bottom they are able to discriminate between different water masses passing over. When the smell of inland water is perceived they can orient themselves to the source by heading into the current. This simple design seems to be very common in nature; the orientation of adult salmon to their parent stream is based on the same principle (WISBY and HASLER, 1954). The combination of smell and water currents in the orientation of marine slugs may also be mentioned here (VAN HAAFTEN and VERWEY, 1960). Good examples of the role of an air current which contains a specifically stimulating odour in the orientation of insects and *Triton* are contributed by FLÜGGE (1934), SCHWINCK (1954), DANZER (1956) and CZELOTH (1931). In all these cases we have to discriminate between two factors: a stimulating factor (a specific odour) and a directing factor (a water or air current) which leads the animals to the source of the stimulating factor.

Hitherto, the observations in the circular stream apparatus did not clarify the features of the field observations in the Texelstroom where apparently during flood more elvers are present in higher water levels than during ebb. It should be noted, however, that the current velocity in the apparatus substantially differed from that in nature. In the Texelstroom the tidal streams generally reach a velocity of about 100 to 150 cm per second while the current velocity in the circular stream apparatus was only 10 to 20 cm per second during these experiments. For this reason further experiments were performed with higher current velocities up to 36 and 50 cm per second.

The results of these experiments were very striking. When during the ebb phase the majority of the elvers swam counter current, the stream was accelerated. The number of elvers heading into the current then immediately decreased. Owing to the strong current these hardly managed to keep their positions though swimming with all their might. The greater part, however, gave up the struggle and clung to the bottom in order not to be washed away (Fig. 2). It looked as if the elvers took shelter from the sand-storm behind the sand ridges being formed by the strong current. This category predominated completely; the water column in the apparatus was almost entirely emptied of elvers.

When, on the other hand, during the flood phase the current was accelerated, the picture was totally different. The majority of the elvers which swam with the current or were transported passively rose to higher watei levels (Fig. 3). It looked as if they were whirled up by the current. Sometimes one or two elvers dashed to the surface. The number of elvers observed in the upper half of the water column was as much as about five times the number observed during the strong ebb.

Through the introduction of greater current velocities in the circular stream apparatus it has become clear that the differences between the ebb and flood catches in the Waddensea aie the result of positive and negative rheotactic responses which are more or less modified by the strong current.

Hitherto the story only deals with the mechanism of migration through an estuarine area from the open sea not far from the coast towards inland waters. In the Texelstroom, for instance, this migration takes place in the months of February, March and April. According to SCHMIDT (1909, 1927) the metamorphosis takes place half a year earlier in the months of August and September at the continental slope west of the Atlantic coasts of France and Great Britain. Hence the elvers which reach the Dutch Waddensea have to cover a distance of about 1000 km in the autumn and winter period preceding the inshore migration.

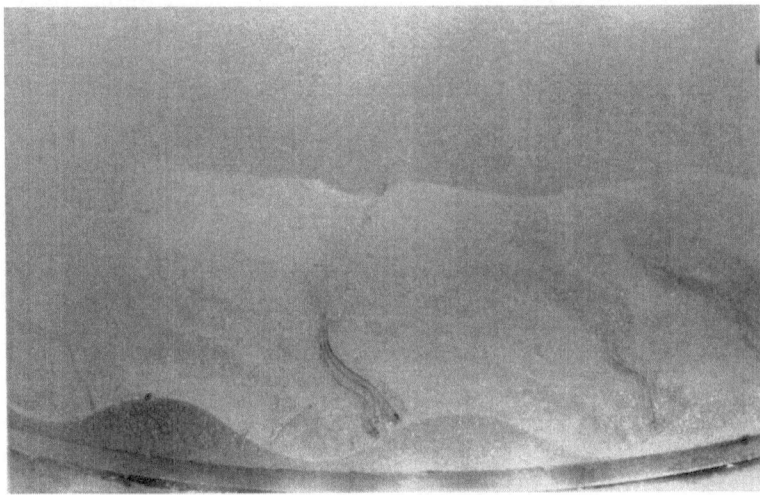

Fig. 2. View of elvers clinging to the bottom during ebb phase (current velocity: 36 cm/sec, clockwise) in the circular stream apparatus, reproduced by increasing the content of inland water

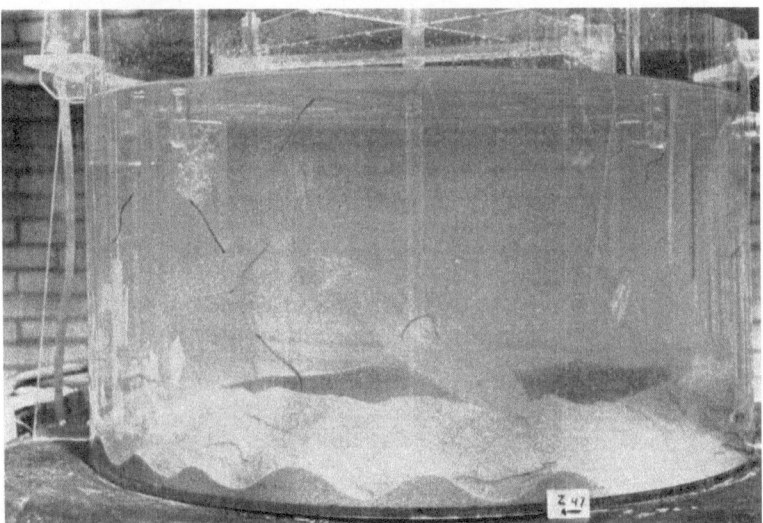

Fig. 3. View of elvers rising to higher water levels during flood phase (current velocity: 36 cm/sec, clockwise) in the circular stream apparatus, reproduced by decreasing the content of inland water

About the pathways followed by the elvers in the open sea and the mechanism of migration and orientation there are still gaps in our knowledge. By reference to the scarce observations in the open sea from our

own work and from that of other workers (SCHMIDT, 1909, 1927; BOW-
MAN, 1912 and GILIS, unpublished) we can form an idea of the trail
followed by the elvers and of the progress they make during those six
months (Fig. 4). It seems unlikely that these movements in the open sea

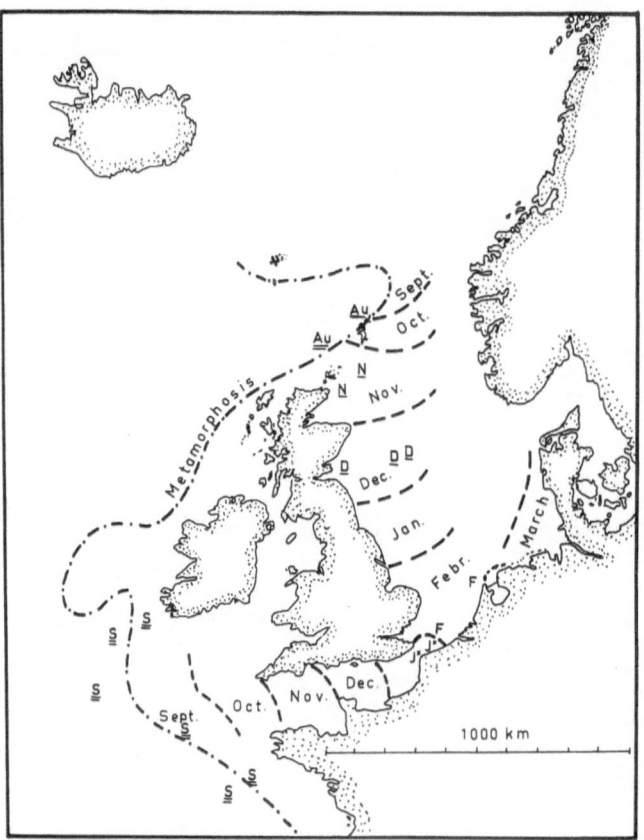

Fig. 4. A tentative representation of the seasonal advance of elvers in the North Sea and the British Channel
based on actual observations in the open sea. Au = August, S = September, N = November, D = December,
J^x = January (GILIS), F = February (own observations). Names of months underscored: BOWMAN (1912),
doublescoring: SCHMIDT (1909, 1927)

should be governed by the same principles as the inward migration in
estuarine areas. It would mean that as a result of their response to certain properties (odour) of inland waters the elvers would tend to be
trapped at a too early stage by the estuarine areas on their way, which
may influence the waters in the open sea at a considerable distance. They
would then accumulate in the estuaries and be barred from further
progress parallel to the coasts.

That indeed inshore migration does not take place during autumn and early winter months may, for instance, be derived from data contributed by MENZIES (1936) and LOWE (1951). They show that also in Ireland inward migration and ascent to inland waters take place as late as in March-April although the distance to the line of metamorphosis might suggest much earlier arrival in coastal water.

This view may be supported by experiments by WEZEMAN (unpublished) with a selection testing apparatus over longer periods. In this apparatus the elvers (one year after metamorphosis) could choose between streamlets of tap water and water from the Yssel Lake by swimming counter current in a plastic tube to a bifurcation and then towards two end containers. The percentage of elvers swimming up to the end containers, however, decreased significantly from July (75%) to November (15%). In January and February this percentage increased again. Undoubtedly internal processes play a role here because this seasonal inactivity could not be correlated with temperature differences. In this connection it may be mentioned that TEICHMANN (1959) observed a remarkably reduced sensitivity of the olfactory sense of one of his young eels (one year after metamorphosis) in about the same period, during the months of October, November, December and January.

Perhaps we may assume that such a seasonal inactivity or reduced sensitivity with respect to streamlets of inland water also occurs in the foregoing autumn and early winter. In this period following metamorphosis elvers can only be observed in open sea and apparently they do not show a positive rheotactic reaction to inland water. However, we have no experimental results as yet to confirm this hypothesis.

On the other hand, as regards the rate of advance in the North Sea off the east coast of England it may be noted that the suggested month by month southern limit of the elvers shows a striking resemblance to the pictures of southward movements of plankton species given by GLOVER (1952, 1955). In both cases there is an advance of about 7 km per day. This fact gives a strong indication that the movements of the elvers in the North Sea can be accounted for in terms of passive transport by residual currents. According to CRAIG (1959) the general southerly drift in this coastal area varies between 2 and 17 km per day, with an average of about 5 km a day. Farther in the North Sea higher velocities may probably be assumed (TAIT, 1937).

Also in the Channel and in the southern North Sea the suggested rate of advance of the elvers (about 6 km per day) is of the same order of magnitude as the velocity of the residual currents (observations on light vessels, British Admiralty, 1946; CARRUTHERS, 1935; DIETRICH, 1950). All these facts indicate a passive transport of elvers with residual currents through the Channel as well as through the large northern entrance of the North Sea round the Shetlands. This passive transport may be considered as a continuation of the transport of the larvae (Leptocephali) before metamorphosis by the Gulf Stream current. Furthermore this passive transport will end in late winter and early spring when the elvers are trapped by the sphere of influence of estuarine waters, where passive transport only occurs during ingoing flood tide, while during ebb positive rheotactic responses to inland water prevent the elvers from being washed back towards the sea.

References

BOWMAN, A.: The distribution of the larvae of the eel in Scottish waters. Sci. Invest. Fish. Scotland 1912, 1—11 (1912).

British Admiralty: Atlas of tides and tidal streams, British Islands and adjacent waters. London 1946.

CARRUTHERS, J. N.: The flow of water through the straits of Dover. Part II. Fish. Invest. Min. Agric. and Fish., Lond., Ser. II, 14, 1—67 (1935).

CRAIG, R. E.: Hydrography of Scottish coastal waters. Mar. Res. Scotland 1959, 1—30 (1959).

CREUTZBERG, F.: Use of tidal streams by migrating elvers *(Anguilla vulgaris* TURT.*)* Nature (Lond.) 181, 857—858 (1958).

— Discrimination between ebb and flood tide in migrating elvers *(Anguilla vulgaris* TURT.) by means of olfactory perception. Nature (Lond.) 184, 1961—1962 (1959).

— On the orientation of migrating elvers *(Anguilla vulgaris* TURT.*)* in a tidal area. Neth. J. Sea Res. 1, 257 — 338 (1961).

CZELOTH, H.: Untersuchungen über die Raumorientierung von *Triton*. Z. vergl. Physiol. 13, 74—163 (1931).

DANZER, A.: Die Duftorientierung von *Geotrupes silvaticus* im natürlichen Biotop. Z. vergl. Physiol. 39, 76—83 (1956).

DIETRICH, G.: Die anomalen Jahresschwankungen des Wärmeinhalts im Englischen Kanal, ihre Ursachen und Auswirkungen. D. hydr. Z. 3 (3—4), 184—201 (1950).

FONTAINE, M., et O. CALLAMAND: Sur l'hydrotropisme des civelles. Bull. Inst. Ocean. Monaco 811, 1—6 (1941).

FLÜGGE, C.: Geruchliche Raumorientierung von *Drosophila melanogaster*. Z. vergl. Physiol. 20, 463—500 (1934).

GILIS, CH.: Personal communication.

GLOVER, R. S.: Continuous plankton records: the Euphausiacea of the north eastern Atlantic and the North Sea, 1946—1948. Hull Bull. Mar. Ecol. 3 (23), 185—214 (1952).

— Science and the herring fishery. Advanc. Sci. 11, 426—434 (1955).

HAAFTEN, J. L. VAN, and J. VERWEY: The role of water currents in the orientation of marine animals. Arch. Néerl. Zool. 13, 493—499 (1960).

HASLER, A. D., and W. J. WISBY: Discrimination of stream odors by fishes and its relation to parent stream behavior. Amer. Naturalist 85 (823), 223—238 (1951).

HEUSDEN, G. P. H. VAN: De trek van den glasaal naar het IJsselmeer. Thesis, Utrecht, 1943.

LOWE, R. H.: Factors influencing the runs of elvers in the River Bann, Northern Ireland. J. Cons. int. Expl. Mer. 17, 299—315 (1951).

MENZIES, W. J. M.: The run of elvers in the river Bann, Northern Ireland. J. Cons. int. Expl. Mer. 11, 249—259 (1936).

SCHMIDT, J.: Remarks on the metamorphosis and distribution of the larvae of the eel *(Anguilla vulgaris* TURT.*)*. Meddel. Komm. Havunders 3, 1—17 (1909).

— Eel larvae in the Faroe Channel. J. Cons. int. Explor. Mer 2, 38—43 (1927).

SCHWINCK, I.: Experimentelle Untersuchungen über Geruchssinn und Strömungswahrnehmungen in der Orientierung bei Nachtschmetterlingen. Z. vergl. Physiol. 37, 19—56 (1954).

TAIT, J. B.: The surface water drift in the northern and middle areas of the North Sea and in the Faroe-Shetland Channel. Sci. Invest. Fisheries. Scotland 1937, 1—60 (1937).

TEICHMANN, H.: Das Riechvermögen des Aales *(Anguilla anguilla* L.*)*. Naturwissenschaften 44, 242 (1957).

— Über die Leistung des Geruchssinnes beim Aal *(Anguilla anguilla* L.*)*. Z. vergl. Physiol. 42, 206—254 (1959).

WEZEMAN, B.: Personal communication.

WISBY, W. J., and A. D. HASLER: Effect of olfactory occlusion on migrating silver salmon *(O. kisutch)*. J. Fish. Res. Board. Canada 11, 472—478 (1954).

Orientation in Three Species of Anuran Amphibians*

By Denzel E. Ferguson**

Department of Zoology, Mississippi State University, State College,
Mississippi (U.S.A.)

With 2 Figures

The present account is a progress report on the initial year of studies concerning the orientation and homing behavior of Fowler's toad *(Bufo fowleri)*, the southern cricket frog *(Acris gryllus)*, and the upland chorus frog *(Pseudacris triseriata)*. These species were selected for study because one or more of them is active at all seasons near State College, Mississippi where the study was conducted. We have attempted to determine if homing skills are exhibited and, if so, to investigate the nature of the orienting mechanisms. Considerable attention has been devoted to possible interspecific differences in method of orientation.

Several investigators have reported homing behavior for frogs and toads. Bogert (1947) displaced 444 Carolina toads *(Bufo terrestris)* and recaptured over 50% of those released at distances of 100—300 yds. He found that 18.6% of the toads returned from a mile displacement. The toads were removed from an area that served as both a home site and breeding place, hence it is not apparent to which of these they were attracted in their return. In discussing possible mechanism of homing, Bogert discounted effects of topography and slope, visual cues (at least in unfamiliar areas) and random searching. A higher percentage of recoveries came from cleared lands, which suggested use of auditory cues, as did failure of toads to return when removed beyond hearing distance from a breeding chorus. The use of visual cues in familiar places was postulated.

Jameson (1957) noted that Pacific tree frogs *(Hyla regilla)* would return to a home pond after being displaced 1000 yds. to a different but occupied pond. He also reported that individuals returned to the same

* Supported by National Science Foundation Grant G-19063.
** Acknowledgement is made to Mr. Claude E. Boyd and Mr. Abner M. Hammond, Department of Zoology, Mississippi State University, U.S.A., who served as field assistants.

place within a pond where they were originally captured. He mentioned olfactory, auditory, and kinaesthetic senses as possible orientation mechanisms but admitted that no one factor would explain his findings.

Other workers have stressed "muscular memory" from repetitive use of certain travel routes (BUYTENDIJK, 1918) and vision in combination with learning (YERKES, 1903; FRANZ, 1927). MARTOF (1962) has studied effects of light and temperature on responses of *Bufo fowleri*.

Experiments and results

Pseudacris triseriata, the upland chorus frog, breeds sporadically from November to March, depending upon temperature and rainfall. The species is primarily nocturnal in habit and utilizes temporary pools for egg deposition. In the summer, individuals are seldom encountered and are presumed to inhabit cracks or holes in the soil.

Chorus frogs were removed from a breeding congress, marked by toe clipping, and released at various distances from the home pond. Among those released were 181 which were placed among another chorus located 1450 ft. away. The intervening obstacles included a veritable tangle of vegetation, a well traveled highway, and a small stream. A total of 409 frogs was moved of which 75 (18%) were recaptured. The results of these experiments are summarized in Table 1. The mechanism of orientation involved in these movements is not presently understood. The seemingly marked attachment to a particular breeding assemblage suggests an auditory response to the home chorus. However, possible reliance upon the night sky, sun, or some other cue cannot presently be discounted.

Table 1. *Recapture data for 409 upland chorus frogs displaced 150—2640 ft. from the home chorus*

Distance Moved (ft.) . . .	150	300	450	500	1350	1450	2640
Number Moved	20	20	95	38	39	181	16
Number Recaptured . . .	9	13	16	15	1	21	0
% Recaptured	45	65	17	39	3	11	—

Bufo fowleri, Fowler's toad, breeds in permanent pools, mainly from March to mid-June. After completion of breeding, individuals establish fairly permanent home ranges for the remainder of the summer. FERGUSON (1960) found a single individual under the same street light 14 times in a two-year period.

That Fowler's toad is less attached to a particular breeding chorus was demonstrated when 14 of 135 marked individuals were later found participating in another breeding assemblage located 1275 ft. away. Also, 2 of 12 marked individuals moved in the opposite direction. When

64 toads were released mid-way between these two choruses, 1 returned to its own pond and 3 went to the wrong pond.

A total of 1162 toads was released at various distances from a breeding pond. Nearly all of these individuals were transported in cloth bags at night. During the next few weeks, a person could clearly hear the home chorus from all release points excepting the two most distant ones (6402 and 8000 ft.). Returns were extremely meager as can be seen in Table 2 where the results of these experiments are summarized. Not indicated in Table 2 is the fact that 1.3 toads moved 425 ft. in the wrong direction to a different pond, 2.2 moved 1500 ft. in the correct direction but stopped at a different pond, 3.3 moved 4026 ft. in a generally correct direction to another pond, 4. several toads were later recaptured at or near points of release.

Table 2. *Recapture data for 1162 Fowler's toads displaced 330—8000 ft. from the home chorus. Specimens denoted by an asterisk (*) differed from others shown in being moved in open containers, in daylight, and at a time when breeding had nearly ceased. The high percent of recapture suggests that the pond may have been their summer home site* (compare with Table 3)

Distance Moved (ft.)	330	450	850	1518	1816	3234	3960	5016	6402	8000
Number Moved . .	25	42*	64	50	50	50	206	175	250	250
Number Recaptured	6	25	1	6	1	—	1	—	—	—
% Recaptured . . .	24	59	2	12	2	—	0.5	—	—	—

After *Bufo fowleri* had moved to summer home sites, they were collected as they came to street lights at night and released at other lights and in unlighted locations. These were displaced both at night and day and in both open and closed containers. The results indicate a much stronger homing tendency than that exhibited by toads removed from a breeding congress (Table 3). It was necessary for many of these toads to pass street lights other than their own in order to return home.

Table 3. *Recapture data for 127 Fowler's toads displaced 150—1100 ft. from summer home sites or home ranges*

Distance Moved (ft.)	150	350	450	500	600	700	800	1000	1100
Number Moved . .	9	5	38	10	23	9	12	6	15
Number Recaptured	7	3	26	7	10	4	6	2	7
% Recaptured . . .	77	60	68	70	43	44	50	33	46

Other studies with *B. fowleri* have yielded the following findings of interest, all of which must be clarified by additional study: 1. By attaching a spool of thread to displaced toads, it has become apparent that movement is usually direct and is accomplished at night; 2. Juveniles placed in a release pen, to be described later, have shown a marked tendency to orient with reference to the sun. Since nocturnal activity commences when a body length of 38—45 mm. is attained (FERGUSON,1960), adults may rely on other cues as well; 3. Adults, when carried away from a familiar home site in view of the sun and then put in a dark container and moved several miles away via a very indirect route, have been observed to travel in an opposite direction to that in which they were initially carried while being moved from the home site. This ability is not displayed when view of the sun is denied the released toads, such as inside a large building.

Southern cricket frogs *(Acris gryllus)* live and breed along the margins of fairly large, permanent bodies of water. When a population was marked to gain an indication of the extent of normal movements, most individuals were found to remain within a few feet of the place where they were first encountered. The maximum movement observed was 660 ft.; the average distance for those which moved 30 or more feet was 62 ft. Even these movements were partially attributable to a response to drying conditions causing the frogs to seek more favorable home sites.

In preliminary displacement studies, 25 of 64 cricket frogs returned 300 ft.; 15 moved 150—270 ft. in the correct direction. None moved more than 60—90 ft. in the wrong direction. When 309 cricket frogs were moved 640 ft. from a home pond to another one supporting its own population, returns were scanty. This was expected in view of the fact that the experiment was conducted in August, at which time these small frogs desiccate in only a few minutes if taken away from the water's edge. Extensive overland movements are therefore presumed to be possible only following rains. The results were as follows: 12 returned; 3 moved 590 ft. in the correct direction, stopping at a small intervening pond; 17 were recaptured at the release pond.

In an attempt to determine if the frogs could return when prevented by some obstacle from taking a direct route (compass direction), 529 frogs were moved 390—1020 ft. around an inlet to the opposite shore. Actually, most were displaced only about 360 ft., if direct distance across the water is considered. The experiment was based on the observation that *A. gryllus* will not voluntarily enter deep water. The survival value of this behavior pattern was illustrated when 25 frogs were thrown 20—25 ft. out into a small pond and fish devoured 19 before they could return to shore. Only 28 of the 529 displaced frogs were recaptured, all being on the same side where they were released. Some had moved 165 ft. in the correct direction and others as much as 330 ft. in the wrong direction, however, the majority remained within 30 ft. of the release points.

Acris thrown into water were observed to swim to the shore where captured rather than to a nearer but unfamiliar one. In order to test possible use of sun orientation, a circular pen 60 ft. in diameter was built in the water of a shallow lake inlet. The walls of the pen were 8 ft. high, constructed of black plastic. Only sky was visible from all points within the pen. Cricket frogs were collected and transported in open containers along a direct line to the center of the pen. The distance of displacement varied from 300—1500 ft. The container of frogs was handed over the wall rather than deviate from a direct course. The frogs were placed in the center of the pen in a large funnel which was designed to sink very slowly during which time observers could leave the test pen. Frogs remained on the funnel until it submerged, then swam to the plastic

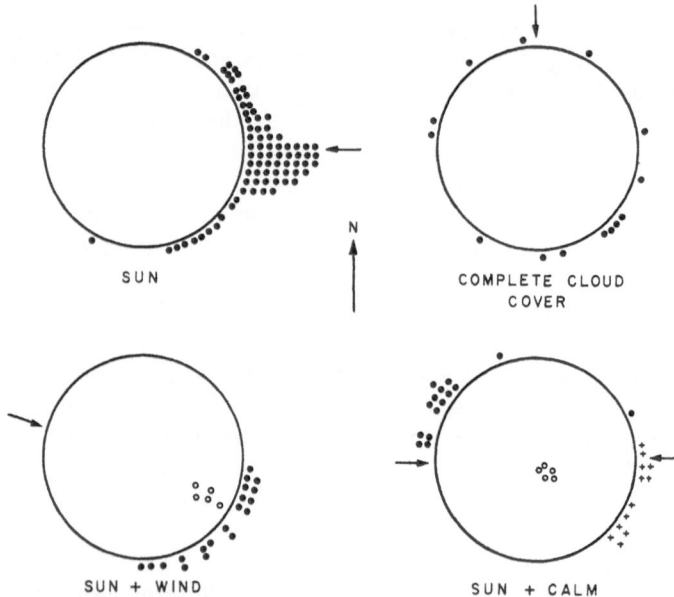

Fig. 1. Results of experiments testing sun orientation in the southern cricket frog under different test conditions. Arrows indicate direction from which frogs were introduced into the test pen (in full view of the sun). Solid dots and plus signs (+) indicate individual frog scores for direction selected. Small open circles denote performance of corks released with frogs to test wind action

wall, climbed out on the plastic and remained in place where they were easily scored for performance in regard to direction selected. Fig. 1 shows the results of several such tests conducted under a variety of conditions.

Both individual releases and simultaneous releases of frogs collected from opposite directions, enabled us to rule out the possibility that the animals were following each other. In strong wind, the frogs generally scored as a group but in an incorrect direction. Small corks released with the test specimens behaved similarly, indicating influence of wind inside the test pen. On these occasions, individuals were noted to make an effort to go in the correct direction but were unable to do so. They would eventually appear to tire and allow themselves to drift with the wind.

Releases made on windless days with complete cloud cover yielded a fairly random dispersion.

In a second series of experiments, cricket frogs were collected and denied all view of light during transport to the test pen. On the way to the test pen the frogs were disoriented by being carried in a confusing pattern of directions, rolled and tossed about like a ball. They were then released in the pen as described above and found to make a proper choice of direction toward the point of original capture. Fig. 2 depicts results of two such tests.

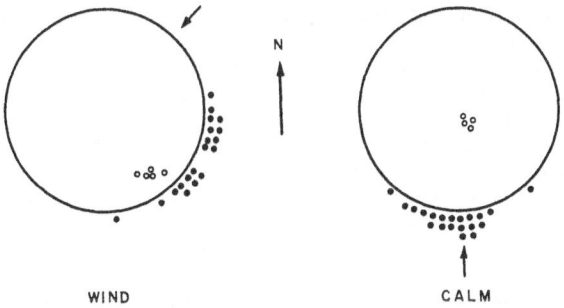

Fig. 2. Tests of the ability of southern cricket frogs to orient toward home under the sun after being displaced in darkened containers, over an indirect route. Arrows indicate direction of the home site in relation to the test pen. Solid dots represent performance of individual frogs upon being released in the pen. Small open circles denote the performance of corks released with frogs to test influence of wind currents

Conclusions

The ability to home has been demonstrated in the upland chorus frog, southern cricket frog and Fowler's toad. Results obtained indicate that the orienting mechanism for homing operations may differ among the species studied. Auditory cues are apparently more important to the chorus frog which breeds in temporary pools. Toads and cricket frogs appear to show little attachment to the breeding assemblage but return with considerable accuracy to home ranges when displaced.

Use of the sun as a compass has been demonstrated for the cricket frog and appears probable for toads as well. Cricket frogs appear able to use the sun azimuth, position, and a biological clock mechanism to properly orient when displaced distances as small as 300 ft. The importance of auditory cues has heretofore been overemphasized as a factor in anuran homing. Many additional experiments are needed to clarify problems posed in these studies.

Summary

When 409 chorus frogs were moved 150—2640 ft. from a home pond, 18% returned. A total of 1162 Fowler's toads was displaced up to 8000 ft. from a home pond. Few were recaptured. However, when marked individuals were removed from summer home sites, 56% returned. Cricket frogs were able to home in a straight line but were unable to do so when forced to use an indirect route. In a release pen with only the sun and sky visible, cricket frogs were able to choose a correct direction toward home. This was true both when they were transported to the release pen in full view of the sun and when transported in containers excluding all light.

References

BOGERT, C. M.: A field study of homing in the Carolina toad. Amer. Mus. Novitates 1355, 1—24 (1947).

BUYTENDIJK, F. J. J.: Instinct de la recherche du nid et experience chez les crapauds *(Bufo vulgaris* et *Bufo calamita)*. Arch. Néer. Physiol. Hom. Anim., Ser. IIIc, II, 1—50 (1918).

FERGUSON, D. E.: Movements and behavior of *Bufo fowleri* in residential areas. Herpetologica 16, 112—114 (1960).

FRANZ, V.: Zur tierpsychologischen Stellung von *Rana temporaria* und *Bufo calamita*. Biol. Zbl. 67, 1—12 (1927).

JAMESON, D. L.: Population structure and homing responses in the Pacific treefrog. Copeia 1957 (3), 221—228 (1957).

MARTOF, B. S.: The behavior of Fowler's toad under various conditions of light and temperature. Physiol. Zool. 35, 38—46 (1962).

YERKES, R. M.: The instincts, habits and reactions of the frog. Psych. Rev. Monog. 4, 579—638 (1903).

Les migrations orientées du Hanneton commun Melolontha melolontha L.
(Coléoptère Scarabeidae)

Par P. Robert

Institut National de la Recherche Agronomique, Colmar (France)

Avec 4 Figures

Sommaire

I. Introduction

Le Hanneton commun est un insecte répandu en Europe dans les régions à climat tempéré. Les adultes s'alimentent surtout aux dépens des arbres forestiers. Les larves ou vers blancs s'attaquent aux racines d'un grand nombre de plantes cultivées et constituent un fléau pour l'agriculture européenne.

Les femelles du Hanneton effectuent au cours de leur vie aérienne trois catégories de vols ou migrations (Fig. 1 et 4) qui ont lieu soit au crépuscule après le coucher de soleil, soit pendant la journée, en particulier le matin. Au printemps elles sortent avec les mâles des terrains cultivés ou enherbés et se dirigent généralement vers un massif boisé: c'est la *migration préalimentaire*. Après une période d'alimentation de 10 à 20 jours, les mères pleines d'oeufs quittent la forêt et pondent dans les champs voisins; au cours de cette *migration de ponte*, elles suivent des trajectoires qui vont en sens inverse de celles du vol préalimentaire. Après l'oviposition, la *migration après ponte* conduit de nouveau les femelles des champs vers la forêt selon la même direction que celle suivie au moment des vols préalimentaires. L'amplitude de ces déplacements est variable, selon les localités et les individus, de 200 m à 2.500 m environ.

Une faculté d'orientation a été mise en évidence chez le Hanneton commun *M. melolontha* L. au cours de nombreux travaux effectués sous la direction de M. COUTURIER, Directeur de la Station de Zoologie agricole de Colmar. Le déterminisme et le rôle de l'orientation ont pu être précisés. Des recherches sont poursuivies sur les guides qui permettent aux Hannetons de retrouver leur direction.

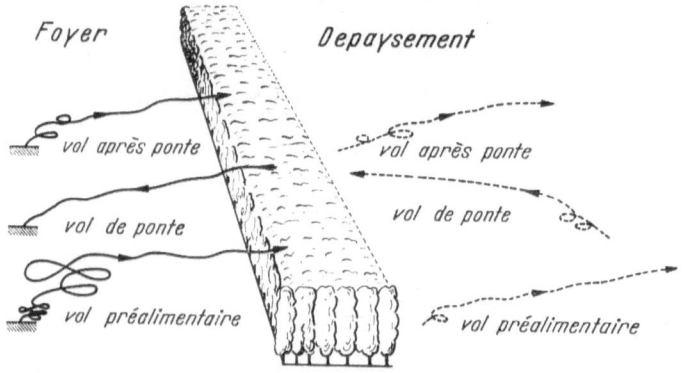

Fig. 1. A gauche, en traits pleins: Diagramme schématique montrant, dans le foyer d'origine, les 3 types de migrations effectuées par une femelle sortie à proximité d'un site attractif constitué par une forêt. La direction prise au cours du vol préalimentaire détermine celle des autres migrations (ponte et après ponte). A droite, en traits interrompus: trajectoires prises lors des dépaysements de femelles capturées au cours des diverses migrations

II. Une faculté d'orientation

Dans un site donné, les Hannetons suivent au cours de leurs migrations des trajets particuliers que HURPIN (1962) appelle des «voies de vols» dont la direction est constante.

Des insectes, capturés au cours des diverses migrations, sont relâchés soit immédiatement dans une enceinte masquant les alentours et laissant voir seulement le firmament, soit quelques minutes plus tard dans un paysage éloigné et inconnu: les Hannetons s'envolent aussitôt et repartent selon les directions suivies au moment de la capture, indépendamment de l'aspect des lieux. De plus, les insectes arrêtés au cours des vols crépusculaires après le coucher du soleil, retrouvent encore leur direction quelques jours plus tard, le matin, dans des paysages inconnus même éloignés de plusieurs centaines de kilomètres du lieu de capture.

De tels comportements impliquent l'existence d'une faculté d'orientation qui se manifeste indépendamment des repères visuels terrestres. La mémoire de la direction est retenue pendant plusieurs jours (COUTURIER et ROBERT, 1955, 1958, 1960).

III. Déterminisme et rôle de l'orientation

1. Migration préalimentaire. Ce déplacement est effectué par les mâles et les femelles, mais la conduite au cours du premier envol printanier, au sortir du sol, n'est pas la même chez les deux sexes. Les femelles effectuent d'abord lentement à 50 cm du sol, au-dessus du lieu de départ, plusieurs spires serrées de 0,5 à 2 m de diamètre. Ensuite, les mères s'élèvent à 2 et 3 m pour effectuer quelques tours plus amples et se dirigent d'abord en zigzags, puis en ligne droite, vers un même point de l'horizon. Les mâles prennent de suite de la hauteur et partent vers le même site après 3 ou 4 tours. Ainsi, tous les adultes, mâles et femelles, issus d'un territoire, vont vers un point particulier ou site attractif qui est une masse faisant contraste sur l'horizon (COUTURIER et ROBERT, 1951). C'est souvent la silhouette qui paraît la plus haute à l'endroit où s'envolent les Hannetons, c'est-à-dire celle qui est vue sous le plus grand angle (SCHNEIDER, 1952).

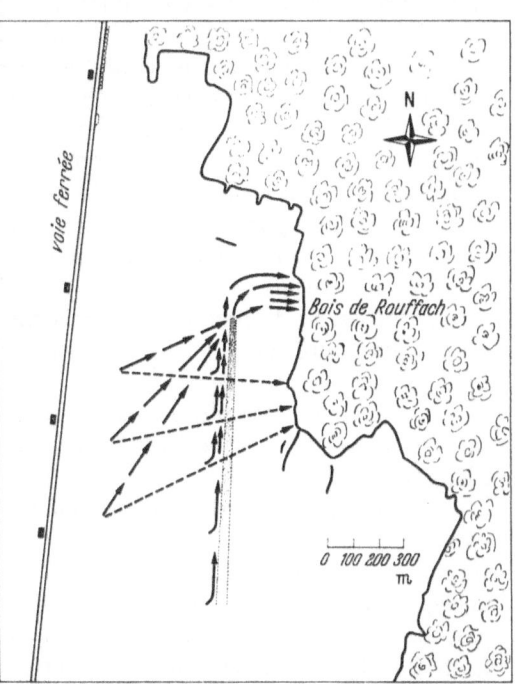

Fig. 2. Comportement des Hannetons devant un écran de fumée cachant le site attractif au cours d'une migration vers la forêt. Les insectes sont déroutés par l'obstacle placé entre eux et le site vers lequel ils allaient. Ils modifient leur trajectoire. Un très grand nombre longent ainsi le nuage, mais parvenus au lieu d'émission ils changent encore leur direction pour aller vers la lisière la plus proche. (En pointillé: trajectoires suivies avant la mise en place de l'obstacle)

Ensuite dans son foyer d'origine, pendant le vol de migration préalimentaire, le Hanneton ne reste pas indifférent au paysage, comme le montre l'observation suivante: au cours d'une migration préalimentaire les observateurs masquent le site attractif à l'aide d'un rideau opaque de fumée blanche qui se confond avec le ciel (Fig. 2). Les insectes déjà en vol sont déroutés par l'écran de fumée placé entre eux et le site vers lequel ils allaient. Ils modifient leur trajectoire et vont vers un autre site attractif qui est la partie de la forêt restée visible au-delà du point de naissance de la fumée. Un très grand nombre longent ainsi le nuage, mais parvenus au lieu d'émission, ils modifient à nouveau leur direction pour aller vers un endroit encore plus attractif qui se trouve être la lisière la plus proche. En cours de migration préalimentaire, les bêtes restent en contact visuel avec le paysage et sont capables de modifier leurs trajectoires en fonction du milieu extérieur (COUTURIER et ROBERT, 1951, 1955).

Cependant, la direction suivie au cours de la migration préalimentaire est retenue; elle est reprise par les Hannetons, mâles et femelles, arrêtés déjà après 100 m de vol et relâchés dans un paysage inconnu. Cette direction est aussi reprise par des insectes qui, capturés sur leur trajectoire orientée avant l'émission du nuage de fumée, sont relâchés peu de temps après devant l'écran de fumée; ils partent aussitôt selon leur direction primitive, qui les conduit à buter dans l'obstacle. Ils ne modifient pas l'orientation de leur corps; ils s'élèvent et passent au-dessus. Ainsi apparaît une différence de comportement entre les bêtes en vol capables de modifier leur chemin devant un obstacle et les insectes qui reprennent un vol interrompu. Ces derniers s'orientent, sans tenir compte du paysage, grâce seulement à leur faculté d'orientation.

2. Migration de ponte. Lorsque les femelles ont muri leurs oeufs, elles s'éloignent de la forêt selon des trajectoires rectilignes sur une longueur qui peut varier de 200 à 2500 m selon les individus et selon les localités.

Des différences de conduite apparaissent entre le comportement des femelles immatures et celui des mères pleines d'oeufs. Nous avons observé que les mères gravides prennent leur envol dans la forêt sans effectuer de spires d'orientation. Elles partent immédiatement dans la direction qui les conduit vers les champs où elles sont nées (COUTURIER et ROBERT, 1958).

Le déterminisme de cette orientation a été recherché expérimentalement: des femelles sont capturées pendant la migration préalimentaire et nourries en cage jusqu'à la maturité des oeufs; à ce moment, ces femelles prêtes à pondre sont relâchées dans divers sites inconnus pour elles; elles partent toutes dans la direction opposée à celle suivie lors de la récolte (pendant l'itinéraire préalimentaire) sans tenir compte du milieu environnant. Dans certaines localités, choisies par les expérimentateurs, cette conduite amène les femelles gravides à pénétrer dans une forêt alors que normalement, au moment de la ponte, elles quittent un massif boisé. Ainsi la maturation des oeufs entraîne des modifications physiologiques chez les femelles. Les mères gravides partent selon des trajectoires parallèles à celles suivies au vol préalimentaire 10 à 20 jours auparavant, mais le sens de la direction est spontanément inversé. La direction est trouvée, grâce à la faculté d'orientation, sans le secours de repères visuels terrestres (COUTURIER et ROBERT, 1956, 1960).

Les femelles pleines d'oeufs, surprises en cours de migration de ponte par un écran de fumée qui barre leur chemin, ont aussi une conduite particulière. Elles ne dévient pas latéralement comme les femelles immatures au vol préalimentaire ou comme les mères après la ponte. Elles conservent leur orientation et contournent l'obstacle en s'élevant au-dessus, puis reprennent leur trajectoire de l'autre côté du nuage en perdant de l'altitude (Fig.3). Le maintien de l'orientation en vol se fait grâce à la faculté d'orientation, sans relations visuelles avec le paysage terrestre.

3. Migration après ponte. Les femelles déposent leurs oeufs dans le sol des champs cultivés ou des prés. Beaucoup d'entre elles remontent en surface pour un nouveau départ vers la forêt.

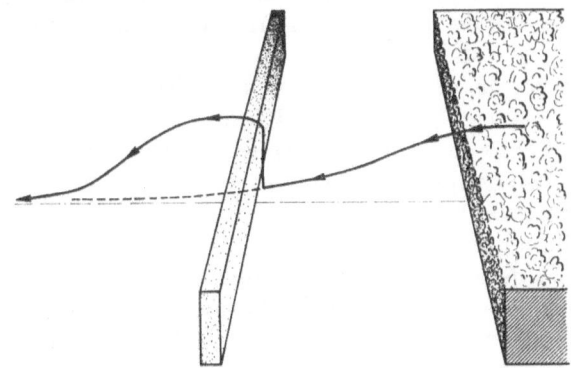

Fig. 3. Trajectoire suivie par les femelles en vol de ponte en présence d'un écran de fumée artificielle. Les mères maintiennent leur orientation et contournent l'obstacle en passant par-dessus, puis reprennent leur trajectoire de l'autre côté du nuage en perdant de l'altitude. (En pointillé: parcours effectué avant la mise en place de l'écran)

Nous avons pu montrer que l'orientation prise à l'envol, après la ponte, est, selon une conduite stéréotypée, la même que celle suivie au cours de la première migration préalimentaire: des femelles gravides capturées pendant la migration de ponte, mises sous des filets ou dans des cages, sont prises aussitôt après le dépôt des oeufs avant leur envol. Lâchées dans leur foyer d'origine comme dans différents sites inconnus, toutes ces bêtes partent spontanément, sans tenir compte du milieu géographique environnant, dans la direction et le sens choisis au cours du vol préalimentaire 15 à 25 jours plus tôt.

Ensuite, pendant le vol, les insectes ont un comportement comparable à celui des insectes de la première migration préalimentaire; ils sont en contact visuel avec le paysage et modifient leur orientation devant un écran de fumée. Après un vol interrompu, ils partent aussi selon leur direction propre grâce à la faculté d'orientation (COUTURIER et ROBERT, 1958).

Quoique apparemment semblables, les deux migrations vers une forêt sont à l'origine de nature différente. L'orientation prise après la ponte est déterminée par la mémoire de longue durée et par la faculté d'orientation. Elle obéit au départ au même processus que la migration de ponte. Toutefois, en cours de route, les individus entrent en relation visuelle avec le milieu extérieur et, le cas échéant, il peut se produire une régulation telle que l'attraction d'un site puisse devenir prépondérante comme cela est de règle pour le vol préalimentaire. Ainsi leurs comportements sont semblables en présence d'un écran de fumée qui modifie la ligne de vol sans toutefois influencer l'impression reçue primitivement.

IV. Les facteurs responsables de l'orientation

Toutes les expériences de dépaysement prouvent que *M. melolontha* possède la faculté de retrouver son orientation sans se repérer sur des signes visuels terrestres.

La direction du vent n'influence pas l'orientation des insectes. Un vent trop fort les empêche de voler. Par fort vent de côté les Hannetons maintiennent l'axe de leur corps selon leur propre direction.

Les Hannetons qui s'orientent au cours d'un vol préalimentaire crépusculaire, après le coucher du soleil, retrouvent leur chemin de vol le lendemain ou les jours suivants dans le courant de la matinée. Ils sont donc capables de maintenir leur direction dans le temps. Cette faculté existe encore lorsque le choix de la direction s'est fait le soir après le coucher du soleil par ciel couvert. Nous avions émis l'hypothèse que les Hannetons, comme les Abeilles ou d'autres Arthropodes, s'orientaient grâce à une faculté d'orientation astronomique, c'est-à-dire grâce à la perception des rayons du soleil ou de la lumière polarisée du ciel et à un «sens interne du temps» selon un rythme de 24 heures permettant de tenir compte du mouvement apparent du soleil (MÉDIONI, 1956).

Nous avons montré que le soleil joue un rôle important dans le courant de la matinée. Les insectes arrêtés en vol le matin reprennent quelques minutes plus tard leur direction dans une enceinte s'ils perçoivent les rayons du soleil. Ils peuvent être déroutés de 180° à l'aide d'un miroir (mnémotactisme). Faute de dispositifs suffisamment vastes, l'influence de la lumière polarisée n'a pas pu être testée (COUTURIER et ROBERT, 1958).

Mais jusqu'à maintenant nous n'avons aucune preuve expérimentale de l'existence d'un sens du temps. Dans la nature, les Hannetons manifestent un rythme nycthéméral d'activité. Ils volent durant les matinées ensoleillées jusque vers 12—14 heures. L'après-midi, les insectes restent immobiles puis ils présentent une activité intense au crépuscule pendant une demi-heure environ. Au laboratoire ce rythme se maintient aussi longtemps que les Hannetons reçoivent de la lumière naturelle même si elle est filtrée par plusieurs épaisseurs de papier calque qui dépolarise totalement la lumière et réduit son intensité.

Ce rythme nycthéméral ne se manifeste d'aucune façon en lumière artificielle présentée de diverses façons:

— alternance de lumière et d'obscurité rappelant le jour et la nuit avec crépuscule artificiel,

— lumière artificielle faible (6 lux) maintenue constante[1] (COUTURIER et ANTOINE, 1962 a).

Cependant, malgré des élevages de longue durée, dans des conditions artificielles, sans manifestation apparente de rythme nycthéméral, les Hannetons conservent la mémoire de leur orientation:

a) des femelles orientées capturées au vol préalimentaire crépusculaire sont élevées à des éclairages constants de 100 lux ou de 0,5 lux. 10 jours 1/2 ou 13 jours 1/2 plus tard elles sont pleines d'oeufs; lâchées le matin dans un site inconnu, elles partent comme les témoins élevés à la lumière naturelle, c'est-à-dire en sens inverse de leur première sortie (COUTURIER et ROBERT, 1960).

b) de même, des femelles prêtes à pondre ne sont pas désorientées si elles sont maintenues pendant 4 jours à un rythme d'éclairement artificiel en avance de 6 heures sur le jour naturel (COUTURIER et ROBERT, 1958), contrairement à ce qui a été constaté chez d'autres Arthropodes (PAPI et al., 1957).

La direction du vol est encore maintenue par des femelles récoltées au vol de ponte puis conservées dans les diverses conditions suivantes:

a) femelles pondeuses soumises à un choc de froid pendant 2 jours 3/4 à $-3°$ C ou à $0°$ C suivi d'un séjour à l'obscurité de 2 jours à $15-18°$;

b) femelles pondeuses maintenues 28 jours 1/2 à l'obscurité dont 27 jours 1/2 à $+2°$ C (COUTURIER et ROBERT, 1960);

c) de même, une narcose de 3 heures pendant la matinée dans le peroxyde d'azote (NO_2) n'a pas d'action sur la mémoire de la direction de vol. Les insectes se réveillent vite et prennent un départ sans difficulté une demi-heure après avoir été placés à l'air libre. La direction originelle est reprise rapidement et le comportement ne présente pas de différence significative avec celui du lot témoin. Pourtant, pendant la narcose, le soleil avait tourné de plus de $50°$.

Ces résultats ne permettent pas de conclure à l'utilisation d'une faculté d'orientation astronomique impliquant l'existence d'un sens du temps lié à un rythme de 24 heures[2]. Mais il apparaît que l'activité de vol est sous la dépendance de facteurs naturels externes momentanément

---

[1] Dans des conditions semblables le rythme d'activité est maintenu pendant quelques jours chez d'autres Coléoptères Scarabeidae: *Rhizotrogus aestivus* OL. et *Amphimallon solstitialis* L. (COUTURIER et ANTOINE, 1962 b).

[2] Des expériences de dépaysement à longues distances, avec décalage horaire important, comparables à celles effectuées avec *Talitrus saltator* (PAPI, 1955) ou avec l'Abeille (RENNER, 1959) apporteraient des précisions intéressantes. La réalisation de telles expériences est impossible, car il faudrait introduire le Hanneton commun, ravageur grave des cultures, dans des pays où il n'existe pas.

perçus durant la matinée et le soir au crépuscule[1]; un mécanisme particulier permet alors au Hanneton de faire le point et d'apprécier sa direction. Le soleil peut servir de guide, d'autres repères utilisés sont encore inconnus; mais la vue du ciel, même couvert de nuages, est obligatoire.

Dans des conditions de laboratoire, SCHNEIDER (1961) a mis en évidence, chez le Hanneton commun, une sensibilité à un champ magnétique. Cependant, nous n'avons pas pu perturber, dans la nature, la direction d'insectes relâchés en plein air le matin ou le soir après le coucher du soleil dans des dispositifs modifiant la direction du champ terrestre. Les insectes ne sont pas non plus désorientés si on colle sur leur thorax de petits aimants produisant à 10 millimètres de leur centre un champ au moins égal à 2 gauss c'est-à-dire 10 fois supérieur au champ terrestre (COUTURIER et ROBERT, 1960).

Signalons, pour terminer, que des insectes privés de leurs antennes retrouvent leur orientation comme les bêtes témoins (COUTURIER, communication personnelle).

V. Conclusions

Il paraît intéressant de dégager deux caractères des déplacements des Hannetons femelles:

1° Ces déplacements doivent être considérés comme des *migrations* selon la définition de VIAUD (1953) «Une migration est un déplacement momentané ou définitif d'une population d'une espèce animale, déplacement dans lequel les individus jouent un rôle actif, et qui intéresse des régions ou territoires différant par leur milieu écologique, climatologique ou géographique». Les Hannetons effectuent des mouvements de populations aller — retour entre deux aires bien distinctes (Fig. 4): *le vol aller*, ou migration préalimentaire, conduit les insectes, qui sortent de terre, depuis une aire de développement larvaire constituée par des champs cultivés ou enherbés vers une aire d'alimentation des adultes représentée le plus souvent par un massif forestier. *Le retour*, ou migration de ponte, est effectué en sens inverse par les femelles gravides. L'amplitude des déplacements est faible, de l'ordre de 200 à 2500 m.

Les migrations de *M. melolontha* sont liées au comportement reproducteur de l'espèce. Le déclenchement de chaque migration se produit pour des états physiologiques caractéristiques. Les vols allers vers la forêt sont effectués par des mères dont l'ovogenèse n'est pas ou peu développée. Au contraire, l'activité migratoire de retour vers les champs se produit au moment où le tractus génital est plein d'oeufs.

2° Au cours de ces migrations, les Hannetons suivent des trajectoires orientées. Au départ de la *migration aller* chaque individu se dirige vers

---

[1] La qualité des radiations lumineuses a une grande influence sur l'activité des Hannetons (COUTURIER et ANTOINE, 1962a).

un objet particulier du paysage: une silhouette faisant contraste sur l'horizon. Ce comportement d'orientation, au début de la vie aérienne, ne se réduit pas à des tropismes. Il apparaît comme une conduite instinctive au cours de laquelle le choix d'une direction est une réponse innée à un objet spécifique perçu visuellement. Cette direction est fixée dans la mémoire de l'insecte qui, grâce à une faculté d'orientation, peut ensuite la retrouver indépendemment de l'heure et de l'aspect des lieux. L'orientation des migrations suivantes est prise sans avoir secours à des repères visuels terrestres. La direction reste toujours la même mais le sens est

Fig. 4. A droite, une aire de développement larvaire de *M. melolontha* constituée par une prairie. Les grandes taches claires représentent des zones ou les larves souterraines (les vers blancs) ont détruit entièrement le gazon. — A gauche, une partie avancée de la forêt qui constitue l'aire d'alimentation des adultes. Les migrations des femelles s'effectuent entre ces deux aires. A l'aller, les insectes qui sortent des prés prennent, grâce à la vue, la direction de la forêt. Ils se dirigent vers les arbres selon une trajectoire rectiligne et orientée. La direction suivie est retenue. A la migration de retour, les femelles pleines d'oeufs partent, depuis l'intérieur de la forêt, sans l'aide de repères visuels terrestres, dans le sens inverse de la direction suivie au vol aller. Elles prennent ainsi, grâce à leur faculté d'orientation, la direction de la prairie

lié à l'état physiologique. Au départ du *voyage de retour* il y a chez les femelles gravides inversion spontanée du sens de la direction. Après le dépôt des oeufs, les femelles effectuent un deuxième aller vers la forêt ou migration après ponte; le sens primitif de la direction réapparaît spontanément. Le deuxième vol de retour, qui succède, présente les mêmes caractéristiques d'inversion que le premier. Le Hanneton commun constitue ainsi un exemple remarquable d'un *migrateur* dont le comportement est conditionné pendant toute son existence par une impression recue dans des circonstances bien déterminées.

Le maintien de la direction, au moment de chaque envol après un vol interrompu et pendant toute la migration de ponte, est si fort que les insectes volent par-dessus un obstacle plutôt que de le contourner. Des phénomènes analogues ont déjà été observés chez d'autres insectes migrateurs (WILLIAMS, 1958). Mais pendant les déplacements vers la forêt, l'insecte a la possibilité d'entrer en relation visuelle avec le paysage, de sorte que, si le site vers lequel il se dirige est caché par un obstacle artificiel, le Hanneton modifie sa trajectoire vers un autre but resté visible. Ces changements n'affectent cependant pas le souvenir de la première orientation.

La direction est toujours retrouvée sans le secours de repères visuels terrestres. Jusqu'ici il n'a pas pu être montré que les Hannetons étaient doués d'une faculté d'orientation astronomique liée à un sens interne du temps. D'autres mécanismes peuvent intervenir. Les Hannetons sont capables de faire le point et d'apprécier leur direction au moment de l'envol grâce à certains repères cosmiques. Le soleil joue un rôle, mais il existe d'autres guides encore inconnus. Les phénomènes d'orientation du Hanneton posent encore des problèmes irritants qui doivent être résolus en poursuivant les recherches et en les étendant à d'autres insectes migrateurs qui sont très nombreux dans la nature (WILLIAMS, 1958; AUBERT, 1961).

Les migrations orientées des Hannetons limitent l'extension des foyers du ravageur. Sur le plan agronomique, il résulte des allers et retours, orientés entre deux territoires, une faible dissémination des pondeuses, qui déposent toujours leurs oeufs dans les mêmes champs où l'on note de grosses concentrations de larves; la compétition alimentaire y est très vive, elle entraîne de fortes mortalités des Vers blancs et freine la multiplication de l'espèce (ROBERT, 1952; COUTURIER et ROBERT, 1956).

VI. Résumé

Le Hanneton commun *(Melolontha melolontha L.)* est un insecte européen. Les larves souterraines ou Vers blancs dévorent les racines des plantes cultivées ou de l'herbe. Les adultes s'alimentent aux dépens des arbres fruitiers ou forestiers. Les femelles ailées effectuent au cours de leur vie aérienne des déplacements ou migrations entre une aire de développement larvaire et une aire d'alimentation des adultes.

Une faculté d'orientation a été mise en évidence grâce à des élevages et des lâchers dans des paysages inconnus.

A la première sortie de terre, les Hannetons effectuent une migration aller ou migration préalimentaire; ils partent vers une silhouette qui fait contraste sur l'horizon et qui est perçue visuellement; la direction suivie au cours de ce vol est retenue. Au moment de la migration de retour ou migration de ponte les mères gravides partent, sans tenir compte de repères visuels terrestres, dans une direction diamétralement opposée à celle suivie lors de la migration aller. De même, après la ponte, les femelles repartent dans la direction prise 20 à 30 jours plus tôt à la première migration aller, sans tenir compte du paysage, grâce seulement à la faculté d'orientation.

Le comportement des insectes devant un obstacle artificiel (nuage de fumée) a été aussi étudié au cours des diverses migrations.

Les facteurs cosmiques, permettant aux Hannetons de retrouver leur direction, ont été recherchés. Le soleil joue un rôle important, mais d'autres guides utilisés sont encore inconnus. Une faculté d'orientation astronomique, impliquant un sens du temps lié à un rythme de 24 heures, n'a pas pu être mise en évidence.

Zusammenfassung

Der Feldmaikäfer ist ein eingeborenes, europäisches Insekt. Die Larven leben im Boden und fressen an den Wurzeln der Kulturpflanzen oder des Grases. Die Käfer ernähren sich an Obst- und Waldbäumen. Die Weibchen führen Wanderungen zwischen dem Larvenbrutgebiet und dem Käferfraßgebiet aus.

Durch Züchtungen und Versetzungsversuche konnte eine Orientierungsfähigkeit erwiesen werden.

Nach dem Schlüpfen findet beim Maikäfer eine Hinwanderung (Fraßwanderung) statt; der Käfer strebt eine vom Horizont abstechende, optisch wahrgenommene Silhouette an; diese Flugrichtung wird im Gedächtnis festgehalten. Bei der Rückwanderung (Legewanderung) ziehen die trächtigen Weibchen, ohne Rücksichtnahme auf optische Landmarken, in der dem Hinflug entgegengesetzten Richtung. Desgleichen schlagen nach der Eiablage die Weibchen dieselbe Richtung ein, die sie 20—30 Tage eher, beim ersten Hinflug, eingenommen hatten; dies geschieht ohne Berücksichtigung der Landschaft, dank dem alleinigen Vorhandensein einer Orientierungsfähigkeit.

Das Verhalten der Käfer vor einem künstlichen Hindernis (künstliche Nebelwand) während seiner verschiedenen Wanderungen wurde ebenfalls bearbeitet.

Die kosmischen Marken, welche dem Maikäfer das Herausfinden seiner Orientierung ermöglichen, wurden erforscht. Die Sonne spielt eine wichtige Rolle, doch verwendet der Käfer zusätzliche, noch unbekannte, Bezugsmarken. Eine astronomische Orientierungsfähigkeit, die einen an einen 24-Stunden-Rhythmus gebundenen Zeitsinn voraussetzt, konnte nicht erwiesen werden.

Bibliographie

Aubert, J.: Observations sur des migrations d'insectes au col de Bretolet (Alpes valaisannes, 1923 m). Note préliminaire. Bull. Soc. Entom. Suisse 35, 130—138 (1962).

Couturier, A., et F. Antoine: Nouvelles observations sur le déterminisme de l'envol du Hanneton commun *(Melolontha melolontha* Linné) (Coléopt. Scarabeidae). C. R. Acad. Sci. (Paris) 254, 159—161 (1962a).

— — Observations sur le déterminisme de l'envol dans les genres *Rhizotrogus* et *Amphimallon* (Coléopt. Scarabeidae). C. R. Acad. Sci. (Paris) 254, 1875—1877 (1962b).

Couturier, A., et P. Robert: Observations préliminaires sur le déterminisme de l'orientation des vols crépusculaires du M. melolontha. IXe Congr. Intern. Entom., Amsterdam (1951).
— — Maintien de la direction de vol chez Melolontha melolontha L. (Coléopt. Scarabeidae). C. R. Acad. Sci. (Paris) 240, 2561—2563 (1955a).
— — Recherches sur le comportement du Hanneton commun (Melolontha melolontha L.) au cours de sa vie aérienne. Ann. Epiphyties 1, 19—60 (1955b).
— — Orientation »astronomique« et déterminisme de la direction des grands vols chez Melolontha melolontha L. (Coléopt. Scarabeidae). C. R. Acad. Sci. (Paris) 242, 3121—3124 (1956a).
— — Observations sur Melolontha hippocastani F. Ann. Epiphyties 3, 431—450 (1956b).
— — Signification au point de vue agricole de l'orientation astronomique du Hanneton commun. C. R. Acad. Agric. Fr. 6, 350—352 (1957a).
— — Recherches sur la faculté d'orientation du Hanneton commun (Melolontha melolontha L.) (Coléopt. Scarabeidae). C. R. Acad. Sci. (Paris) 245, 2399—2401 (1957b).
— — Recherches sur les migrations du Hanneton commun (Melolontha melolontha L.). Ann. Epiphyties 3, 257—328 (1958).
— — Caractères particuliers des migrations du Hanneton commun (Melolontha melolontha L.). XI. Congr. Intern. d'Entom., Vienne (sous presse) (1960).
Hurpin, B.: Super-famille des Scarabaeoidae. In A. S. Balachowsky, Entomologie appliquée à l'Agriculture, T. 1, 24—204. Paris: Masson et Cie 1962.
Médioni, J.: L'orientation »astronomique« des arthropodes et des oiseaux. Ann. Biol. 32, 37—67 (1956).
Papi, F.: Experiments on the sense of time in Talitrus saltator (Montagu) (Crustacea-Amphipoda). Experientia (Basel) 11, 201 (1955).
— L. Serretti e S. Parrini: Nuove ricerche sull'orientamento e il senso del tempo di Arctosa perita Latr. (Araneae Lycosidae). Z. vergl. Physiol. 39, 531—561 (1957).
Renner, M.: Über ein weiteres Versetzungsexperiment zur Analyse des Zeitsinnes und der Sonnenorientierung der Honigbiene. Z. vergl. Physiol. 42, 449—483 (1959).
Robert, P.: L'évolution d'une population de Hannetons communs (Melolontha melolontha L.) dans un foyer simple à Rouffach (Haut-Rhin). Ann. Epiphyties 2, 257—281 (1953).
Schneider, F.: Untersuchungen über die optische Orientierung der Maikäfer (Melolontha vulgaris und M. hippocastani F.) sowie über die Entstehung von Schwärmbahnen und Befallskonzentrationen. Mitt. Schweiz. Ent. Ges. 29, 269—340 (1952).
— Zur Orientierung des Maikäfers beim Rückflug. Mitt. Schweiz. Ent. Ges. 25, 269—340 (1956).
— Beeinflussung der Aktivität des Maikäfers durch Veränderung der gegenseitigen Lage magnetischer und elektrischer Felder. Mitt. Schweiz. Ent. Ges. 33, 223—237 (1961).
Viaud, G.: Problèmes psycho-physiologiques posés par les migrations animales. J. Psychol. 46, 12—48 (1953).
Williams, C. B.: Insect migration, 235 pp., London, Collins 1958.

Ultraoptische Orientierung
des Maikäfers (*Melolontha vulgaris* F.)
in künstlichen elektrischen und magnetischen
Feldern

Von F. Schneider

Eidgenössische Versuchsanstalt Wädenswil (Zürich)

Mit 6 Abbildungen

Unter „Ultraoptischer Orientierung" oder „Feld-Orientierung" werden hier alle jene Orientierungsphänomene zusammengefaßt, welche sich auf statische oder wechselnde physikalische Felder unter Ausschluß der optischen Orientierung und der rein mechanisch erklärbaren Reaktionen auf akustische Felder und das vertikale Gravitationsfeld der Erde beziehen (SCHNEIDER, 1961). Die Empfindlichkeit der Tiere auf ultraoptische Einwirkungen muß nicht nur im Zusammenhang mit Unter-

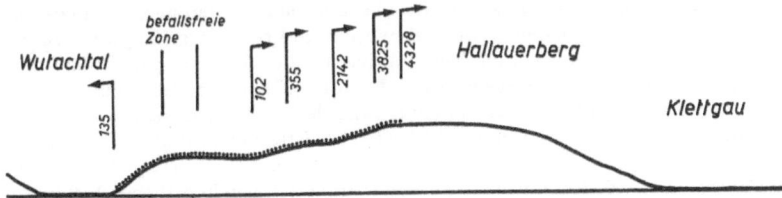

Abb. 1. Rückkehr legereifer Maikäfer in ihr altes Brutgebiet unabhängig von der topographischen Situation. Pfeile: Flugrichtungen mit den entsprechenden Zählresultaten. Punkte: Wald. Zwischen den von links und rechts beflogenen Befallsgürteln bleibt eine befallsfreie Zone, welche die beiden Populationen isoliert (24. 5. 1954)

suchungen über die räumliche Orientierung berücksichtigt werden (SCHNEIDER, 1957, 1960, 1961, 1962), sondern auch beim Studium von Aktivitätsschwankungen und anderen rhythmischen Erscheinungen (SCHNEIDER, 1960, 1961, 1963) und vor allem beim Problem der sog. „Inneren Uhr" (BROWN, 1960).

Die Untersuchungen am Maikäfer sind durch eine Beobachtung im Freiland ausgelöst worden. Nachdem die Bildung von Schwärmbahnen und Befallskonzentrationen an Waldrändern auf Grund der hypsotaktischen Orientierung (SCHNEIDER, 1952) rein optisch erklärt werden konnte, standen wir vor der Frage, wie sich der Käfer beim Rückflug vom Fraßplatz zum Brutplatz orientiert (SCHNEIDER, 1956). Wir fanden ein günstiges Versuchsgelände im Kanton Schaffhausen bei Hallau (Abb. 1).

10*

Beim ersten Ausflug werden die Tiere, welche sich in der Ebene rechts entwickelt haben, durch die überragende Silhouette des Hallauerberges angezogen; sie finden jedoch erst am bewaldeten Hang gegen das Wutachtal hinunter geeignete Fraßbäume. Der gleiche Wald wird auch vom Wutachtal her beflogen. Die beiden Befallsgürtel sind durch eine breite befallsfreie Zone voneinander getrennt. Zur Zeit des massivsten Rückfluges zur Eiablage (24. Mai 1954) richteten wir längs einer Waldschneise am ganzen Abhang gegen die Wutach verteilt sieben Kontrollposten mit Zähluhren ein; sie hatten die Aufgabe, Zahl und Richtung fliegender Käfer in einem begrenzten Himmelsabschnitt festzustellen. Das Ergebnis (Abb. 1) beweist, daß die legereifen Käfer in ihr altes Brutgebiet zurückkehren, auch wenn sie ganz regelwidrig zuerst bergwärts fliegen müssen, und daß die beiden Populationen im Klettgau und im Wutachtal sich nicht vermischen. Die Käfer offenbaren beim Rückflug zur Eiablage einen präzisen Richtungssinn und eine Gedächtnisleistung. Unabhängig von topographischen Gegebenheiten ist die Rückflugrichtung gegenüber der hypsotaktisch gesteuerten Ausflugrichtung um 180° gedreht, was Couturier und Robert (1956) mit Dislokationsversuchen bestätigen konnten.

Zur Analyse der Bezugssysteme, welche der Rückflug-Orientierung dienen, verwendeten wir vom Mai 1957 an folgende Methoden: Flugreife, jedoch durch Licht noch immobilisierte Tiere wurden während der abendlichen Dämmerung in einer bestimmten Himmelsrichtung an einem Aluminiumdraht im Freien aufgehängt. Im Verlaufe dieser „Exposition" wurden die Tiere aktiviert und schwirrten am Draht; damit ahmten wir den natürlichen Ausflug nach. Nach 1—2 Tagen prüfte man ihre Orientierung beim Abflug im Laboratorium unter variablen, mehr oder weniger künstlichen Bedingungen. Dazu diente anfänglich eine in zwei Etagen unterteilte Kiste mit einer Grundfläche von 36 × 36 cm. Der obere Teil erlaubte eine gleichmäßige Ausleuchtung des Flugraumes. Die Käfer wurden im Flugraum unter dem künstlichen Himmel auf einen Kork gesetzt und durch allmähliche Herabsetzung der Beleuchtung aktiviert. Sie flogen dann nach einigen Wendungen auf dem Kork in einer bestimmten Richtung gegen die Kistenwand; solche Flüge konnten mit den gleichen Käfern unmittelbar nacheinander mehrmals wiederholt werden.

Die Abflüge in der Kiste waren wie erwartet nach der Expositionsachse ausgerichtet, wenn vor dem künstlichen Himmel eine Polarisationsfolie befestigt wurde, die Schwingungsrichtung des Lichtes mit den natürlichen Verhältnissen nach Sonnenuntergang übereinstimmte und der Versuch mit der natürlichen abendlichen Dämmerung zusammenfiel. Doch bald zeigte es sich, daß die Käfer auch in diffusem Licht ganz bestimmte Himmelsrichtungen bevorzugen und das Orientierungsverhalten mit der Tageszeit oft sehr charakteristische Änderungen erleidet.

Vorversuche mit künstlichen magnetischen Feldern verliefen positiv. Wenn ein exponierter Käfer in der Kiste zum Abflug gebracht wurde, und zwar abwechslungsweise ohne künstliches magnetisches Feld und in der Nähe eines kleinen Hufeisenmagneten in ganz bestimmter Stellung,

erzielte man Unterschiede in den entsprechenden Flugdiagrammen. Wir
beschränken uns hier auf ein einziges Beispiel (Abb. 2), weil mit einer
anderen Methode der Nachweis einer magnetischen Orientierung leichter
gelingt. Oder wenn man eine Reihe gleichgerichteter Käfer im Freien in
der Nähe eines Hufeisenmagneten exponierte, zeigten die Abflugrichtungen in der Versuchskiste später Abweichungen, welche mit der
ursprünglichen Lage im künstlichen magnetischen Feld in Beziehung
gesetzt werden konnten. Aus den Versuchen des Jahres 1957 durfte
gesamthaft geschlossen werden, daß die Käfer bei der Wahl ihrer Abflug-

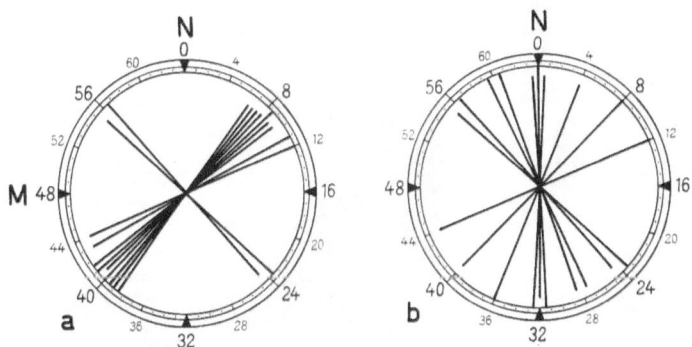

Abb. 2a u. b. Ein nach Himmelsrichtung 8 exponierter Käfer fliegt in der Versuchskiste abwechslungsweise
neben einem bei 48 außerhalb der Kistenwand angebrachten Hufeisenmagnet (a) und ohne künstliches
magnetisches Feld (b). N = Norden. Das künstliche magnetische Feld bewirkt eine straffe Flugorientierung
in der NO/SW-Achse und senkrecht dazu (4. 7. 1957)

richtung oft schon auf geringe Richtungsänderungen magnetischer Felder
reagieren (SCHNEIDER, 1957), was seither mit bedeutend exakteren
Versuchsdispositionen mehrfach bestätigt werden konnte. 1958 ersetzte
ich die starre Flugkiste durch einen auf einem Glaskugellager drehbaren
Flugraum, dessen Lage zur geographischen Nordsüdachse an einer Skala
abgelesen werden kann. Im Innern wurde ein zylindrischer, mit einem
Messingring beschwerter Vorhang aus Samt aufgehängt mit einer durch
einen schwarzen Streifen getarnten Beobachtungsöffnung von 9×9 cm
(Abb. 3). Die Helligkeit des künstlichen Himmels aus Milchglas läßt
sich auch hier mit einem Schiebewiderstand regulieren. Dieser Flugraum
aus Samt ist optisch indifferenter als die vier Kistenwände, die Käfer
werden nicht beschädigt, wenn sie in geradlinigem Flug aufprallen, sie
können sich auf dem Samt festkrallen und sind doch leicht davon ab-
zuheben. Die Richtungskreise des Drehtisches und des Flugraumes sind
in 64 Sektoren unterteilt. Die Käfer werden in die Mitte des Flugraumes
auf einen Kork gesetzt und durch künstliche Dämmerung in Flugstim-
mung gebracht. Oft vollführen sie vor dem Abflug einen Rundlauf oder
drehen sich um die Vertikalachse, um dann nach links und rechts kor-

rigierend oder erregt tanzend die Abflugrichtung auszuwählen. Die Flugrichtung wird auf dem inneren Skalenkreis abgelesen. Nach jedem Flug
wird der Tisch um beispielsweise 10 Teilstriche gedreht. Man erhält auf
diese Weise von jedem Tier zwei Flugdiagramme; das eine bezieht sich
auf den Flugraum und enthält eventuelle optische Komponenten der
Orientierung, das andere auf die geographische Nordsüdachse und gibt

Auskunft über ultraoptische Orientierungstendenzen (Schneider, 1960). Stark streu-
ende Diagramme lassen sich auswerten,
indem man die Symmetrieachse bezüglich
aller vom Käfer gewählten Flugrichtungen
berechnet.

Abb. 3a u. b. Drehbare Flugkiste zum Nachweis einer ultraoptischen Orientierung des Maikäfers. a Gesamtansicht; Samtzylinder mit Guckloch; der Beobachter wirft ein schwarzes Tuch über sich, um das Innere der
Kiste vor seitlichem Lichteinfall zu schützen. b Samtzylinder gehoben, um die Abflugrampe (Kork) und
die innere Richtungsskala zur Bestimmung der Flugrichtung sichtbar zu machen

Der Wechsel vom starren zum richtungsvariablen Flugraum hatte
eine vermehrte Streuung in der Richtungswahl der Käfer zur Folge. Die
wiederholte Drehung des optischen gegenüber dem ultraoptischen Bezugssystem führt den Käfer scheinbar in ein Dilemma. Oft zeigen die Käfer
eine präzise ultraoptische geographische Orientierung, ohne sich um die
Expositionsrichtung zu kümmern. Am 3. September 1958 sind beispielsweise 7 Käfer in den Himmelsrichtungen 6, 15, 26, 36, 39, 46 und 59
exponiert worden. Der Flugversuch zwei Tage später ließ keine Beziehung
zwischen Expositionsrichtung und Lagen der Symmetrieachsen der Flugdiagramme erkennen, dagegen eine Korrelation zwischen Flugzeit und
Orientierung. Von 14.34 bis 15.42 liegen die Symmetrieachsen in einem

lockeren Bündel bei $4^1/_2/36^1/_2$, $7/39$, $7^1/_2/39^1/_2$ und $11/43$ von 16.15 bis 17.10 in einem dichten Bündel bei $20/52$, $20^1/_2/52^1/_2$ und $21/53$. Ein neun-stündiger Versuch am 9. September 1958 mit nicht exponiertem Material zeigte schließlich, daß sich der Maikäfer spontan, ohne jede Flug-erfahrung, ultraoptisch orientieren kann und daß die Richtungswahl

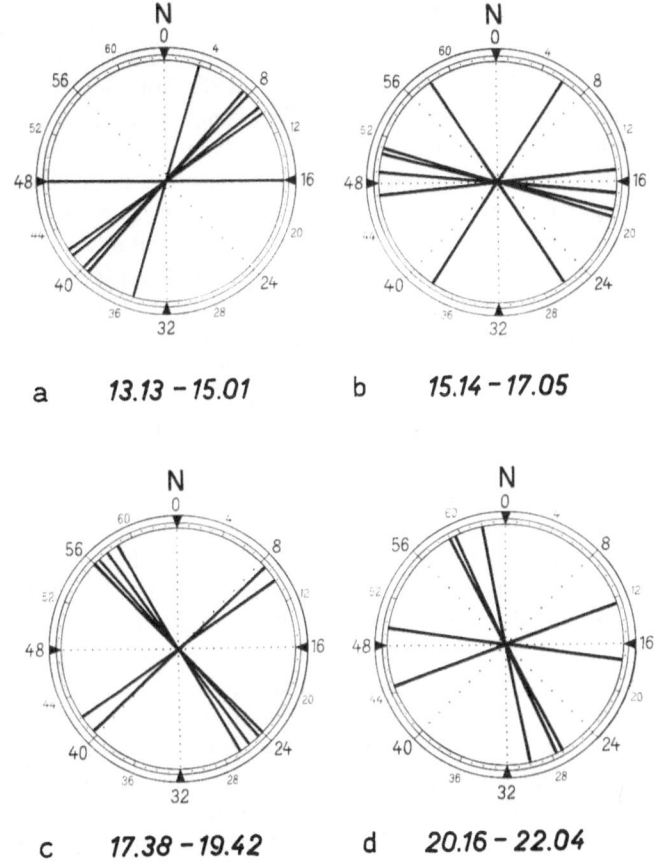

a   *13.13 – 15.01*   b   *15.14 – 17.05*

c   *17.38 – 19.42*   d   *20.16 – 22.04*

Abb. 4a—d. Ultraoptische, mit der Zeit veränderliche Orientierung nicht exponierter Käfer;
N = Norden; die Strahlen sind Symmetrieachsen von Flugdiagrammen der nacheinander zum
Abflug gebrachten Versuchstiere (9. 9. 1958)

zeitlichen Variationen unterworfen ist. In Abb. 4 sind für vier Zeit-phasen (notiert sind die Zeiten der ersten Flüge) die Symmetrieachsen der ultraoptischen Flugdiagramme eingetragen. Jeder Strahl entspricht einem Versuchstier und 6 bis 47, durchschnittlich etwa 23 Flugrichtungen, wobei der Beobachtungsraum nach jedem Abflug um 10 Teilstriche gedreht wurde; in den 4 Diagrammen sind also 520 Flugbeobachtungen verarbeitet. Man sieht recht deutlich, wie sich die Achsenbündel drehen;

von 17.38 bis 19.42 ist die Streuung sehr gering, und die beiden Bündel stehen genau senkrecht aufeinander. Auffällig ist in der ersten und dritten Versuchsphase die Vorliebe für das geographische Achsenkreuz 9/41 — 25/57. Wenn man die optischen Flugdiagramme, d. h. die Flugrichtungen in bezug auf den richtungsvariablen Flugraum zeichnet und ihre Symmetrieachsen berechnet, läßt sich in keiner Phase des Versuchs eine einheitliche Orientierung feststellen.

Dieser Versuch beweist, daß der Maikäfer trotz Ausklammerung des optischen Bezugssystems über einen Sinn für die Himmelsrichtung verfügt. Die zeitlichen Variationen in der Richtungswahl zeigen uns ferner, daß diese ultraoptische Orientierung nicht allein auf einer Ausrichtung nach dem stationären erdmagnetischen Feld beruhen kann, sondern sich wenigstens partiell auf ein sehr wirksames zweites Bezugssystem stützen muß. Das vorläufig unberechenbare Auftreten einer mit der Zeit variablen Spontanorientierung neben der durch die Exposition geprägten Richtungswahl ist ein schweres Hindernis für eine systematische Untersuchung der mnemotaktisch gesteuerten, natürlichen Rückflugsorientierung.

Das folgende und letzte Beispiel eines Drehtischversuches soll zeigen, wie sich unter günstigen Voraussetzungen auch mit dieser Methode eine magnetische Orientierung nachweisen läßt. In zwei Bohrungen der Säule des Drehtisches (Abb. 3) wurden starke Stabmagnete derart angebracht, daß die Nadel eines Kompasses am Abflugort der Käfer nach Richtung 55 wies. Der künstliche Nordpol (KN) war somit gegenüber dem natürlichen um 9 Teilstriche nach links verschoben. Durch Verlagerung des einen Magnetstabes in eine dritte Bohrung ließ sich der künstliche Nordpol leicht nach 48 abdrehen. Am 5. 12. 1959 von 10.12—11.32, am 9. 12. von 10.26—11.24, am 10. 12. von 20.06—21.25 und am 12. 12. von 10.23—11.46 wurden nun bei KN-Stellung 55 in der Versuchskiste 15 Käfer Richtung 58 kurz exponiert, sofort vom Aluminiumdraht gelöst und auf den Kork gesetzt. Abwechslungsweise blieb für einen ganzen Flugversuch mit richtungsvariabler Kiste der KN bei 55 oder 48. An jedem Tag flogen somit Käfer vergleichsweise bei KN 55 und 48 trotz gleicher Exposition bei KN 55. In den beiden Diagrammen a und b der Abb. 5 sind die entsprechenden Symmetrieachsen der einzelnen Flugdiagramme eingetragen. Bei KN 55 bilden sie zwei senkrecht aufeinander stehende Bündel Richtung 9/41 und 25/57. Die Ähnlichkeit mit dem dritten Diagramm in Abb. 4 ist sehr auffallend. Die Käfer haben ihre Flüge symmetrisch zu den immer wieder in solchen Versuchen hervortretenden Vorzugsachsen NE/SW und SE/NW ausgerichtet oder parallel und senkrecht zur Expositionsrichtung. Bei Drehung des künstlichen Nordpols von 55 nach 48 (b) drehen auch die Symmetrieachsen um beinahe denselben Winkel und die Streuung wird eher größer. Der Einfluß des künstlichen magnetischen Feldes auf die Abflugsorientierung ist eindeutig erkennbar. Doch ist es nicht erwiesen, ob die Käfer sich bei der Exposition die Stellung des magnetischen Feldes eingeprägt haben und beim Abflug der Drehung dieses Feldes folgen oder ob es sich wie in Abb. 4 bloß um Spontanorientierungen unter zwei verschiedenen Konstellationen der ultraoptischen Bezugssysteme handelt.

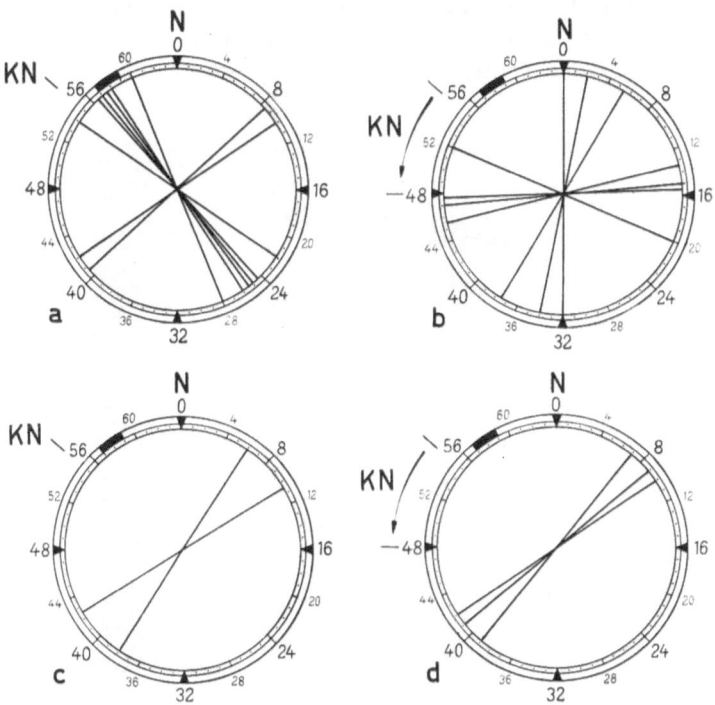

Abb. 5a—d. Gelungener und mißlungener Versuch, die ultraoptische Orientierung exponierter Käfer durch Drehung eines künstlichen magnetischen Feldes zu beeinflussen. N = geographischer Nordpol, KN = künstlicher Nordpol; die Strahlen sind Symmetrieachsen durch die einzelnen Flugdiagramme. a Ausrichtung der Symmetrieachsen nach der Expositionsrichtung und senkrecht dazu, wenn der KN unverändert bleibt, b Drehung der Symmetrieachsen, wenn der KN von 55 nach 48 gedreht wird. c und d trotz Drehung des KN bleiben die Symmetrieachsen senkrecht zur Expositionsachse ausgerichtet (Dezember 1959)

Nach 8 Tagen wurde dieser Versuch wiederholt, d. h. am 16. 12. von 10.50—11.15 und um 19.35 und am 17. 12. von 10.53—11.50 (Abb. 5c, d). Die Symmetrieachsen liegen auf oder in unmittelbarer Nachbarschaft der Vorzugsachse 9/41 wie in der ersten Versuchsserie bei KN 55. Von einer Ablenkung durch die veränderte Lage des künstlichen Nordpols ist nichts zu sehen, obwohl auch hier eine ultraoptische Orientierung vorliegt. Der Nachweis einer durch magnetische Felder beeinflußten Orientierung gelingt manchmal sehr eindrücklich, manchmal überhaupt nicht. Außer den magnetischen Feldern müssen dem Maikäfer demnach noch weitere, in ihrer Wirksamkeit variable ultraoptische Bezugssysteme zur Verfügung stehen.

Im Jahre 1960 ersetzte ich das Experiment mit dem Drehtisch durch den Schalenversuch. Die Versuchsbedingungen entfernen sich hier noch weiter von den natürlichen, indem die Käfer überhaupt nicht mehr fliegen, sondern sich nur noch zu Fuß bewegen; doch ist das zu diesem Zweck verwendete Versuchsmaterial physiologisch einheitlicher, die Ausbeute an verwertbaren Ergebnissen pro Zeiteinheit kann gesteigert werden und es ist technisch einfacher, die Käfer präzis definierten elektrischen und magnetischen Feldern auszusetzen.

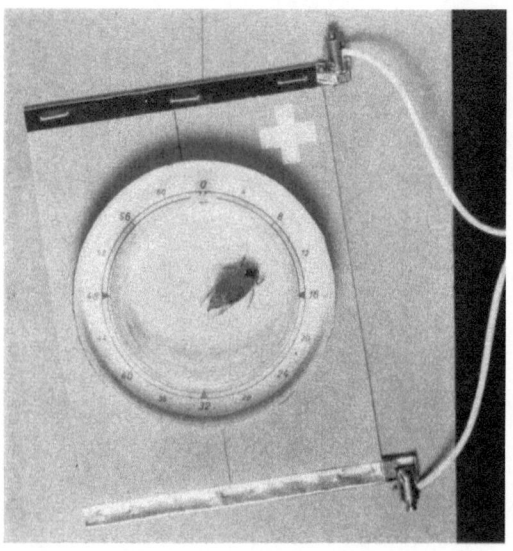

Abb. 6. Schalenversuch zur Untersuchung der Wirkung künstlicher elektrischer Felder auf Orientierung und Aktivität des Maikäfers. Kartondose geöffnet. Der Käfer läuft in der horizontalen Glasdoppelschale und kommt in einer Vorzugsrichtung zur Ruhe, welche mit Hilfe des linierten und mit Zeiger ausgerüsteten Glasdeckels und der Richtungsskala (O = Norden) bestimmt werden kann. Die beiden mit einer Batterie verbundenen Aluminiumstreifen auf der drehbaren Kartonmanschette erzeugen ein elektrisches Feld, dessen Vektor ebenfalls auf der Richtungsskala abgelesen werden kann

Der Schalenversuch geht auf eine merkwürdige Beobachtung im Sommer 1959 zurück. Um Käfer für Drehkistenversuche vorzubereiten, setzte ich sie einzeln ohne Futter in kleine Glasdoppelschalen. Nach anfänglichen Aktivitätsphasen fielen sie in Starre und stellten sich dabei schief und in ganz verschiedenen Winkeln zum einfallenden indirekten Sonnenlicht ein. Die Körperachsen aller dieser zur Ruhe gekommenen Tiere waren jedoch parallel oder standen senkrecht aufeinander. Diese parallele oder rechtwinklige Ausrichtung wiederholte sich später mehrmals auch nach vorübergehenden Aktivitätsphasen, wobei sich die Himmelsrichtung der Körperachsen allmählich etwas drehte.

Die heute gebräuchliche Versuchsanordnung ist aus Abb. 6 ersichtlich; sie ist bereits früher kurz beschrieben worden (SCHNEIDER, 1960, 1961). Ein kältestarres Tier wird in die Mitte der in einem Richtungskreis fixierten horizontalen Petrischale gesetzt, Kopf nach Norden gerichtet. Die Doppelschale läßt sich mit dem Deckel einer Kartondose verdunkeln. Unter der Wirkung der relativ hohen Temperatur des Versuchsraums (20° C) beginnt der Käfer umherzulaufen. In der glatten horizontalen und verdunkelten Schale ist keine Himmelsrichtung in taktiler oder optischer Hinsicht irgendwie ausgezeichnet. Ultraoptische Bezugssysteme treten gegenüber trivialen in den Vordergrund. Alle 10 min wird die Dose einige Sekunden geöffnet, um den Aktivitätsgrad und die Körperstellung des Versuchstiers festzustellen. Wenn der Käfer in zwei aufeinanderfolgenden Kontrollen ruhig sitzt und die gleiche Körperstellung $\pm \frac{1}{64}$ des Kreisumfangs einnimmt, wird die Richtung der Körperachse als definitive Ruhelage notiert und der Käfer durch einen frischen ersetzt. Man kann gleichzeitig mehrere Schalen mit Käfern beschicken, und die Versuche können ohne Unterbrechung stundenlang fortgesetzt werden. Mit einfachen technischen Mitteln werden die Schalen elektrischen und magnetischen Feldern ausgesetzt mit genau definierter Stärke und Vektorenrichtung. Die Richtung des elektrischen Feldes zwischen den beiden Aluminiumstreifen in Abb. 6 läßt sich durch Drehung der Kartonmanschette beliebig verändern. Künstliche magnetische Felder werden durch Stabmagnete erzeugt, welche in horizontaler Lage unter der Tischplatte auf einer drehbaren, mit Richtungsskala ausgerüsteten Scheibe fixiert sind. Mit Hilfe eines geeigneten Verschiebungsschemas können die Orientierungstendenzen nach künstlichen Feldern und nach der Himmelsrichtung getrennt dargestellt und analysiert werden. Die Versuche werden heute in einem fensterlosen, unterirdischen, temperaturkonstanten, schwach und diffus beleuchteten Kellerraum ausgeführt, der Versuchsort ist zudem durch einen feinmaschigen Faradaykäfig vor Radiowellen geschützt.

In den Jahren 1960 bis 1962 sind mit Schalenversuchen[1] unter anderen folgende Ergebnisse erzielt worden:

1. Die Maikäfer bevorzugen oder meiden oft bestimmte Richtungen oder Sektoren in bezug auf die geographische Nordsüdachse und auf die Vektoren künstlicher elektrischer und magnetischer Felder (SCHNEIDER, 1960, 1962).

2. Drehungen künstlicher Felder in bezug auf die geographische Nordsüdachse können deutliche Effekte auf die Orientierung ausüben.

3. Auch in einem künstlichen, richtungsvariablen, horizontalen, magnetischen Feld, das etwa 50mal stärker ist als die Horizontalkomponente des erdmagnetischen Feldes, bewahren die Käfer ihren Sinn für Himmelsrichtungen. Sie stehen offenbar auch unter dem Einfluß natürlicher, jedoch nicht erdmagnetischer, ultraoptischer Bezugssysteme.

4. Einzelne natürliche ultraoptische Bezugssysteme sind veränderlich. Sie beeinflussen damit nicht nur die Orientierung nach der geographischen Nordsüdachse, sondern auch nach künstlichen elektrischen

---

[1] Ein paar Beispiele sind in Form von Diagrammen vorgewiesen worden; sie werden an anderer Stelle ausführlich und mit statistischem Kommentar veröffentlicht.

und magnetischen Feldern, weil alle diese ultraoptischen Einflüsse als Komplex auf den Käfer einwirken (Schneider, 1960, 1963).

5. Die gegenseitige Lage der Vektoren natürlicher ultraoptischer Bezugssysteme und künstlicher elektrischer und magnetischer Felder beeinflußt die Laufzeit und somit die Aktivität der Käfer in den Versuchsschalen (Schneider, 1960, 1961).

6. Die Käfer sind überindividuellen rhythmischen Aktivitäts- und Orientierungsänderungen unterworfen, welche mit der Variabilität ultraoptischer Bezugssysteme zusammenhängen (Schneider, 1963).

7. Viele der bisherigen Versuchsergebnisse lassen sich am besten in folgendem Modell unterbringen: Die Käfer sind mit spezifischen, funktionell radiären Sinnesorganen ausgerüstet zur Wahrnehmung der Vektoren ultraoptischer Bezugssysteme. Die Zahl der funktionellen Einheiten oder „Facetten" und damit der „Öffnungswinkel" eines Elements ist bei den einzelnen Organen (z. B. zur Wahrnehmung magnetischer und elektrischer Felder) verschieden. Bei der Orientierung vor der definitiven Ruhestellung suchen die Käfer besonders disponierte „Facetten" oder „Facettengruppen" nach den Vektoren adäquater physikalischer Bezugssysteme auszurichten. Je nach der gegenseitigen Stellung und Wirksamkeit der Vektoren natürlicher und künstlicher Felder und der gegenseitigen Lage der funktionellen Elemente dieser ultraoptischen Sinnesorgane gelingt eine solche Ausrichtung leicht, nur kurzfristig und bei ganz bestimmten Körperstellungen oder überhaupt nicht.

Im ersten Fall kommen die Käfer relativ rasch zur Ruhe; da sie unter mehreren möglichen Körperstellungen auswählen können, ist die Streuung in der Orientierung verhältnismäßig groß; im zweiten Fall wird der Käfer erst nach mehreren Rundläufen und Wendungen in eine geeignete Körperstellung einklinken, Laufzeit und Aktivität sind erhöht und die Ausrichtung beschränkt sich auf wenige Sektoren oder Richtungen; im letzten Fall handelt es sich um Dauerläufer, welche während der zweistündigen Beobachtungszeit kein Ergebnis liefern. Mit diesem Modell ließe sich auch erklären, wie nicht nur periodisch pendelnde sondern auch kontinuierlich sich drehende Vektoren natürlicher ultraoptischer Bezugssysteme rhythmische Änderungen in der Aktivität und in der Orientierung erzeugen, wie sie im Schalenversuch jederzeit nachgewiesen werden können.

Literatur

Brown, F. A.: Response to pervasive geophysical factors and the biological clock problem. Cold Spr. Harb. Symp. quant. Biol. **25**, 57—71 (1960).

Couturier, A., et P. Robert: Orientation «astronomique» et déterminisme de la direction des grands vols chez *Melolontha melolontha* L. C. R. Acad. Sci. (Paris) **242**, 3121—3124 (1956).

SCHNEIDER, F.: Untersuchungen über die optische Orientierung der Maikäfer (*Melolontha vulgaris* F. und *M. hippocastani* F.) sowie über die Entstehung von Schwärmbahnen und Befallskonzentrationen. Mitt. Schweiz. Entomol. Ges. 25, 269—340 (1952).

— Zur Orientierung des Maikäfers beim Rückflug. Mitt. Schweiz. Entomol. Ges. 29, 69—70 (1956).

— Die Fernorientierung des Maikäfers während seiner ersten Fraßperiode und beim Rückflug in das alte Brutgebiet. Verh. Schweiz. Naturforsch. Ges., Neuenburg 1957, 95—96.

— Der experimentelle Nachweis einer magnetischen und elektrischen Orientierung des Maikäfers. Verh. Schweiz. Naturforsch. Ges., Aarau 1960, 132—134.

— Beeinflussung der Aktivität des Maikäfers durch Veränderung der gegenseitigen Lage magnetischer und elektrischer Felder. Mitt. Schweiz. Entomol. Ges. 33, 223—237 (1961).

— Die Feinorientierung des Maikäfers nach physikalischen Feldern als Indiz für Leistungsfähigkeit und Bau entsprechender Sinnesorgane. Verh. Schweiz. Naturforsch. Ges., Schuls (im Druck) 1962.

— Orientierung und Aktivität des Maikäfers unter dem Einfluß richtungsvariabler künstlicher elektrischer Felder und weiterer ultraoptischer Bezugssysteme. Mitt. Schweiz. Entomol. Ges. 36 (im Druck) 1963.

— Systematische Variationen in der elektrischen, magnetischen und geographisch-ultraoptischen Orientierung des Maikäfers. Vierteljahrsschr. Naturforsch. Ges. Zürich 108 (im Druck) 1963.

Kompaßorientierung

Von MARTIN LINDAUER

Zoologisches Institut der Universität München

Mit 8 Abbildungen

Inhaltsübersicht

I. Einleitung

Wenn ein Orientierungsziel vom Ausgangspunkt so weit entfernt liegt, daß kein direkter sinnesmäßiger Kontakt mit diesem besteht, dann gibt es für ein Tier nur 2 Möglichkeiten, dieses Ziel zu finden:

1. Auf einem Erkundungsflug werden Geländemarken als *Zwischenziele* eingeprägt; sie ermöglichen, daß der Endpunkt bei der Wiederorientierung etappenweise angesteuert wird. Gelegentlich setzen die Tiere *selbständig* solche Hilfsmarken, wie Ameisen oder Säugetiere, wenn sie Duftspuren am Boden auslegen.

2. Eine solche Orientierungsmethode ist umständlich, und je weiter das Ziel entfernt ist, um so höhere Forderungen werden an die Aufmerksamkeit und das Gedächtnis gestellt. Es bedeutet eine wesentliche Erleichterung, wenn auf einem langen Anmarschweg lediglich die *Zug-Richtung* zu merken ist. An Stelle vieler Orientierungsmarken benötigt jetzt das Tier nur einen einzigen Bezugspunkt; zu ihm hält es menotaktisch den festgelegten Wanderkurs, bis das Ziel in „Sichtweite" ist.

Damit der Wanderkurs aber nicht zur Manegebewegung wird, müssen entweder Richtmarken Verwendung finden, die sich in großer Entfernung befinden, wie die Himmelsgestirne, oder es erfolgt eine menotaktische Einstellung zu einem Reizfeld, das über weite Horizontal- bzw. Vertikalwanderungen seine Richtungskonstante nicht ändert — zum erdmagnetischen Feld und zur Schwerkraft. Bei der *astronomischen Orientierung*, auf die sich unser Bericht beschränken wird, begegnen die Tiere jedoch der Schwierigkeit, daß die Bezugspunkte keine feststehenden Marken sind, sondern ihre Lage am Himmelsgewölbe in festgelegter Bahn verändern. Es genügt demnach nicht allein, winkeltreu zum Bezugspunkt zu wandern; erst ein komplizierter Verrechnungsmechanismus, der Tageszeit, Jahreszeit und Azimutwinkelgeschwindigkeit ortsgerecht einbezieht, sichert die kompaßtreue Festlegung und Beibehaltung des Wanderkurses. Eine andere Schwierigkeit entsteht dadurch, daß diese Himmelsmarken tageszeitlich nur beschränkt zur Verfügung stehen und daß sie bei bedecktem Himmel als Kompaß versagen.

Es ergeben sich somit 3 Grundfragen der Kompaßorientierung:

I. Wie und wann wird der Kompaßkurs — unter Konkurrenz von Landmarken — primär festgelegt und in welchen Situationen wird er für die Orientierung zu Hilfe genommen?

II. Wie wird die Beibehaltung eines Kompaßkurses vom Startpunkt bis zum Ziel unter erschwerten natürlichen Bedingungen (s. S. 165) gewährleistet?

III. Welcher Ersatz steht bereit, wenn Sonne, Mond oder Sterne als Kompaß versagen?

II. Festlegen des Kompaßkurses

Wir kennen 2 Prinzipien zur Festlegung des Kompaßkurses:

a) Die Zugrichtung ist angeboren und ohne jede Erkundung fixiert: Bei Zugvögeln, Uferspinnen, beim Strandkrebs, bei *Velia*, vielleicht bei Schildkröten und wandernden Fischen. Beweise hierfür liefern Tiere, die unter Ausschluß von Himmelslicht aufgezogen wurden, so wie das von PARDI (1960), SAUER (1956), HOFFMANN (1960) BIRUKOW (1956) u. a. geschehen ist (vgl. hierzu die folgenden Berichte von BIRUKOW, PAPI, BRAEMER und SAUER). Es scheint die Regel zu sein, daß immer da, wo der Zugweg bzw. die Fluchtrichtung ein ganzes Leben lang und für viele Generationen in gleichem Sinne beibehalten werden, diese angeborenermaßen festliegen.

b) Wenn Ziele und Wanderwege im Laufe des Lebens ständig wechseln, wie bei Bienen und Ameisen, dann muß jeweils ein Erkundungsflug den neuen Kompaßkurs festlegen. Dabei ist zu bedenken, daß schon der erste erfolgreiche Anflug — auch wenn er auf Umwegen zum Endpunkt geführt hatte — die Lage eines Zieles so gut dem Gedächtnis einprägen muß, daß das Tier bei der Wiederorientierung den direkten Weg dorthin findet. Es ist weiter zu bedenken, daß die bei der Erkundung gewonnenen Informationen erst *rückwirkend* als Signale ihren

Wert erhalten, dann nämlich, wenn der Suchflug mit einem *Erfolg* endet. Zweifellos werden auf diesem ersten Erkundungsflug die weitaus größten Anforderungen an das Orientierungsvermögen gestellt. Um die dabei gezeigten Leistungen genauer zu analysieren, stellen wir folgende Fragen:

1. Welche Signale haben für die Wiederorientierung Primärrang: Landmarken oder Himmelsmarken?

2. Welche Bedeutung kommt dem Hinflug, welche dem Rückflug für die Wiederorientierung zu?

3. Erfolgt auf späteren Ausflügen eine Umwertung der Orientierungssignale, etwa dergestalt, daß zunächst Landmarken als Zwischenziele gemerkt und erst später diese durch den Himmelskompaß ersetzt werden?

4. Für Bienen bleibt noch ein Sonderproblem: Neulinge erhalten durch Tänzerinnen Anweisung über Richtung und Entfernung des Zieles. Inwieweit ist für die Wiederorientierung die Anweisung der Stockgenossen maßgebend, inwieweit die eigene Erfahrung auf dem ersten Erkundungsflug?

Folgende Versuche sollten einen Teil dieser Fragen beantworten:

a) Die Orientierungsleistung des ersten Erkundungsfluges

Bienen, die auf ihrem *ersten* Erkundungsflug als alarmierte Neulinge an unsere Futterstelle kamen, wurden in andere Richtung und Entfernung versetzt. Die Versetzung erfolgte immer nur mit einer einzelnen Biene, während diese am Zuckerwasser saugte, und sie erfolgte nur ein einziges Mal. Die Bienen waren also nur einmal aus bestimmter Richtung und Entfernung angeflogen, und *der Rückflug führte sie aus anderer Richtung bzw. anderer Entfernung zum Stock zurück.* Um nun herauszufinden, ob die Sinneseindrücke des Hinfluges oder jene des Rückfluges entscheidend für die Wiederorientierung waren, standen für den zweiten Anflug je ein Futtertischchen am ursprünglichen Anflugplatz, ein zweites am Abflugplatz zur Wahl. Tab. 1 faßt das Ergebnis — zunächst für die Richtungsversetzung — zusammen; die Werte für „passive" Versetzung (s. u.) seien zuerst diskutiert (Tab. 1, Säulen 2, 4 u. 5):

Es zeigt sich, daß der zweite Anflug unterschiedlich auf Abflugplatz und Anflugplatz verteilt ist: Bei *Nah*versetzung wird vorwiegend der *Abflugplatz* aufgesucht; wenn die Futterstelle weiter als 50 m vom Stock wegrückt, dann bekommt der Hinflugplatz das Übergewicht. Das Gesamtverhältnis der Zweitanflüge am Anflug- bzw. Abflugplatz ist 5:25 (bei Nahversetzung) und 28:5 (bei Fernversetzung). (Werte bei 50 m sind als Grenzwerte unberücksichtigt.) Auch OPFINGER (1931) hatte ähnliche Versetzungen durchgeführt und gefunden, daß — entsprechend dem Ergebnis meiner Nahversetzung — zu etwa 80% der Abflugplatz beim zweiten Ausflug gewählt wird. Bei allen diesen Versuchen war jedoch der Futterplatz sehr nahe beim Stock (in etwa 20 m Entfernung) aufgestellt. Fernversetzungen hatte OPFINGER nicht durchgeführt.

Tabelle 1. *Wiederorientierung von Neulingen, die passiv oder aktiv in andere Richtung versetzt worden waren — bei abgestufter Entfernung des ersten Anflugortes.* Versetzung 45°, 90° und 180°

| Erster Anflug erfolgte bei m | Zahl der versetzten Bienen | | Zweiter Anflug nach | | | |
| | | | passiver Versetzung | | aktiver Versetzung | |
	passiv	aktiv	Anflugplatz	Abflugplatz	Anflugplatz	Abflugplatz
5	20	8	3	17	0	8
20	10	10	2	8	1	9
50	10	10	4	6	0	10
150	20	12	15	5	0	12
300	8	3	7	0	0	3
500	6	—	6	0	—	—

Aus obigen Ergebnissen lag die Folgerung nahe, *daß mit zunehmender Entfernung die Richtungsweisung der Tänzerinnen,* die ja den Anflugplatz anzeigen und durch die unsere Neulinge verständigt worden waren, *ausschlaggebend für die Wahl des zweiten Ausfluges sein könnte.* Dies um so mehr, als der Umschlag gerade beim Übergang vom Rundtanzgebiet in das des Schwänzeltanzes erfolgte (vgl. Tab. 1 bei 50 m).

Wieweit die Anweisung durch Tänzerinnen einerseits und eine davon abweichende subjektive Erfahrung andererseits für die Wiederorientierung ins Gewicht fielen, mußte durch eine neue Methode entschieden werden. Bisher wurde die Versetzung *passiv* durchgeführt, also so, daß die Biene am Zuckerschälchen saugend von einem Ort zum anderen getragen wurde. Wir änderten diese Methode so, daß die Biene selber *aktiv bei der Versetzung beteiligt war:* Unmittelbar nach der Landung am Anflugplatz A — nachdem er eine kurze Kostprobe genommen hatte — wurde der Neuling kurz aufgescheucht und — während das Tischchen langsam wanderte — hinten drein gelockt; zwischendurch durfte er immer wieder kurz am Futterschälchen nippen, aber erst am endgültigen Versetzungsplatz B konnte er sich vollsaugen. Damit ist der Ablauf eines normalen Erkundungsfluges weitgehend nachgeahmt: die Biene besucht da eine Blüte und dort, und findet schließlich nach längerem Herumsuchen das endgültige Ziel, das reichlich Futter bietet. Mit einer einzigen Ausnahme aus 43 Versuchen wurde bei dieser aktiven Versetzung beim zweiten Ausflug stets der *Abflugplatz* angeflogen (s. Tab. 1 die beiden letzten Säulen).

Damit ist erwiesen:

1. Die Information von seiten einer Tänzerin wird zwar beim ersten Erkundungsflug befolgt, sie verliert jedoch mit dem ersten Ausflug ihren Einfluß, sofern auf Grund einer abweichenden subjektiven Erfahrung ein lohnendes Ziel gefunden wurde.

11

2. Für die Wiederorientierung auf größere Entfernung (über 50 m) ist der *Hinflug* entscheidend. Wenn auch bei aktiver Versetzung stets der Abflugplatz aufgesucht wurde, so muß man diesen für die Bienen trotzdem als Anflugsort werten, denn das aktive Mitfliegen bis zum Endziel bedeutet nichts weiter als einen *verlängerten Hinflug*. Bei Nahversetzung liegen die Verhältnisse anders: eine dem Hinflug entsprechende Richtungsorientierung kommt nicht zum Durchbruch; offenbar, weil beim Abflug von der Futterstelle durch Orientierungsschleifen (OPFINGER, 1931) die optische Konstellation des Zieles eingeprägt wird; sie kommt beim zweiten Abflug vom Stock sofort in Sichtweite und lockt die Biene unmittelbar an.

3. Die angeführten Versuche lassen noch keine Aussage darüber zu, ob beim ersten Erkundungsflug wirklich schon der Kompaßkurs zum Ziel festgelegt wird, oder ob zunächst nur Landmarken für die Wiederorientierung benutzt werden. Bei 19 der insgesamt 117 in Tab. 1 aufgeführten Versetzungen haben die Sammlerinnen schon nach dem ersten Heimflug getanzt. (Dies läßt sich nur bei akutem Futtermangel erreichen, da normalerweise mehrere erfolgreiche Sammelflüge dem ersten Tanz vorausgehen müssen; vgl. LINDAUER, 1948.)

Alle diese Ersttänze waren in der Richtungs- und Entfernungsangabe sauber orientiert, und sie zeigten ausnahmslos jene Himmelsrichtung an, die beim zweiten Ausflug gewählt wurde. Der Kompaßkurs mußte also — ohne Zuhilfenahme von Landmarken — bereits beim ersten Ausflug — und zwar, gemäß Tab. 1, beim Hinflug zum Ziel — festgelegt worden sein.

OTTO (1959) ist es in seinen Versetzungsversuchen gelungen, auch dem Rückflug ein bestimmtes Gewicht zu verleihen. Dies gelang dadurch, daß er in einer *Dauerdressur* seine Bienen tagelang vom Anflugplatz in andere Richtung versetzte. Solche Bienen lernten — ganz im Gegensatz zu den einmalig passiv versetzten Bienen in unserer Tab. 1 — in direktem Flug vom Versetzungsplatz zum Stock zurückzukehren. Die Tänzerinnen zeigten dann die Winkelhalbierende zwischen Hinflug- und Rückflugschenkel. Zusätzliche optische Markierung konnte die Tanzrichtung zugunsten des markierten Flugschenkels verlagern. Das sind jedoch Ergebnisse, die erst durch einen nachträglichen Lernprozeß im Zuge einer anhaltenden Dressur gewonnen wurden. Sie offenbaren aber, wenn wir sie unseren Daten gegenüberstellen, eine Orientierungsleistung der Bienen, die grundlegend für jede höhere Orientierungsform ist: die höheren Instanzen im Zentralnervensystem haben nicht nur die Fähigkeit zu Assoziation, Verrechnung (in Raum und Zeit) und Integration, sie haben auch die „Vollmacht", *die über die Sinnesorgane einlaufenden Informationen verschieden zu bewerten.* Wir werden noch in anderem Zusammenhang (s. S. 173) auf diese *selektive Tätigkeit* des Zentralnervensystems im Orientierungsgeschehen hinzuweisen haben.

Wird die Versetzung vom Anflugplatz aus in andere *Entfernung* durchgeführt, dann sind auch da die Ergebnisse unterschiedlich: Versetzen wir so, daß der Abflugplatz weiter vom Stock weg liegt als der Anflugplatz, dann wird ausnahmslos der Abflugort aufgesucht; versetzen wir näher zum Stock hin, hat der Anflugplatz den Vorrang; letzteres kommt besonders darin zum Ausdruck, daß in 3 Fällen sogar bei aktiver Versetzung (Tab. 2, Versetzung von 300 auf 50 m) von fern auf nah der *ferne*

Tabelle 2. *Wiederorientierung von Neulingen, die passiv oder aktiv auf andere Entfernung versetzt worden waren*

Erster Anflug erfolgte bei m	versetzt auf m	Zahl der versetzten Bienen	Zweiter Anflug bei m	Art der Versetzung
1	150	5	150	passiv
1	1000	3	1000	passiv
15	150	3	150	passiv
50	300	8	300	5 mal passiv
				3 mal aktiv
100	1000	16	850	passiv
100	2000	8	2000	passiv
200	50	6	50 (3 Bienen)	aktiv
			200 (3 Bienen)	passiv
300	50	14	50 (6 Bienen)	aktiv
			300 (8 Bienen)	5 mal passiv
				3 mal aktiv

(Die erste Gruppe: *Versetzung von nah auf fern*; die zweite Gruppe: *Versetzung von fern auf nah*)

Platz für die Wiederorientierung gewählt wurde. Um dieses Verhalten zu verstehen, sei daran erinnert, daß die Bienen unter normalen Bedingungen ihre Weidegründe immer vom Stock ausgehend stufenweise auf einen zunehmend größeren Radius ausdehnen; niemals erkunden sie zunächst ferne Futterstellen und dann erst nahe gelegene.

Wir haben diese Richtungs- und Entfernungsversetzung auch in unbekanntem Gelände ausgeführt; ein Volk wurde in einer fremden Landschaft aufgestellt und ein Futterplatz — der einigen wenigen Bienen durch Vordressur bekannt war — sofort bei 50 m Süd aufgestellt. In wenigen Minuten hatten wir dann den ersten Neuling. Wenn wir diesen nach Süden in gleicher Richtung 300 m, 500 m, ja sogar 1 km versetzten, suchten die Bienen beim zweiten Ausflug stets den Abflugplatz wieder auf; gewiß eine erstaunliche Orientierungsleistung! Wenn aber nicht nur in andere Entfernung, sondern auch auf andere Richtung versetzt wurde, also z. B. von 50 m Süd auf 500 m Ost, dann blieb etwa die Hälfte der versetzten Bienen verschollen, die übrigen blieben lange Zeit — bis zu einer Stunde und noch länger — aus und sie kamen stets zum *Anflugplatz* bei 50 m wieder. Das beweist, daß zur Zeit der Versetzung das Gelände noch — in bezug auf Landmarken — fremd war, und es wird uns noch einmal eindrucksvoll demonstriert, wie rasch und nachhaltig die Prägung auf einen Kompaßkurs stattfindet: der einmalige Flug in Südrichtung in fremder Gegend genügt, um der Biene ausreichende Kursrichtung beim Heimfinden zum Stock und beim Wiederfinden des Zieles zu bieten. Landmarken haben wenig Bedeutung und die Entfernung kann entlang des festgelegten Kompaßkurses weitgehend variiert werden.

b) Bedeutung des ersten Anfluges gegenüber späteren Ausflügen

Um zu klären, ob spätere Ausflüge eine Bedeutung für die weitere Festigung des Kompaßkurses bzw. für eine Änderung desselben erwirken, wurden die Versetzungen im gleichen Sinne nach dem zweiten, dritten und vierten Anflug wiederholt, etwa in der Weise, daß beim ersten Anflug von 100 m auf 1000 m versetzt wurde; wenn dann die Biene beim zweiten Ausflug wie erwartet bei 1000 m wieder angeflogen war, wurde sie auf 1500 m gebracht; jetzt hatte die Versetzung keine Wirkung mehr. Die Biene kam wieder bei 1000 m an und man konnte die gleiche Versetzung zehnmal und öfter wiederholen: immer wieder erfolgte der Anflug bei 1000 m. Auch bei der Richtungsversetzung, ganz gleich ob diese aktiv oder passiv erfolgte, wurde wieder jene Futterstelle aufgesucht, die durch den *ersten* Anflug bzw. Abflug festgelegt war. Der erste Ausflug also, der erfolgreich endet, liefert die entscheidenden Sinneseindrücke für die Prägung des Kompaßkurses und für die Festlegung der Entfernung zum Ziel. Solange der Anflug belohnt wird, haben spätere Versetzungen keine Wirkung mehr. Damit sind frühere Befunde von OPFINGER bestätigt, die ebenfalls fand, daß der erste An- bzw. Abflug entscheidend für die Wiederorientierung ist. Auch BAERENDS (1941) berichtet über *Ammophila*, daß lediglich die kurzen Orientierungsflüge während des Nestgrabens für das Einprägen der Nestumgebung von Bedeutung seien. Eine Umdressur auf neue Wegmarken gelingt bei den späteren Verproviantierungsflügen nicht mehr. Wir haben damit wiederum einen ausgeprägten Auslesevorgang höherer Instanzen des ZNS festgestellt: nur was der erste erfolgreiche Erkundungsflug an Information beibringt, wird gespeichert und der Wiederorientierung zur Verfügung gestellt. Zweifellos wird damit auf späteren Ausflügen eine Entlastung der höheren Zentren zugunsten anderer, neu auftretender Orientierungsaufgaben (s. S. 175) ermöglicht.

Zusammengefaßt:

1. Schon beim ersten Erkundungsflug auf größere Entfernung wird der Kompaßkurs für die Wiederorientierung festgelegt — auch wenn die Biene auf Umwegen zum Ziel gekommen war. Nur bei nahe gelegenen Zielpunkten spielt die Richtung eine untergeordnete Rolle und die optische Konstellation der unmittelbaren Umgebung dient der Wiederorientierung als Leitsignal.

2. Spätere Ausflüge sind für die Festlegung bzw. Änderung des Kompaßkurses und für die gedächtnismäßige Prägung der Lage des Zieles ohne Bedeutung.

3. Die von einer Tänzerin weitergegebene Information verliert gegenüber der subjektiven Erfahrung beim ersten Erkundungsflug für die Wiederorientierung ihr Gewicht.

III. Die Konstanthaltung des Kompaßkurses

Ist die Marschroute festgelegt, dann obliegt dem Tier die Aufgabe, bei allen späteren Ausflügen den gleichen Kurs erneut anzupeilen und vom Heimatort zum Ziel kompaßrichtig beizubehalten.

Wir müssen noch einmal zu Anfang die Frage stellen, inwieweit hierbei irdische Landmarken zu Hilfe genommen werden; diesmal bietet sich uns die Möglichkeit, Himmelskompaß und Landmarken experimentell in Konkurrenz zu stellen: Eine Bienenschar wird in einem fremden Gelände 200 m nach Süden dressiert. Mitten auf dem Weg zum Ziel steht eine auffallende Landmarke — ein hoher Baum, eine größere Baumgruppe oder dgl. Die Bienen haben die Wahl, die Himmelsrichtung ihres Flugweges sich einzuprägen oder die irdische Geländemarke als Leitsignal zu verwenden. Um zu prüfen, welche der beiden Alternativen den Vorrang hat, versetzen wir das Volk über Nacht in eine andere fremde Landschaft und stellen in 200 m Entfernung 4 Futtertischchen zur Wahl — im Süden, Osten, Westen und Norden. Wir bieten auch die Landmarke vom Vortag wieder, aber diesmal steht sie auf dem Flugweg, der zum *östlichen* Futterplatz führt. Die Bienen folgen dem Sonnenkompaß und lassen sich nicht durch die für uns so auffallende Baumgruppe in die falsche Richtung verleiten. Nur solche irdische Orientierungsmarken, die als *durchgehende Leitlinien* vom Stock bis zum Ziel geboten werden — eine Straße, ein Seeufer, ein Waldrand —, können mit der Sonne konkurrieren. Wenn sie nach der Versetzung so geboten werden, daß sie um 90° verdreht das Gelände durchqueren, folgen die Bienen dieser irdischen Leitlinie. Dies sind aber Ausnahmefälle, die sich in der Natur wohl kaum einmal den Bienen anbieten (v. FRISCH und LINDAUER, 1954).

Damit ist der Vorrang der Himmelsgestirne für die Orientierung erneut erwiesen. Die Konstanthaltung des Kompaßkurses ohne Hilfestellung von Landmarken erfordert, wie schon eingangs angedeutet, Orientierungsleistungen besonderer Art. Wir haben speziell zu prüfen:

1. Inwieweit wird die tageszeitlich, jahreszeitlich und ortsmäßig festgelegte *Azimutänderung* von Sonne, Mond und Sternen einberechnet?

2. Inwieweit wird ein aktiver bzw. passiver *Umweg*, der die Tiere oft (s. u.) vom direkten Kurs ablenkt, integriert?

3. Wie erfolgt bei freifliegenden Tieren die Kompensation einer *Seitenwindabtrift*?

1. Einkalkulieren der Sonnenwanderung

Die ersten Beweise für das Einkalkulieren der Sonnenwanderung haben v. FRISCH (1950, 1952) und KRAMER (1952a/b) unabhängig voneinander an Bienen und Vögeln erbracht. KRAMER gelang es, Stare in einem Rundkäfig auf Himmelsrichtung zu dressieren; sie orientierten sich nach der natürlichen Sonne kompaßgerecht, nach einer feststehenden künstlichen Sonne azimutgerecht.

Für Bienen hat das Versetzungsexperiment den entscheidenden Nachweis geliefert: Eine Bienenschar wird wieder in einer unbekannten Gegend nach Süden dressiert; sie darf die Futterstelle aber nur an einem einzigen Nachmittag besuchen. Über Nacht wird das Volk in eine andere Gegend verfrachtet, und den süddressierten Bienen stehen am nächsten Morgen wieder 4 Futtertischchen in den vier Himmelsrichtungen zur Wahl. Sie kommen spontan und einmütig nach Süden und zeigen damit, daß sie die veränderte Sonnenstellung am Morgen in Rechnung gestellt haben: am vorhergehenden Nachmittag waren sie stets links zur Sonne geflogen; jetzt am Morgen müssen sie sich rechts von der Sonne halten, um zum Ziel zu kommen (v. FRISCH, 1952, v. FRISCH und LINDAUER, 1954).

Einen zweiten Beweis für das Einkalkulieren der scheinbaren Sonnenwanderung liefern die Dauertänzerinnen: Man kann durch einen Kunstgriff Sammelbienen oder Schwarmspurbienen zu beliebiger Zeit veranlassen, durch Tänze ihre Ziele anzuzeigen, ohne daß sie unmittelbar vorher das ihnen bekannte Ziel besucht hatten; diese Tänzerinnen zeigen den Winkel zwischen Ziel und Sonne immer in bezug auf den Sonnenstand, der für die Zeit des Tanzes gilt — auch wenn die Sicht zum Himmel während des Tanzes tagelang, ja wochenlang vorher verwehrt war. Sogar bei Nacht beziehen sich solche Dauer-Tänzerinnen auf den Sonnenstand und demonstrieren damit die Fähigkeit, daß sie nicht nur die Sonnenbahn bei Tag einberechnen — die ihnen aus dem Gedächtnis bekannt sein dürfte —, sondern daß sie auch den nächtlichen Sonnenbogen aus seinem Tagesverlauf extrapolieren können (LINDAUER, 1954, 1957).

MEDER (1958) hat einen weiteren Beleg erbracht, indem er Bienen, die auf einem Flugplatzgelände richtungsdressiert waren, am Futterplatz abfing und für mehrere Stunden in ein Kästchen sperrte. Während der Gefangenschaft war die Sicht zur Sonne verwehrt. Nach der Freilassung flogen die Bienen nicht winkeltreu zur Sonne, sondern azimutgetreu zurück. Versetzte er seine Bienen während der Dunkelhaft vom südlich gelegenen Futterplatz nach Osten, dann flogen sie nach der Freilassung in der gewohnten Kompaßrichtung nach Norden ab und nicht etwa in direktem Flug nach Hause — ein Beweis, daß in dem weiten flachen Gelände irdische Landmarken nicht für den kompaßgerechten Flug im ersten Versuch zu Hilfe genommen worden waren.

Einen unerwarteten Beweis für das Einkalkulieren der Sonnenbahn lieferten neuerdings solche Bienen, die wir darauf dressiert hatten, eine *wandernde* Futterstelle zu besuchen. Um den Tagesgang der Restmiß-weisung (s. v. FRISCH u. LINDAUER, 1961) auszuschalten, wurde einer Sammelschar ein Futterplatz den ganzen Tag über in Richtung zur Sonne geboten. Jede Minute wurde das Futtertischchen neu zur Sonne einvisiert und wanderte also im Minutenabstand bei einer Entfernung von 200 m ständig weiter gegen West. Das Versuchsgelände war flach

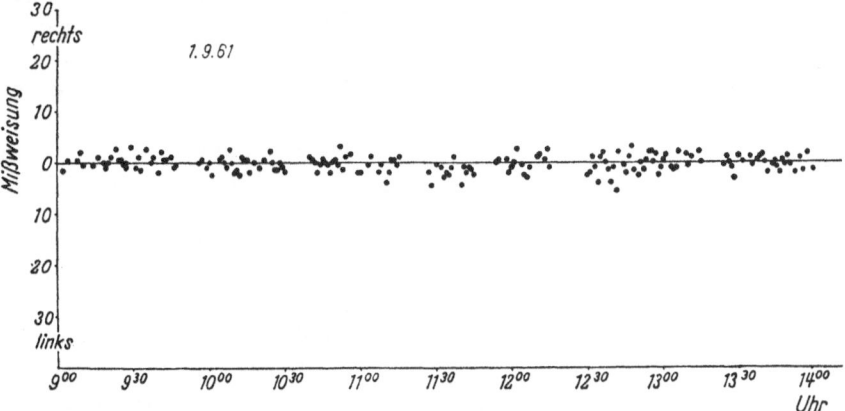

Abb. 1. Mißweisung von solchen Bienen, deren Futterplatz immer mit der Sonne mitwanderte. Die Bienen waren ohne Sonnenerfahrung aufgezogen worden. Punkte auf der Null-Linie zeigen einen Schwänzellauf senkrecht nach oben an; sie gelten für alle Tänze mit Mißweisung 0°. Werte unter dieser Linie bedeuten „Links"-Mißweisung, d. h. Abweichung der Richtungsangabe gegen Ost; Werte über der Null-Linie zeigen entsprechend Rechts-Mißweisung, d. h. Abweichung gegen West an. Die Tänze schwanken gleichmäßig den ganzen Tag über um den Nullpunkt

und offen, das Tischchen war optisch wie auch geruchlich auffallend markiert, und so folgten die Bienen ohne Schwierigkeit diesem wandern-den Ziel. Das erwartete Ergebnis, die Mißweisung würde jetzt unter Wegfall jeder Tagesperiode durchgehend um die Null-Linie liegen, bekamen wir aber nur in einem Ausnahmefall; dann nämlich, wenn wir sonnenlos aufgezogene Jungbienen bei der Dressur verwendeten (Abb. 1). In allen anderen Fällen traten erhebliche und gesetzmäßige Fehler auf, die sogar über den Durchschnitt einer normalen Restmißweisung hinausgingen. Dabei zeigten die Tänzerinnen *konstant eine Linksmiß-weisung*, d. h. sie hinkten 30, 60 oder gar 120 min in ihrer Richtungs-weisung nach (Abb. 2); sie zeigten immerzu eine Futterstelle an, die sie gar nicht beim Ausflug, der dem Tanz unmittelbar vorausging, besucht hatten, sondern eine längere Zeitspanne vorher. Dieser Tatbestand wurde noch auffälliger, wenn das Futtertischchen nur alle 30 min oder jede Stunde zur Sonne hin versetzt wurde (Abb. 3), die Richtungs-weisung blieb am alten Anflugsziel hängen und näherte sich erst langsam

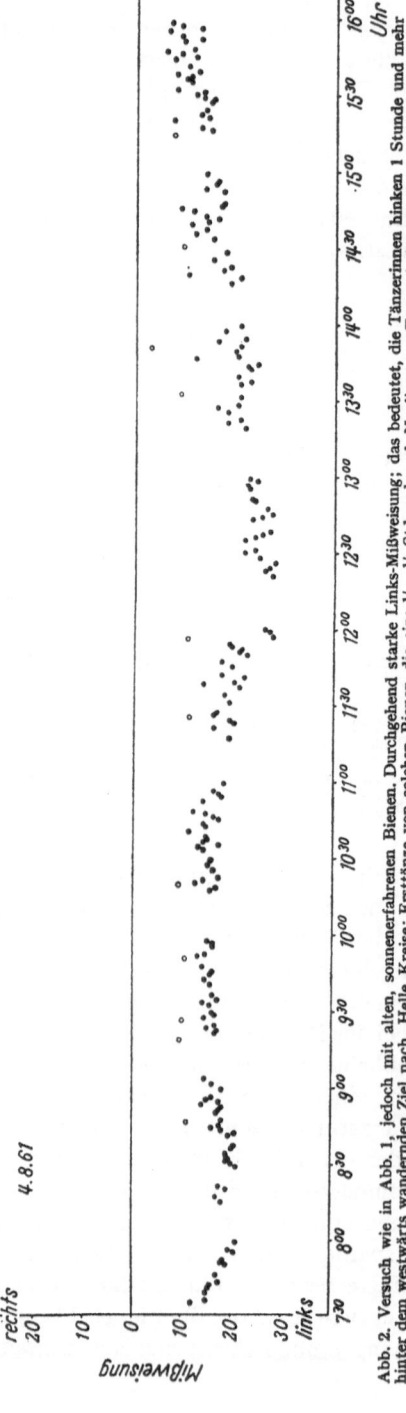

Abb. 2. Versuch wie in Abb. 1, jedoch mit alten, sonnenerfahrenen Bienen. Durchgehend starke Links-Mißweisung; das bedeutet, die Tänzerinnen hinken 1 Stunde und mehr hinter dem westwärts wandernden Ziel nach. Helle Kreise: Ersttänze von solchen Bienen, die eine $^1/_4$—$^1/_2$ Std vorher als Neulinge am Futtertisch markiert worden waren. Ihre Mißweisung ist am Anfang kleiner, da sie den Ort ihres noch nicht lang zurückliegenden ersten Zielanfluges anzeigen

dem neuversetzten Futtertisch. Das Ergebnis war sinngemäß das gleiche, wenn das Futtertischchen 30° links von der Sonne oder 30° rechts von der Sonne weiterwanderte.

Wie ist dieses Ergebnis zu deuten? Es ist wichtig, zunächst darauf hinzuweisen, daß Bienen nicht nur blütenstete, sondern auch extrem *ortsstete* Sammlerinnen sind: Wenn eine ergiebige Futterstelle entdeckt ist, dann begrenzen sie ihren Sammelbereich auf wenige Quadratmeter, z. T. sogar auf Bruchteile eines solchen (MINDERHOUD, 1931). Man muß aus den obigen Ergebnissen folgern, daß sich die Bienen während eines *ersten erfolgreichen* Sammelfluges die Richtung zum Ziel einprägen und von jetzt ab automatisch für diesen Punkt durch einen zentralen Rechenmechanismus die Azimutänderung der Sonne einberechnen. Spätere Informationen von seiten der Peripherie werden nicht oder kaum mehr beachtet. Es läßt sich aus den vorliegenden Beobachtungen nicht entnehmen, ob nun jeder neue Ausflug, der in eine andere Richtung führt, mit der im Gedächtnis liegenden Information *graduell* in Konkurrenz tritt und so die Mißweisung nachhinkend immer wieder in die neue Richtung verschiebt, oder ob ein bestimmter späterer Ausflug, dem die Biene wieder besondere Aufmerksamkeit widmet, die gesamte Information, die im Gedächtnis gespeichert liegt, auslöscht und in einem einmaligen Prägungsvorgang eine neue Ausgangsposition schafft. Die Ergebnisse in Abb. 3 sprechen für die erste Möglichkeit.

Einen letzten Beweis haben vor kurzem NEW (1961) und NEW, D., u. J. NEW (NEW, D. A. T., and J. K. NEW, 1962) vorgelegt: Sie berichten, daß selbst bei *Zenitdurchgang* die Bienen den Sonnenverlauf einkalkulieren können. Wie Ref. selbst vor Jahren zeigen konnte (LINDAUER, 1957), sind die Tänzerinnen etwa $2^1/_2°$ vor und nach Zenitdurchgang

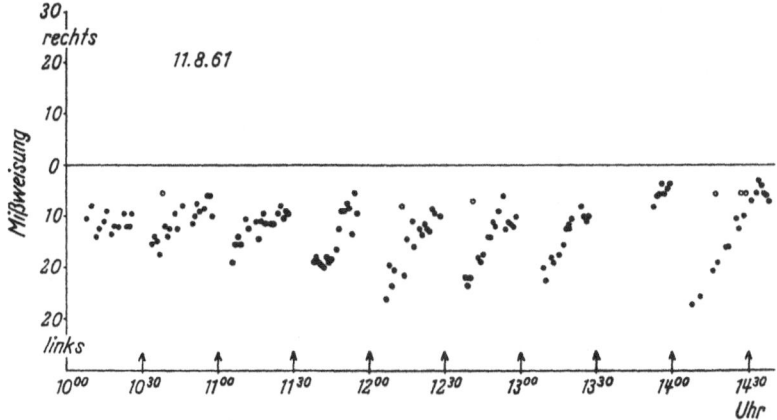

Abb. 3. Bei halbstündiger Versetzung des Futtertisches — die Versetzung ist durch Pfeile auf der Abszisse angezeigt — ergibt sich eine sprunghafte Zunahme der Mißweisung nach jeder Versetzung: wieder zeigen die Tänzerinnen die alte Futterstelle an und verrechnen aus dem Gedächtnis den zugehörigen Sonnenwinkel. Erst nach und nach erfolgt eine Angleichung an das versetzte Ziel

desorientiert; sie können den Sonnenwinkel nicht mehr ablesen. Das wird dann besonders klar, wenn man den Stock horizontal legt. Bei vertikal gestelltem Stock gibt es zu dieser Zeit neben den desorientierten Tänzen zwar auch einige gerichtete Schwänzelläufe, ihre Richtungsangabe aber ist unsauber, und da seinerzeit keine Kontrolle möglich war, zu welchem Zeitpunkt genau die Tänzerinnen am Futtertisch gewesen waren, mußten solche Werte ausgeschieden werden. NEW, D., u. J. NEW haben nun auf diese gerichteten Tänze besonderes Augenmerk gelegt und gefunden, daß manche Bienen um die Mittagszeit allein durch Extrapolation den Sonnenstand festlegen können; sie bestätigen allerdings, daß erhebliche Fehler auftreten und daß der Mechanismus der Extrapolation in verschiedener Hinsicht uneinheitlich ist: die einen Tänze springen zu früh in die Gegenrichtung um, andere zu spät; auch die Umkehr in dem Sinne, daß man zunächst *im* Uhrzeigersinn verrechnet und dann über Nacht *entgegen* dem Uhrzeigersinn — wenn die Sonne von Süd über den Zenit nach Nord wandert — macht Schwierigkeiten; Völker und einzelne Bienen verhalten sich subjektiv verschieden

(Abb. 4). Diese Befunde sind im einzelnen noch schwer zu deuten, aber sie weisen darauf hin, daß das Einkalkulieren der Sonnenwanderung zu einem zentralen Automatismus geworden ist und daß die Tiere dabei auf periphere Kontrolle der Sinnesorgane zeitweise verzichten können.

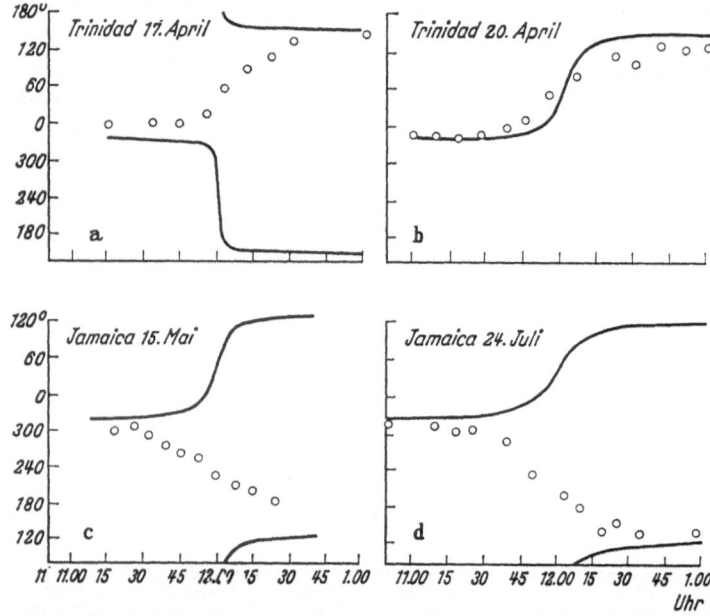

Abb. 4a—d. Tänze von Sammelbienen auf vertikaler Wabenfläche bei Zenitstand der Sonne bzw. zenitnahem Sonnenstand. Ausgezogene Linie: Winkel zwischen Sonnenazimut und Azimut des Futterplatzes, also: „erwarteter" Tanzwinkel. Helle Kreise: tatsächlich registrierte Tänze. In a nehmen die Tänzerinnen bereits auf eine nördliche Sonnenbahn in ihrer Richtungsweisung Bezug, noch ehe die Sonne von Süd nach Nord hinübergewechselt war. b Die Tänze folgen zwar sinngemäß dem tatsächlich gegebenen nördlichen Sonnenbogen, aber sie gleichen den jähen Umsprung von Ost nach West dadurch aus, daß sie dem Einberechnen der scheinbaren Sonnenbahn aus dem Gedächtnis eine flachere Azimutkurve zugrunde legen. c Die Sonne zieht ihre Bahn bereits von Ost über Nord nach West, die Tänzerinnen aber halten sich noch an den südlichen Sonnenbogen, der einige Tage vorher noch Gültigkeit hatte. d Noch ehe der Sonnenbogen von Norden herkommend dem Zenit sich nähert und am Südhimmel erscheint, hat sich eine Biene bereits im vorhinein auf den Südbogen eingestellt (nach NEW, D. A. T., and J. K. NEW, 1962)

2. Integration des aktiven Umweges

Wenn eine Biene vom Stock zum Futterplatz fliegt, hält sie nicht schnurgerade ihren Kurs bei, sondern beschreibt Kurvenlinien, die sie 20° und mehr von der direkten "Bee Line" abbringen. Gelegentlich muß eine Sammelbiene auch einen größeren Umweg machen — um einen Häuserblock, eine Baumgruppe oder eine Bergnase; v. FRISCH hat als Erster die Frage gestellt, ob die Sammlerinnen die Richtungsabweichungen auf solchen Umwegen zugunsten des direkten Kompaßkurses einberechnen können. Diese Frage ist ja hinsichtlich der Richtungsweisung im nachfolgenden Schwänzeltanz von größtem Interesse. Der klassische Umwegversuch (v. FRISCH, 1948) hat die erstaunliche Fähigkeit der

Bienen zu solcher Integrationsleistung unter Beweis gestellt. Physiologisch ließe sich diese Leistung so deuten, daß die Meldungen jener beiden Ommatidiengruppen, von denen die eine auf dem ersten und die andere auf dem zweiten Flugschenkel den Einfallswinkel der Sonne festhält, gemittelt werden. Das ergibt aber nur dann die direkte Zugrichtung, wenn beide Flugschenkel *gleich lang* sind (vgl. Abb. 5); sind sie es nicht, dann muß auf umständlicherem Weg der direkte Kurswinkel bestimmt werden. Wie die Bienen diese Berechnung durchführen, wissen wir nicht. Da sie schon nach einem ersten Umwegflug, der zum Ziel geführt hatte, die direkte Richtung zurückfinden und anschließend diesen direkten Kurs durch einen Tanz anzeigen können (s. S. 162), wird man annehmen dürfen, daß sie nach dem System der vektoriellen Summation und Subtraktion die direkte Fluglinie festlegen. JANDER (1957) hat marschierenden Ameisen durch wechselweises Aufleuchten zweier Lämpchen einen Umweg vorgetäuscht; die Tiere liefen in der Resultanten zwischen den beiden Dressurreizrichtungen, wobei auch die Reizzeiten je Lämpchen mit einkalkuliert wurden. Für die Festlegung

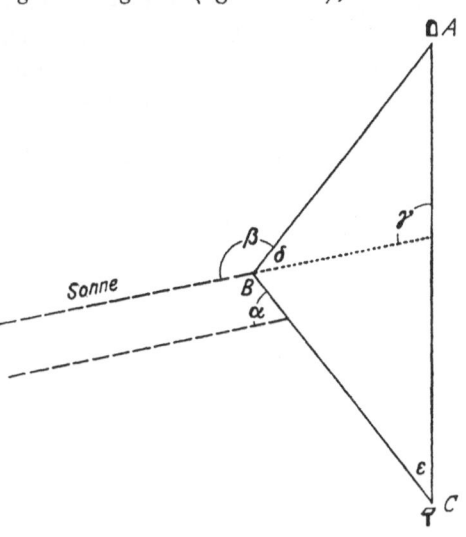

Abb. 5. Eine Biene fliegt von A auf einem Umweg über B nach C. Auf dem Flugschenkel AB sieht sie die Sonne unter dem Winkel β, auf dem Flugschenkel BC unter dem Winkel α. Sind beide Flugschenkel gleich lang, so errechnet sich der Sonneneinfallswinkel γ für die direkte Fluglinie AC wie folgt:

$$\gamma = 180° - \delta - \varepsilon = 180° - 180° + \beta - \frac{\beta - \alpha}{2}$$

$$\boxed{\gamma = \frac{\alpha + \beta}{2}}$$

Die Tänzerinnen könnten also für die Richtungsweisung den Winkel der direkten Fluglinie durch einfaches Mitteln der beiden Umwegschenkel gewinnen

der Luftlinie eines unter natürlichen Verhältnissen gelaufenen Umweges fordert JANDER eine Integrationsleistung nach folgender Gleichung:

$$\alpha_r = \frac{1}{t_z - t_N} \int_N^Z \alpha_i \, dt;$$

(α_r = Luftlinie; α_i = Gesamtzahl der Richtungswinkel vom Nest zum Ziel; $t_z - t_N$ = Gesamtreizzeit; dt = zerlegte Zeitabschnitte der Gesamtreizzeit.)

Apis indica, die schon auf 2 m Entfernung durch gerichtete Schwänzelläufe das Ziel anzeigen kann, bot uns günstige Gelegenheit, den Integrationsvorgang im Umwegversuch genauer zu analysieren.

Es wurde vor dem Flugloch eine Glasgalerie angebracht (Abb. 6), die die ausfliegenden Bienen zunächst einige Meter in Südrichtung lotste. Im rechten Winkel zur Galerie stand — in gleicher oder verschiedener Entfernung — der Futterplatz. Der Plan war, durch Abdunkeln oder durch Verlängern des einen Flugschenkels quantitativ die Teilhabe beider Flugschenkel an der Festlegung des direkten Kompaßkurses zu messen. Dabei gab es eine Überraschung: die Tänzerinnen haben *den Flugschenkel, der sie in der Galerie nach Süd führte, bei ihrer Richtungsweisung nicht berücksichtigt.* Stand der Futtertisch wie in Abb. 6 östlich vom Galerieausgang, dann zeigten die Tänze genau nach Ost und nicht nach Südost, wie es dem direkten Flug, vom Flugloch an gerechnet, entsprochen hätte (Tab. 3, Versuch 1—3). Um zu prüfen, ob

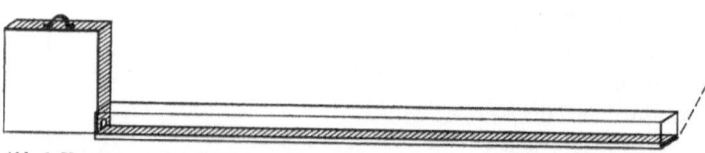

Abb. 6. Umwegversuch mit *Apis indica* F.: eine Glasgalerie zwingt die Sammlerinnen zunächst in Richtung Süd zu fliegen, nach einer 90°-Schwenkung gegen Ost kommen sie dann zum Futtertisch. Die Entfernung der beiden Flugschenkel kann beliebig gleich- oder ungleichmäßig variiert und der eine Flugschenkel abgedeckt werden. Bei der obigen Anordnung werten die Tänzerinnen jedoch bei ihrer Richtungsweisung nur den freien Flug gegen Ost. Die Glasgalerie ist noch Stockrevier. Flugloch ist der Ausgang dieser Galerie

die Galerie selbst — infolge des etwas behinderten Fluges — der Grund für diese Ausschaltung des einen Flugschenkels war, wurde die gleiche

Tabelle 3. *Richtungsweisung von Bienen, die auf einem Umweg ihre Futterstelle erreichen. Beide Flugschenkel gleich lang*

Versuch Nr.	Versuchsanordnung	Abweichung des Tanzwinkels von der		Zahl der gemittelten Tänze
		direkten Flugrichtung	Flugrichtung Ost	
1	Galerie am Stock, 2 m südwärts. Freier Flug 2 m nach Ost (Abb. 6)	44,07° gegen Nord	0,93° gegen Süd	35
2	Galerie am Stock, 3 m südwärts. Freier Flug 3 m nach Ost	55,87° gegen Nord	10,87° gegen Nord	20
3	Galerie am Stock, 4 m südwärts. Freier Flug 4 m nach Ost	48,28° gegen Nord	3,88° gegen Nord	50
4	Galerie am Stock, 4 m südwärts (abgedeckt). Freier Flug 4 m nach Ost	49,68° gegen Nord	4,68° gegen Nord	19
5	Freier Flug vom Stock 4 m südwärts, Galerie 4 m, ostwärts (Abb. 7)	6,72° gegen Nord	38,28° gegen Süd	30

Galerie am anderen Flugschenkel aufgestellt (Abb. 7a). Jetzt wurde die *direkte* Flugrichtung angegeben, so wie es v. FRISCH bei einem normalen Umwegversuch gefunden hatte (Tab. 3, Versuch 5). Kein Zweifel, die Bienen berechneten in der ersten Anordnung den Abflug erst vom Ende der Galerie ab und *betrachteten die Galerie selber noch als ihr Heim*. Das war auch dadurch offenkundig, daß stets Wächterbienen in der Galerie postiert waren und diese von Unrat sauber gehalten wurde. Überdies zeigte der Kontrollversuch mit abgedunkelter Galerie die gleichen Tanzwinkel (Tab. 3, Versuch 4) wie bei offener Galerie. Dieser unerwartete

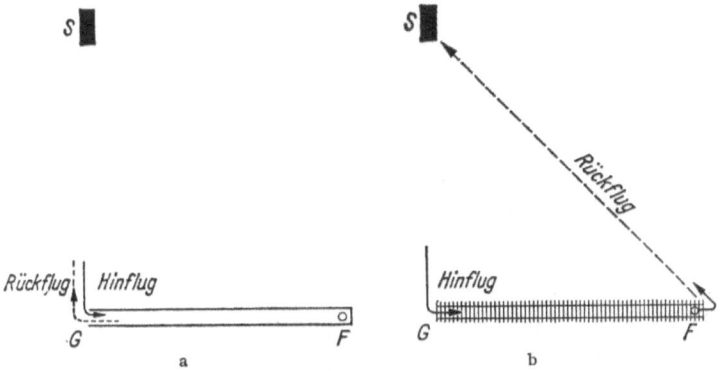

Abb. 7a u. b. Die Glasgalerie ist vom Stock abgerückt und leitet die Bienen im rechten Winkel gegen Ost. a Die Galerie bietet freien Blick zum Himmel; Hinflug und Rückflug erfolgen über den Westausgang. b Die Glasgalerie ist abgedeckt, der Hinflug wird auf dem Umweg geleitet, für den Rückflug aber wird der Ostausgang der Galerie geöffnet, so daß die Sammlerinnen auf direktem Weg den Stock erreichen. Die Tänzerinnen weisen nach Süden, der Rückflug wird nicht für die Richtungsweisung berücksichtigt

Nebenbefund hat uns wieder die selektive Leistung des ZNS im Orientierungsgeschehen kundgetan: die Biene kann willkürlich und subjektiv Sinneseindrücke *während des Fluges* so lange in übergeordneten Zentren abschalten, bis eine bestimmte Raum- oder Zeitmarke, hier die Grenze des Heimbereiches, überschritten wird. Die willkürliche Filterung der afferenten Information tritt in diesem Versuch besonders deutlich an den Tag.

Tab. 4 gibt Auskunft, welche Richtung die Tänzerinnen anzeigen, wenn einer der beiden Flugschenkel verlängert (Versuch Nr. 6) bzw. der Ostschenkel abgedunkelt (Versuch Nr. 7 und 8) ist. Die Berechnung des direkten Flugweges gelingt demnach auch, wenn die Flugschenkel ungleich lang sind. Ob die Abweichung der Tanzwinkel gegen Ost (Versuch 6) eine echte Überbewertung des kürzeren Schenkels, d. h. der Endstrecke bedeutet, so wie das für die Entfernungsweisung angenommen wird (v. FRISCH u. KRATKY, 1962), läßt sich statistisch aus den vorhandenen Daten nicht sichern (vgl. S. 174, Versuche von BISETZKY).

In den Versuchen Nr. 7 und 8 fällt auf, daß die Tänze nicht genau nach Süd, d. h. in jene Flugrichtung mit freier Sicht zeigen, sondern

stark zur Seite des abgedunkelten Schenkels abgelenkt werden. Die Abweichung gegen Ost ist im Versuch 7 gegenüber der Alternative Abweichung gegen Süd sehr gut gesichert, im Versuch 8 gut gesichert. Daraus muß man folgern, daß die Bienen die 90°-Wendung, die sie vor Ein-
flug in die verdunkelte Galerie noch unter freiem Himmel vollführen,
zugunsten der direkten Fluglinie berücksichtigen.

Die Anordnung in Versuch 9 der Tab. 4 (vgl. Abb. 7b) erlaubte
zudem eine nochmalige Prüfung der Frage, inwieweit Hinflug und Rückflug für die Richtungsweisung Bedeutung haben: wurde der Hinflug

Tabelle 4. *Richtungsweisung von Bienen, die auf einem Umweg ihre Futterstelle erreichen. Flugschenkel verschieden lang oder teilweise abgedeckt*

Versuch Nr.	Versuchsanordnung	Fehler (Mittelwert) bezogen auf			Zahl der gemittelten Tänze
		direkte Flugrichtung	Flugschenkel Süd	Flugschenkel Ost	
6	Freier Flug vom Stock 8 m; Galerie 4 m ost- wärts	13,6° gegen Ost	39,6° gegen Ost	51,4° gegen Süd	20
7	Freier Flug vom Stock 3 m Süd; Galerie 3 m ost- wärts (abge- deckt, diffuses Licht)	25,6° gegen Süd	19,4° gegen Ost	70,6° gegen Süd	20
8	Freier Flug vom Stock 3 m Süd; Galerie 3 m ost- wärts (abge- deckt, völlig dunkel)	32,9° gegen Süd	12,1° gegen Ost	77,9° gegen Süd	20
9	Freier Flug vom Stock 3 m Süd, Galerie 3 m ost- wärts (völlig dunkel), Rück- flug auf direk- ter Flugbahn (Abb. 7b)	31,0° gegen Süd	14,0° gegen Ost	76,0° gegen Süd	20

über die abgedunkelte Ostgalerie geleitet, der Rückflug aber auf der
direkten Luftlinie, also vom Ostausgang der Galerie zum Flugloch, dann
wurde die Südrichtung angezeigt, d. h. die Flugrichtung des Hinfluges
und nicht die des Rückfluges.

Bisetzky (1957) hatte Sammelbienen so dressiert, daß sie unter einer
Glasplatte auf dem Fußweg zum Ziel marschieren mußten. Wenn die
Bienen in einem rechtwinklig geknickten Umweg oder in einem Halb-
bogen den Endpunkt erreichten, gaben sie beim Tanz ebenfalls in etwa
die direkte Wanderrichtung an. Abweichungen traten insofern auf, als
in beiden Fällen die Endstrecke „überbewertet" wurde; einer solchen
Deutung würde auch unser Ergebnis in Versuch 6 entsprechen. Wurde

ein Teil des Zugweges abgedeckt, so wurde — entsprechend unseren Ergebnissen — nur jener Marschweg für die Richtungsweisung verwertet, der freien Ausblick zum Himmel hatte.

3. Kompensation der Seitenwindabtrift

Die Konstanthaltung des Kompaßkurses erfordert bei fliegenden Insekten eine zusätzliche Orientierungsleistung: wenn Seitenwind die Tiere von ihrem Kurs ablenkt, muß die Abtrift kompensiert werden. Die direkte Beobachtung an Bienen zeigt, wie sie das machen: Wie der Fährmann sein Boot, so stellen sie ihren Körper schräg gegen den Wind und kommen so auf direkter Bahn zum Ziel. Erstaunlicherweise können nun Sammelbienen Richtung und Stärke der Abtrift für die Richtungsweisung im Tanz sehr genau einkalkulieren: Obwohl die entsprechenden Ommatidien der Biene einen anderen Sonnenwinkel zwischen Ziel und Sonne melden, als er dem direkten Flugkurs entspricht, zeigt die Tänzerin im Stock die direkte Flugrichtung zwischen Stock und Ziel an (v. FRISCH u. LINDAUER, 1955).

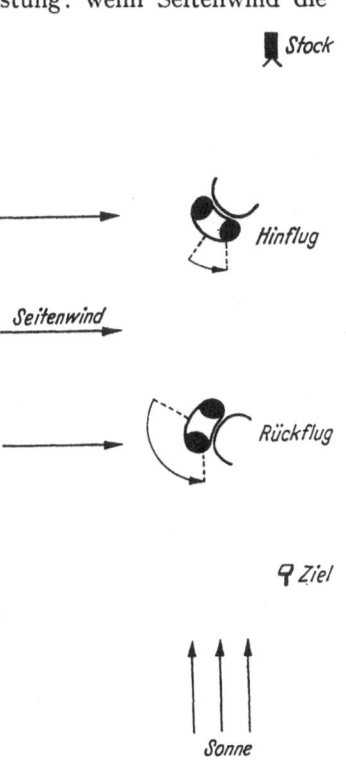

Das ist eine Integrationsleistung, die vielleicht noch über jene beim Umwegversuch zu stellen ist: Bei der Schrägstellung gegen den Seitenwind kann die Biene die peripheren Informationen nicht mehr mitteln und nicht mehr vektoriell summieren, denn auf dem gesamten Hinweg erhält sie ein-

Abb. 8. Schematische Darstellung der falschen Kompaßeinstellung von solchen Bienen, die unter Seitenwind ihre Sammelflüge ausführen. Sowohl beim Hinflug wie beim Rückflug ist das linke Auge aus der direkten Flugrichtung heraus sonnenwärts gedreht

sinnige „Falschmeldungen"; und selbst wenn man für diesen Sonderfall voraussetzen wollte, daß die Informationen des Rückfluges mit einbezogen würden, gäbe dies keinen Ausgleich zum realen Mittelwert. Wie Abb. 8 zeigt, werden bei der Schrägstellung auf dem Rückflug nicht die entsprechenden Ommatidien der Gegenseite, wie es einem symmetrisch um 180° verdrehten Rückflug entsprechen würde, von den Sonnenstrahlen getroffen. Das wäre nur dann der Fall, wenn der Wind beim Rückflug auf die Gegenseite umschlagen und damit die Abtrift wieder ausgleichen würde. Am einfachsten wäre das Problem mit der Annahme gelöst, die Biene würde sich stets auf einen früheren Flug bei Windstille beziehen und dann wie Dauertänzerinnen den Sonnenwinkel errechnen. Das kann aber nicht sein, da ja auch *Neulinge* bei Seitenwind ans Ziel finden. Die Frage nach dem Mechanismus der zentralen Verrechnung bleibt also wiederum offen. Voraussetzung ist, daß die Abtrift über die Sinnesorgane sehr exakt kontrolliert werden kann. Hierfür gibt es zwei Möglichkeiten:

1. Die Biene „erfliegt" sich, bevor sie zum Ziel abfliegt, in Stocknähe den Seitenwind. Sie benutzt den Stock oder seine nächste Umgebung als feststehende optische Marke und errechnet in einem Rundflug Stärke und Richtung der Abtrift. Diese Methode würde aber nur wirksam sein, *wenn auf dem gesamten Flug vom Stock zum Ziel der Wind völlig gleichmäßig Richtung und Stärke beibehielte.* Das ist praktisch niemals der Fall, und wir müssen es den Bienen als besondere Leistung zuerkennen, daß sie während des Fluges Änderungen der Abtrift ebenfalls kompensieren und in Rechnung stellen können und daß sie diese vielfachen Schwankungen auf einen gemeinsamen Nenner für die Richtungsweisung bringen. Sinnesphysiologisch steht hierfür zunächst die optische Kontrolle zur Verfügung: durch Beobachtung des Untergrundes, etwa des Schattenwinkels, den Grashalme, Steine usw. werfen, ließe sich die Abtrift*richtung* feststellen; gleichzeitige Kontrolle der Flughöhe würde die Voraussetzung schaffen, daß die *Stärke* der Abtrift wahrgenommen werden kann.

2. Zusammen mit dem optischen System können aber auch Mechanoreceptoren die schwankende Abtrift kontrollieren. HERAN (1956, 1958, 1959) konnte nachweisen, daß die Antennen als Anzeiger für die Eigenfluggeschwindigkeit dienen, indem sie die Anströmungsgeschwindigkeit messen. Sie reagieren auf erhöhte Anströmung von vorne durch stufenweises Zusammenführen der Antennenspitzen; gleichzeitig verkleinert sich die Flügelschlagamplitude. Als Strömungsreceptor ist von HERAN einwandfrei das Johnstonsche Organ ausgewiesen.

Mein Schüler NEESE konnte Belege erbringen (unveröffentlicht), daß auch die *Augenborsten* der Bienen bei der Kontrolle der Windabtrift beteiligt sind. Im histologischen Bild weisen sich diese Augenborsten, die in der Regel an den Berührungspunkten dreier Facettensechsecke inserieren, als Sensilla chaetica aus. NEESE ist es gelungen, diese Borsten einseitig oder beidseitig vollständig zu entfernen und hat daraufhin festgestellt, daß bei Seitenwind jene Bienen, denen auf der Luvseite die Borsten genommen waren, abgetrieben wurden. Beidseitig borstenlose Tiere ließen sich bei Rückenwind — im Gegensatz zu Normalbienen — treiben; sie kamen schneller ans Ziel. Bei Gegenwind hingegen war ihre Flugzeit bis ums Fünffache verlängert.

Inwieweit nun die Bienen das Hauptgewicht auf die optische Kontrolle oder auf ihre mechanischen Meßinstrumente legen, wie ferner Antennen und Augenborsten in ihrer Funktion sich ergänzen, können wir nicht sagen. Sicher sind mehrere Regelkreise zusammen geschaltet (vgl. HERAN, 1959, S. 157), die sowohl die Kompensation der Windstärke wie der Windrichtung — entsprechend der im Flugverkehr bekannten Koppelnavigation — garantieren.

IV. Kompaßorientierung bei unsichtbarer Sonne

Der hohe Rang der Kompaßorientierung nach Himmelsmarken wird dadurch unterstrichen, daß selbst dann, wenn das maßgebliche Himmelsgestirn unsichtbar ist — bei Nacht, im abgedunkelten Bienenstock, bei bewölktem Himmel — die astronomische Orientierung nicht aufgegeben wird. Im Falle einer Sonnenkompaßorientierung wird dabei folgender Ersatz gewählt:

1. Bei Nacht benutzen die Tiere an Stelle der Sonne den *Mond* als Kompaß.

2. Wenn die Sonne durch Wolken verdeckt ist, wird das *polarisierte Licht des blauen Himmels* als Anzeiger des Sonnenstandes benutzt.

3. Im finsteren Stock, wo die Tänzerinnen ebenfalls die Sonne als Bezugspunkt für die Richtungsweisung verwenden, wird der optische Kompaßkurs ins Schwerefeld *transponiert*.

Zu 1: *Talitrus*, der auch bei Nacht die Fluchtrichtung zum Meer finden muß, kann sich in den Nachtstunden auf den Mondkompaß umstellen (PAPI und PARDI, 1953, 1954, 1959, PAPI, 1960). Das gleiche konnte ENRIGHT (1961) für einen Verwandten, *Orchestoidea corniculata*, bestätigen. Im letzteren Fall war es jedoch nötig, daß die Tiere mindestens eine Stunde vor Sonnenuntergang und zwei Stunden vor einem anschließenden Mondaufgang unter freiem Himmel gehalten wurden. Anderenfalls orientierten sie sich winkeltreu zum Mond. Die schwierige Frage, die hier auftaucht: Umstellung auf den Mondazimut einerseits, andererseits das Vermögen, auch bei Nacht sich nach der Azimutwinkelgeschwindigkeit der Sonne zu richten — wenn man ein Lämpchen als künstliche Sonne bietet — versuchen die genannten Forscher mit der Annahme zu erklären, daß zwei Orientierungsrhythmen vorhanden seien, die unabhängig voneinander arbeiteten; der zweite Orientierungscyclus müsse jeweils in den Intervallen zwischen Sonnenuntergang und Mondaufgang in Gang gesetzt werden.

Zu 2: Erstmalig hat v. FRISCH (1948, 1949, 1950) für Bienen die Fähigkeit nachgewiesen, aus dem Polarisationsmuster eines blauen Himmelfleckes den Sonnenstand abzulesen. In der Folge wurde bei zahlreichen anderen Arten unter den Insekten, Krebsen, Spinnen und neuerdings auch bei Mollusken diese Fähigkeit, die Schwingungsrichtung des polarisierten Lichtes zu analysieren, nachgewiesen (Überblick über die umfangreiche Literatur bei v. FRISCH, LINDAUER und DAUMER, 1960). Überraschenderweise können sich die Bienen aber auch dann noch nach der Sonne orientieren, wenn die Wolkendecke — die freilich nicht zu dicht sein darf — völlig geschlossen ist (v. FRISCH, 1953). v. FRISCH wies nach, daß für die Wahrnehmung der Sonnenscheibe durch die Wolkendecke nur der UV-Bereich wirksam ist; wenn die Wolkendecke nicht zu dicht ist, dringt in unmittelbarer Umgebung der Sonne etwas mehr UV (bis zu 10%) durch die Wolkenschicht, und nur dann sind Tänze auf horizontal gestellter Wabe orientiert (v. FRISCH, LINDAUER u. SCHMEIDLER, 1960).

Zu 3: Eine einzigartige Orientierungsleistung hat v. FRISCH (1946) an seinen Bienen darin entdeckt, daß sie einen Kompaßkurs vom optischen Bereich ins Gebiet der Schwerkraft transponieren können. Damit ist die Möglichkeit gegeben, bei Ausfall der optischen Orientierung andere Sinnesorgane zu beanspruchen. Erst dieses Transponierungsvermögen hat es den Bienen ermöglicht, ihre Bauten in schützenden Höhlen anzulegen. Die primitiveren Zwerghonigbienen in Süd-Ost-Asien können den optischen Sonnenwinkel nicht ins Schwerefeld auf der vertikalen Wabe übertragen; sie sind gezwungen, ungeschützt im Freien ihre Nester anzulegen; denn nur so können sie auf horizontaler Tanzfläche unter freiem Himmel ihre optisch ausgerichteten Tänze aufführen (LINDAUER, 1956).

Eine ähnliche Transponierung wurde seitdem von vielen anderen Insekten berichtet: von Ameisen, vom Mistkäfer, vom Marienkäferchen usw. (VOWLES, 1954a/b, BIRUKOW, 1954, 1956). Dabei interessieren zusätzlich zwei Fragen: 1. nach welchem Schlüssel wird transponiert? 2. wie stimmen optischer Kompaßkurs und jener im Schwerefeld quantitativ überein?

Zu 1: Die Übersetzung kann mehrdeutig oder eindeutig sein, sie kann so erfolgen, daß Lauf zum Licht = Lauf nach oben und umgekehrt bedeutet.

Zu 2: JANDER (1960) hat den wichtigen Hinweis erbracht, daß Köcherfliegen nicht winkeltreu, sondern proportionstreu transponieren: der Lichtwinkel verhält sich zum Schwerewinkel wie 3:2. Vielleicht liegt auch der Mißweisung der Bienen ein ähnliches Prinzip zugrunde: je weiter die Tanzwinkel von der Lotrechten und der Horizontalen im Schwerefeld abweichen, um so größer ist die Mißweisung. Die Verhältnisse sind hier aber sicher komplizierter als bei den Köcherfliegen und warten noch auf endgültige Klärung (v. FRISCH u. LINDAUER, 1961).

Schließlich sei noch vermerkt, daß man den tanzenden Bienen Schwerkraft und Licht in Konkurrenz setzen kann. Bietet man der Tänzerin auf vertikaler Fläche Sicht zum blauen Himmel, nicht aber Sicht zur Sonne, dann stellt sie ihren Schwänzellauf sehr genau in die Resultante zwischen der erwarteten Tanzrichtung bezogen auf Schwerkraft einerseits und Licht andererseits. Das gilt auch, wenn die Sonne gerade unter dem

Bauch der Tänzerin steht, also in einer Situation, die die Biene bei ihren Ausflügen nie vorfindet. Ist der Blick zur Sonne frei, dann stellt die Tänzerin ihre Richtungsweisung voll und ganz auf den optischen Bereich um — die Schwerkraft hat ihre Richtungswirkung verloren (v. FRISCH, 1962).

V. Angeboren oder erlernt?

Zum Abschluß müssen wir noch die Frage erörtern, inwieweit das Vermögen, sich nach Himmelsgestirnen kompaßgerecht zu orientieren, auf angeborenen Fähigkeiten beruht und inwieweit Erfahrung und Lernvorgänge beteiligt sind. Es wäre vorstellbar, daß ein grobes Raum-Zeit-Modell des Sonnenbogens im Erbgut vorliegt; ein solches müßte jedoch abwandelbar sein, da sich einesteils erhebliche *jahreszeitliche* Unterschiede als auch solche, die sich mit dem *geographischen Ort* verändern, einstellen; diese kommen z. T. schon für das gleiche Individuum, mindestens aber für eine Generationenfolge oder eine weiträumige Population zur Geltung. Für Bienen, die nur eine Lebenszeit von wenigen Wochen haben und deren Urheimat in Äquatornähe ist, wäre es schwer verständlich, daß ein starres Sonnenbogenmodell jedem Tier angeboren sein sollte; zwischen den Wendekreisen ändert sich nicht nur die Azimutwinkelgeschwindigkeit, sondern auch der Richtungssinn der scheinbaren Sonnenwanderung so radikal, daß mit einem einzigen solchen Modell nichts anzufangen wäre. Entgegen den Befunden von KALMUS (1954, 1956) konnten wir nachweisen, daß die Bienen den Verlauf des Sonnenbogens ortsgerecht und der Jahreszeit entsprechend individuell erlernen können. Bienen, die ohne Sonnenerfahrung aufgezogen wurden, können, nachdem sie unter freien Himmel gebracht sind, in den ersten Tagen die Sonnenwanderung noch nicht einkalkulieren und solche Bienen, die man von der Südhalbkugel zur Nordhalbkugel verfrachtet, stellen sich nach einiger Zeit auf den neuen Sonnenkompaß um (LINDAUER, 1959).

Für Ameisen konnte JANDER (1957) ebenfalls nachweisen, daß ein Lernvorgang bei der Sonnenorientierung beteiligt ist: Überwinterte Tiere orientieren sich zunächst winkeltreu zur Sonne, erst nach und nach gehen sie zur kompaßgerechten Orientierung über. Eine offene Frage bleibt noch, in welchen Stufen dieser Lernvorgang abläuft: ist es im wesentlichen ein kurzzeitiger oder gar einmaliger Prägungsvorgang auf einen jahreszeitlichen und örtlichen Sonnenbogen, oder sind wiederholte Ausflüge zu verschiedener Tageszeit nötig, bis Rhythmus, Richtungssinn und Geschwindigkeit der scheinbaren Sonnenbahn dem Gedächtnis eingeprägt sind? Nach LINDAUER (1959) genügt es jedenfalls, wenn sonnenunerfahrene Bienen nur den Nachmittagsbogen der Sonnenbahn erlernen. Das fehlende Stück des Bogens können sie dann daraus extrapolieren. Über weitere Einzelheiten solcher Lernvorgänge hätten wir aber gerne in Zukunft noch Näheres erfahren.

12*

Literatur

Baerends, G. P.: Fortpflanzungsverhalten und Orientierung der Grabwespe *Ammophila campestris* Jur. T. Entomol. **84**, 68—275 (1941).

Birukow, G.: Photo-Geomenotaxis bei *Geotrupes silvaticus* Panz. und ihre zentralnervöse Koordination. Z. vergl. Physiol. **36**, 176—211 (1954).

— Lichtkompaßorientierung beim Wasserläufer *Velia currens* F. (Heteroptera) am Tage und zur Nachtzeit. Z. Tierpsychol. **13**, 463—484 (1956).

Bisetzky, A. R.: Die Tänze der Bienen nach einem Fußweg zum Futterplatz. Unter besonderer Berücksichtigung vom Umwegversuchen. Z. vergl. Physiol. **40**, 264—288 (1957).

Enright, J. T.: Lunar orientation of *Orchestidea corniculata* Stoub. (Amphipoda). Biol. Bull. **120**, 148—156 (1961).

Frisch, K. v.: Die Tänze der Bienen. Österr. Zool. Z. **1**, 1—48 (1946).

— Gelöste und ungelöste Rätsel der Bienensprache. Naturwissenschaften **35**, 12—23 (1948).

— Die Polarisation des Himmelslichtes als orientierender Faktor bei den Tänzen der Bienen. Experientia (Basel) **5**, 142—148 (1949).

— Die Sonne als Kompaß im Leben der Bienen. Experientia (Basel) **6**, 210—221 (1950).

— Die Richtungsorientierung der Bienen. Verh. Dtsch. Zool. Ges. Freiburg, 59—71 (1952).

— Die Fähigkeit der Bienen, die Sonne durch die Wolken wahrzunehmen. S. B. Bayer. Akad. Wiss. math.-nat. Kl. **17**, 197—199 (1953).

— Über die durch Licht bedingte „Mißweisung" bei den Tänzen im Bienenstock. Experientia (Basel) **18**, 49—53 (1962).

— u. O. Kratky: Über die Beziehung zwischen Flugweite und Tanztempo bei der Entfernungsmeldung der Bienen. Naturwissenschaften **49**, 409—417 (1962).

— u. M. Lindauer: Himmel und Erde in Konkurrenz bei der Orientierung der Bienen. Naturwissenschaften **41**, 245—253 (1954).

— — Über die Fluggeschwindigkeit der Bienen und über ihre Richtungsweisung bei Seitenwind. Naturwissenschaften **42**, 377—385 (1955).

— — Über die „Mißweisung" bei den richtungsweisenden Tänzen der Bienen. Naturwissenschaften **48**, 585—594 (1961).

— — u. K. Daumer: Über die Wahrnehmung polarisierten Lichtes durch das Bienenauge. Experientia (Basel) **16**, 289—302 (1960).

— — u. F. Schmeidler: Wie erkennt die Biene den Sonnenstand bei geschlossener Wolkendecke? Naturwiss. Rdsch. H. 5, 169—172 (1960).

Heran, H.: Ein Beitrag zur Frage der Wahrnehmungsgrundlage der Entfernungsweisung der Bienen *(Apis mellifica* L.*)*. Z. vergl. Physiol. **38**, 168—218 (1956).

— Fluggeschwindigkeitswahrnehmung bei der Honigbiene. Verh. Dtsch. Zool. Ges. Graz 1957. Zool. Anz. Suppl. **21**, 331—338 (1958).

— Wahrnehmung und Regelung der Flugeigengeschwindigkeit bei *Apis mellifica* L. Z. vergl. Physiol. **42**, 103—163 (1959).

Hoffmann, K.: Experimental manipulation of the orientational clock in birds. Cold Spr. Harb. Symp. quant. Biol. **25**, 379—387 (1960).

Jander, R.: Die optische Richtungsorientierung der roten Waldameise *(Formica rufa* L.*)*. Z. vergl. Physiol. **40**, 162—238 (1957).

Kalmus, H.: Sun navigation by animals. Nature (Lond.) **173**, 657 (1954).

— Sun navigation of *Apis mellifica* L. in the southern hemisphere. J. exp. Biol. **33**, 554—565 (1956).

Kramer, G.: Die Sonnenorientierung der Vögel. Verh. Dtsch. Zool. Ges. Freiburg, 72—84 (1952a).

KRAMER, G.: Experiments on bird orientation. Ibis 94, 265—285 (1952b).

LINDAUER, M.: Über die Einwirkung von Duft- und Geschmackstoffen sowie anderer Faktoren auf die Tänze der Bienen. Z. vergl. Physiol. 31, 348—412 (1948).

— Dauertänze im Bienenstock und ihre Beziehung zur Sonnenbahn. Naturwissenschaften 41, 506 (1954).

— Über die Verständigung bei indischen Bienen. Z. vergl. Physiol. 38, 521—557 (1956).

— Sonnenorientierung der Bienen unter der Äquatorsonne und zur Nachtzeit. Naturwissenschaften 44, 1—6 (1957).

— Angeborene und erlernte Komponenten in der Sonnenorientierung der Bienen. Bemerkungen und Versuche zu einer Arbeit von KALMUS. Z. vergl. Physiol. 42, 43—62 (195).

MEDER, J.: Über die Einberechnung der Sonnenwanderung bei der Orientierung der Honigbiene. Z. vergl. Physiol. 40, 610—641 (1958).

MINDERHOUD, A.: Untersuchungen über das Betragen der Honigbiene als Blütenbestäuberin. Gartenwiss. 4, 342 (1931).

NEW, D. A. T.: Effects of small zenith distances of the sun on the communication of honey bees. J. Ins. Physiol. 6, 196—208 (1961).

— and J. K. NEW: The dances of honeybees at small zenith distances of the sun. J. exp. Biol. 39, 271—291 (1962).

OPFINGER, E.: Über die Orientierung der Biene an der Futterquelle. (Die Bedeutung von Anflug und Orientierungsflug für den Lernvorgang bei Farb-, Form- und Ortsdressuren.) Z. vergl. Physiol. 15, 431—487 (1931).

OTTO, F.: Die Bedeutung des Rückfluges für die Richtungs- und Entfernungsangabe der Bienen. Z. vergl. Physiol. 42, 303—333 (1959).

PAPI, F.: Orientation by night: the moon. Cold Spr. Harb. Symp. quant. Biol. 25, 475—480 (1960).

— e L. PARDI: Ricerche sull'orientamento di *Talitrus saltator* (MONTAGU) (Crustacea-Amphipoda). II. Sui fattori che regolario la variazione dell'angolo di orientamento nel corso del giorno. L'orientamento di notte. L'orientamento diruno di altre polplazioni. Z. vergl. Physiol. 35, 490—518 (1953).

— — La luna come fattore di orientamento degli animali. Boll. Ist. Zool. Univ. Torino 4, 1—4 (1954).

— — Nuovi reperti sull'orientamento lunare di *Talitrus saltator* (MONTAGU). Z. vergl. Physiol. 41, 583—596 (1959).

PARDI, L.: Innate components in the solar orientation of littoral Amphipods. Cold Spr. Harb. Symp. quant. Biol. 25, 262—401 (1960).

SAUER, F.: Zugorientierung einer Mönchsgrasmücke *(Sylvia atricapilla L.)* unter künstlichem Sternenhimmel. Naturwissenschaften 43, 231—232 (1956).

VOWLES, D. M.: The orientation of ants. 1. The substitution of stimuli. J. exp. Biol. 31, 341—355 (1954a).

— The orientation of ants. 2. Orientation to light, gravity, and polarized light. J. exp. Biol. 31, 356—375 (1954b).

Vom Rhythmus der Sonnenorientierung am Äquator (bei Fischen)

Von WOLFGANG BRAEMER* und HORST O. SCHWASSMANN**

Max Planck-Institut für Verhaltensphysiologie,
Seewiesen über Starnberg, Obb., Germany
Universität Wisconsin, Hydrobiology Laboratory, Madison 6, Wisconsin, USA

Mit 10 Abbildungen

Inhaltsverzeichnis

I. Einleitung und Fragestellung

Seit 1958 ist bekannt, daß Fische, ebenso wie Vögel, Eidechsen und verschiedene Arthropoden die Sonnenbewegung kompensieren können (HASLER, HORRALL, WISBY und BRAEMER, 1958). Hierdurch können alle diese Tiere bei Sonnensicht am Tage beliebige Himmelsrichtungen einhalten. Ein solcher Kompaß kann im Dienste der verschiedensten Funktionen stehen. Darüber ist bei Fischen allerdings so gut wie nichts bekannt.

* WOLFGANG BRAEMER verstarb kurz nach der Vollendung der vorliegenden Arbeit in Lima, Peru. Seine Frau HELGA BRAEMER führt seine Aufgaben im Max Planck-Institut Seewiesen fort.

** Unterstützt durch Grants der National Science Foundation des Office of Naval Research an Professor A. D. HASLER.

Die Versuche wurden durch großzügige Hilfe des Museu Paraense E. Goeldi, Belém (Brasilien), ermöglicht. Besonders für den Bau der Spiegeleinrichtung, Transport- und Unterbringungsmöglichkeiten und das Tiermaterial sprechen wir dem Direktor, Dr. EDUARDO GALVAO und dem Personal des Museums unseren aufrichtigsten Dank aus.

Wir konnten zeigen, daß dem grünen Sonnenfisch (*Lepomis cyanellus* RAFINESQUE) und dem Portcichliden (*Aequidens portalegrensis* HENSEL) die Fähigkeit zur Einrechnung der Sonnenbewegung *angeboren* ist. Lediglich die *Richtung*, in der sich die Sonne bewegt, muß von den tropischen Cichliden gelernt werden (BRAEMER und SCHWASSMANN, in Vorb., SCHWASSMANN, 1962). Die Phase des Licht-Dunkel-Wechsels stellt bei Fischen, ebenso wie bei Vögeln, Eidechsen und den daraufhin untersuchten Arthropoden die physiologische Uhr auf den jeweiligen Längengrad, das ist die Ortszeit, ein. Bei solchen Versuchen im phasenverschobenen Tag bemerkten wir zum erstenmal einen Einfluß der *Sonnenhöhe*, sowohl auf das Verhalten als auch auf die Richtungswahl der Fische (BRAEMER, 1959, 1960; SCHWASSMANN, 1960, 1962). Diese Arbeit enthält weitere Versuche, die den Einfluß der Sonnenhöhe im Spiegelversuch direkt zeigen.

Um Mißverständnissen vorzubeugen, möchten wir gleich am Anfang die Bemerkung einschalten, daß wir bei unseren Messungen, wie jedermann, der Versuche über die Sonnenorientierung auf der Horizontalen macht, den Winkel angeben, den der flüchtende Fisch zum Azimut der Sonne einschlägt. Dieser Winkel *muß* sich im Laufe des Tages, wenn das Tier Himmelsrichtungen einhält, ebenso ändern wie das Azimut der Sonne, nur mit umgekehrtem Vorzeichen. Aus dieser Art der Auftragung der Meßergebnisse geht aber keinesfalls hervor, daß die jeweilige Höhe der Sonne bei der Bestimmung dieses Winkels auch für das Tier ohne Bedeutung ist. Das muß erst in davon getrennten Experimenten mit abweichenden Sonnenhöhen bewiesen werden.

Das Problem, wie die Sonnenorientierung auf verschiedenen Breitengraden beschaffen ist, und welche Faktoren dabei eine Rolle spielen, ist bei Wirbeltieren ein noch so gut wie völlig unbeackertes Feld. Es gibt darüber nur eine Arbeit an Fischen (HASLER und SCHWASSMANN, 1960). Wir führen hier die Ergebnisse aus dieser Arbeit an, die für unsere jetzt beschriebenen Versuche von Belang sind: Fische können Himmelsrichtungen nicht mehr einhalten, wenn der Zenitabstand der Sonne kleiner als 5° ist. Darüber hinaus wird gezeigt, daß verschiedene Cichliden, die auf der südlichen Hemisphäre beheimatet sind, dort im September auf Himmelsrichtungen dressiert werden können, das heißt, daß sie die mittags im Norden kulminierende Sonnenbahn kompensieren können. Mit diesen Fischen sind allerdings zu wenig Versuche gemacht, um den quantitativen Verlauf der kompensierten Sonnenbewegung beurteilen zu können. Wir beschreiben in der vorliegenden Arbeit weitere Versuche, in denen wir die tägliche Kompensation genau gemessen haben.

Wir hatten Gelegenheit, in den Monaten April bis August 1962 in Belém, Brasilien, (1°27′ S) die Sonnenorientierung von Fischen genauer zu studieren. Auf dem Äquator ist die Situation für jedes Tier, das mit Hilfe der Sonne und einer physiologischen Uhr Himmelsrichtungen einhalten muß, nicht nur wegen des jährlich zweimaligen Zenitdurchganges der

Sonne, sondern auch sonst ganz besonders kompliziert. Die Tageslänge beträgt zwar während des ganzen Jahres 12 Std und fällt deshalb als Faktor, der die quantitative Einrechnung der Sonnenbewegung beeinflussen könnte, aus. Zugleich aber befinden wir uns auf dem Breitengrad, auf dem die Sonnenazimutkurven den denkbar ungleichförmigsten Verlauf sowohl während eines Tages als auch im Laufe des Jahres haben. Das illustriert die Abb. 1. Dort sind die Azimutkurven zu den Äquinoktien (21. 3. und

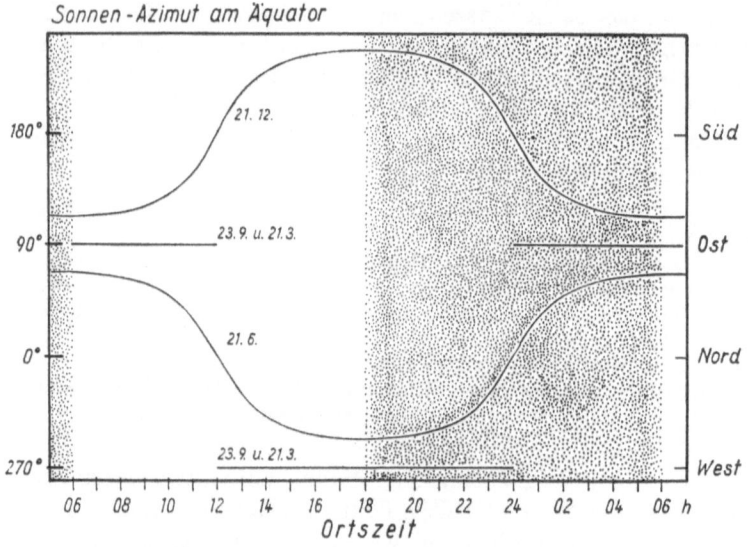

Abb. 1. Die jahreszeitliche Änderung der täglichen Azimutkurve der Sonne am Äquator. Abszisse: Ortszeit, Ordinate: Kompaßgrade. Die Kurve für die beiden Äquinoktien ist wegen des Zenitdurchganges der Sonne unterbrochen

23. 9.) und den Solstitien am 21. 6. und am 21. 12. abgebildet. Die Kurven in den dazwischenliegenden Zeiten kann man sich leicht dazudenken. Die Sonne geht am 21. 3. und am 23. 9. um 6^{00} genau im Osten auf. Sie ändert ihr Azimut bis Mittag nicht. Um 12^{00} geht die Sonne durch den Zenit, hat also für einen Moment kein Azimut und hat danach bis 18^{00} ein Westazimut. Stellt man sich die Erde durchsichtig vor und zeichnet die Kurve für 24 Stunden auf, dann hat die Sonne in der ersten Nachthälfte ein Westazimut und nach 24^{00} ein Ostazimut. Am 21. 6. geht die Sonne bei 66,5° auf, ändert in den Morgen- und Abendstunden ihr Azimut langsam, in den Mittagsstunden erheblich schneller, kulminiert im Norden, läuft also dem Uhrzeigersinn entgegengesetzt und geht bei 293,5° unter. Die Sonnenazimutänderung beträgt in den 12 Tagesstunden 133°. Durchschnittlich müßte demnach von dem Kompaßtier in den 12 Tagesstunden eine Azimutänderung von 11°/h eingerechnet werden und, wenn keine Richtungsumkehr in der Kompensation der

Sonnenbewegung stattfindet, müßte in den 12 Nachtstunden in einer Kompensation von 19°/h der Kreis zu 360° geschlossen werden. In Wirklichkeit durchläuft in der Nacht das Azimut der Sonne denselben Weg wie am Tage in umgekehrter Reihenfolge, ändert also die Verlaufsrichtung nach Sonnenuntergang und vor Sonnenaufgang. Spiegelbildlichsymmetrisch verläuft die Kurve am 21. 12. Eine ähnliche Richtungsumkehr in der 24stündigen Sonnenazimutkurve, d. h. mitternachts dasselbe Azimut wie mittags, finden wir zwischen den Wendekreisen des Krebses und Steinbocks immer dann, wenn die Deklination der Sonne größer ist als die geographische Breite, unabhängig vom Vorzeichen. Weitere Einzelheiten findet man dazu bei SCHWASSMANN (1960).

Tatsächlich ist eine Richtungsumkehr in der Einrechnung der Sonnenbewegung in der Nacht bei verschiedenen Arthropoden (PARDI, 1960; BIRUKOW, 1960) und bei Brieftauben (SCHMIDT-KÖNIG, 1961) entdeckt worden. Diese Versuche sind allerdings in mittleren nördlichen Breiten durchgeführt, wo in der Nacht die Sonnenazimutrichtung nicht eine eben beschriebene Richtungsumkehr aufweist. Von nordamerikanischen Sonnenfischen wissen wir indessen aus Versuchen im Sommer auf dem 43. nördlichen Breitengrad, daß sie in der Nacht fortfahren, die im Uhrzeigersinn umlaufende Sonne zu verrechnen (BRAEMER, 1959, 1960). Gleiches ist von der Honigbiene (LINDAUER, 1954, 1957) und von Staren (HOFFMANN, 1958) nach Versuchen in nördlichen Breiten beschrieben worden. Es ist vorläufig noch nicht genug darüber bekannt, unter welchen Bedingungen diese beiden Verrechnungstypen in der Nacht auftreten, ob sie immer artgebunden sind oder durch Änderung der Außenbedingungen (Tageslänge, Sonnenhöhe oder gesehene Sonnenazimutbewegung) bei der gleichen Art ineinander überführt werden können, oder ob sie von der Methode abhängen, mit der die Sonnenorientierung in der Nacht untersucht wurde. Nur von *Velia currens* F. ist bekannt, daß die Richtungsumkehr sowohl im kurzen Wintertag, als auch im langen Sommertag vorkommt (BIRUKOW, 1956, 1957) und von Fischen wissen wir vorerst nur, daß die Tageslänge einen Einfluß auf die quantitative Verrechnung der Sonnenbewegung am Tage hat (SCHWASSMANN und BRAEMER, 1961). Wir versprachen uns jedenfalls von der Untersuchung der Sonnenorientierung in der Nacht auf dem Äquator weitere Fortschritte in dieser physiologisch so außerordentlich interessanten Frage.

In der vor kurzem veröffentlichten Zweikomponententheorie über "Control Systems of Orientation in Insects", die darüber hinaus aber Allgemeingültigkeit beansprucht (MITTELSTAEDT, 1962), wird für den Rhythmus der Sonnenorientierung der Honigbiene prophezeit, daß am Äquator die oben beschriebene Richtungsumkehr in der Nacht gefunden werden müßte. Die Honigbiene ist für solche Untersuchungen besonders günstig,da sie den Winkel zum Azimut der Sonne auf die Schwerkraftrichtung überträgt, und deshalb bei der Untersuchung der Sonnenorientierung in der Nacht nicht eine unwirkliche Sonne über dem Horizont eingeführt werden muß. Für die Tiere, bei denen diese Umkehr schon in nördlichen Breiten gefunden ist, muß, wie MITTELSTAEDT selbst schon erwähnt, ein anderer als in seiner Theorie geforderter Modus der Sonnenorientierung gelten.

Für die Untersuchung der Sonnenorientierung in der Nacht bei Fischen stehen uns 2 Methoden zur Verfügung:

1. Wir verstellen den auf eine Himmelsrichtung dressierten Versuchstieren durch eine Phasenverschiebung des Hell-Dunkel-Wechsels die

physiologische Uhr und können sie dann in ihrer subjektiven Nacht mit der örtlichen Sonne prüfen.

2. Wir dressieren Fische in der Nacht mit der künstlichen Sonne und messen dann am Tage, in welche Himmelsrichtung diese Fische schwimmen. Wir haben beide Methoden angewandt.

Wir arbeiteten mit den beiden Cichlidenarten *Cichlaurus severus* s. Heckel = *Cichlasoma severum* Regan und *Crenicichla saxatilis* L. Von der ersten Art ist schon durch Versuche auf dem 43. nördlichen Breitengrad bekannt, daß sie dort auf Himmelsrichtungen dressiert werden kann (Schwassmann, 1962).

II. Methodisches

Unsere Versuchstiere fingen wir kurz vor Beginn der Dressur aus den Teichen des Museum Goeldi. Dort hatten die Fische Gelegenheit gehabt, die Sonne tagsüber zu sehen. Die Fische hielten wir in Blechkanistern einzeln im Freien, aber im Sonnenschatten. Zu den Versuchen, in denen die Phase des Licht-Dunkel-Wechsels verstellt wurde, stellten wir die Kanister mit dem dressierten Tier in einen fensterlosen Raum in einen lichtdicht verschlossenen Kasten. Während ihres Tages wurden die Fische mit einer 20 Watt-Neonröhre beleuchtet. Temperaturschwankungen im Wasser waren durch eine geringe Wärmeabgabe der Leuchtröhren mit dem Lichtzyklus synchronisiert und nicht merklich durch die kleinen Schwankungen der Außentemperatur beeinflußt. Das Licht wurde durch eine elektrische Uhr ein- und ausgeschaltet, ohne die Dämmerung zu berücksichtigen.

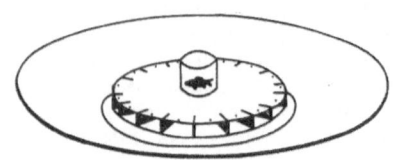

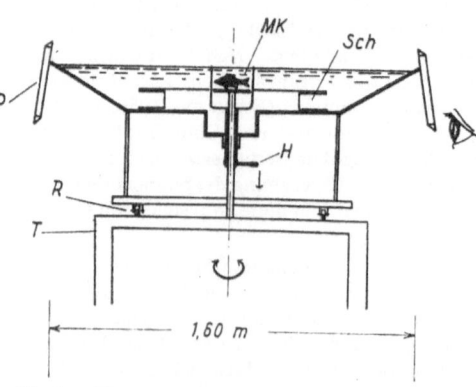

Abb. 2a. Oben: Versuchstank in schräger Aufsicht. Alle sechzehn Schachteln sind offen zum Test. Unten:Querschnitt. *MK* = Hebel zum Heben und Senken des Mittelkäfigs, *Sch* = eine der Schachteln, *P* = eins der vier Periskope, *R* = Rollen, auf denen sich die Apparatur auf dem Tisch *T* um die Mittelachse drehen läßt

Unsere Versuchsanordnung bestand aus einem wassergefüllten, drehbaren Rundtank (Abb. 2a). Er war auf dem flachen Dach eines kleinen Gebäudes aufgestellt. Landmarken der Umgebung konnten von dem Fisch im Versuchstank nicht gesehen werden. In diesem Tank befinden sich 16 kleine Schachteln. Diese sind mit ihren Öffnungen nach außen kreisförmig aneinander gelegt und durch eine runde Platte bedeckt. 15 dieser Schachteln werden während der Dressurperiode durch einen Ring verschlossen: die 16. Schachtel wird in die Himmelsrichtung, auf die der Fisch dressiert werden soll, gedreht. Der Fisch wird in der Mitte der Platte von einem runden Plexiglasgefäß festgehalten. Er wird freigesetzt, indem dieser Zylinder von unterhalb des Tanks versenkt wird. Er lernt nach oft wiederholten Dressuren in die Richtung zu schwimmen, in der sich die offene Schachtel befindet, wo er sich verstecken kann. Vier Periskope am Wannenrand ermöglichen die genaue Beobachtung durch den für den Fisch selbst unsichtbaren Beobachter. Man kann mittels der Periskope sehen, ob der Fisch eine Schachtel auf ihrer linken oder rechten Seite überschwimmt. Wir haben deshalb in der Abb. 2a in jede Schachtelmitte noch einen Punkt eingezeichnet. Wir haben gemessen, an welcher Stelle der Fisch den Plattformrand überschwimmt, und nicht, in welcher Schachtel er sich schließlich versteckt, da viele Fische sich angewöhnen, stereotyp nach dem Überschwimmen des Plattformrandes eine Links- oder Rechtswendung zu machen.

Neu war die Spiegeleinrichtung (Abb. 2b). Einen großen Schattenspender, dessen Höhe und Neigungswinkel verstellbar waren, stellten wir zwischen Sonne und Versuchstank. Man konnte mit ihm in den Morgenstunden den Versuchstank

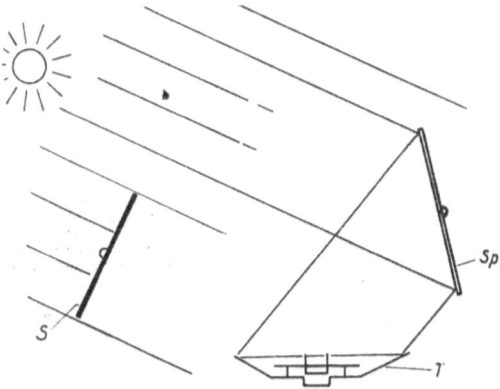

Abb. 2b. Schematische Skizze der Versuchsanordnung mit Schattengeber *S*, Spiegel *SP* und Versuchstank *T*. Das Azimut der Sonne wird in diesem Fall um 180° verdreht, die Höhe um 25° vergrößert

vollständig beschatten, ohne ihn allzu dicht an die Versuchswanne heranstellen zu müssen. Den 2×1 m großen Spiegel stellten wir in allen Versuchen der Sonne genau gegenüber auf. Das Azimut der Sonne war also um 180° verstellt. Der Spiegel war ebenfalls drehbar, so daß auch die Höhe der einfallenden Lichtstrahlen verändert werden konnte. Der Winkel des Spiegels konnte auf $\pm 1°$ genau eingestellt werden. Nach jedem Spiegelhöhenversuch haben wir den Winkel gemessen. Diese Zusatzeinrichtung wurde natürlich nur in einem Teil der Versuche angewendet.

III. Die Sonnenorientierung am Tage

Wir begannen dann, eine Reihe Fische, vor allem *Crenicichla saxatilis*, zu dressieren. Dieser Cichlide ist für die Versteckmethode ganz besonders gut geeignet. In mehreren Fällen konnten wir schon nach zweimaliger Dressur am Vormittag Versuche am Nachmittag machen, in denen diese Fische ihre Himmelsrichtung gut einhielten. Wir haben bei dieser Art niemals bei Dressuren am Tage einen rein azimutwinkelkonstanten Fisch bekommen. Leider ist diese Art ziemlich anfällig gegen Flossenfäule (finrot), und mancher dressierte Fisch ging während der Versuche ein.

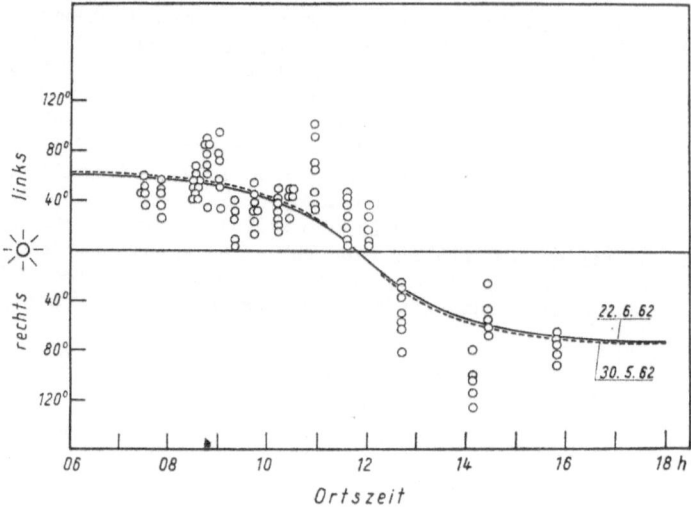

Abb. 3. Die Kompensation der Sonnenbewegung eines Hechtcichliden (*Crenicichla saxatilis* L.). Abszisse: Wahre Ortszeit, Ordinate: Kompaßgrade als Winkel zur Sonne. 0° = Richtung auf die Sonne zu. Die Kurven geben die Wahlrichtungen bei idealer Kompensation der Sonnenbewegung an, sie sind aus den Azimutkurven der Sonne berechnet. Die zwei Extreme während des Zeitraumes der Versuche sind vom 22. 6. 1962, 23° 30′ Deklination und 30. 5. 1962, 21° 30′ Deklination. Wie man sehen kann, kommt eine gewisse Streuung dadurch zustande, daß der Fisch über einen längeren Zeitraum hinweg oder auch von Tag zu Tag die Richtung etwas ändert

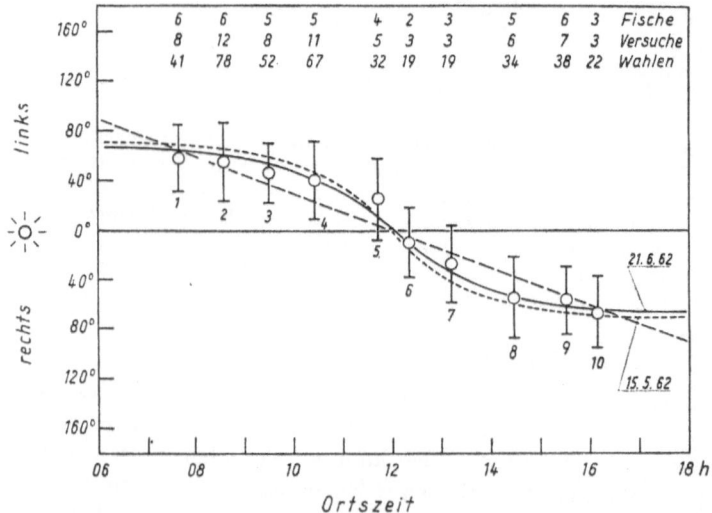

Abb. 4. Die Sonnenorientierung von sechs dressierten Fischen am Äquator, später zu Umstimmungs- und Nachtdressuren verwendet. Diese Fische wurden einige Wochen hindurch zu allen Tagesstunden geprüft. Jeder Punkt gibt den Mittelwert aus der Anzahl von Wahlen, Versuchen und Fischen, die darüber angegeben ist. Senkrechte Striche = Standardabweichung. Die Kurven geben die Wahlrichtungen bei idealer Kompensation der Sonnenazimutbewegung an, die gestrichelte Linie die Wahlrichtungen bei Kompensation einer Sonnenazimutbewegung von 15° pro Stunde. Wie man deutlich sieht, ändern die Fische ihren Schwimmwinkel zum Azimut der Sonne in den Morgen- und Abendstunden langsam, in den Mittagsstunden sehr schnell

Im folgenden ist nur von solchen Fischen die Rede, mit denen wir wochen- oder monatelang Versuche zu allen Tageszeiten gemacht haben, und die wir auch für die später geschilderten Spiegel- und Umstimmungsversuche verwendet haben. In der Abb. 3 sind über der Ortszeit die Winkel zum Azimut der Sonne eingetragen, die ein immer nur kurz am Vormittag nach Norden dressierter Fisch im Laufe des Tages einschlägt. Die Versuche sind in der Zeit vom 31. 5. bis 7. 7. gemacht. Die Extreme der Azimutkurven sind als ausgezogene Linie (22. 6. 1962) und als punktierte Linie (30. 5. 1962) eingezeichnet. In der Abb. 4 haben wir die Richtungswahlen von den sechs Fischen zusammengefaßt, die wir später für Spiegel- und Umstimmungsversuche verwendet haben.

IV. Die Spiegelversuche

In den bisher veröffentlichten Versuchen über die Sonnenorientierung von Fischen haben wir noch niemals mit einem Spiegel direkt gezeigt, daß und ob die Sonne der einzige äußere Bezugspunkt ist, der vom richtungsschwimmenden Fisch verwendet wird. Wir haben viele Versuche gemacht, die kaum anders zu verstehen sind. Es kommt z. B. oft vor, daß Fische azimutwinkelkonstant flüchten. Solche Fische ändern im Laufe des Tages die Himmelsrichtung, in die sie schwimmen, dem jeweiligen Azimutstand der Sonne entsprechend. Bei bedecktem Himmel schwimmen sowohl azimutwinkelkonstante als auch richtungskonstante Fische unorientiert in alle Himmelsrichtungen. Werden Fische zu einer bestimmten Tageszeit mit einer künstlichen Sonne im geschlossenen Raum dressiert, in einem bestimmten Winkel zum Azimut dieser Sonne zu flüchten, dann verhalten sich viele Fische bei Testen mit der natürlichen Sonne so, als wären sie auf die entsprechende sonnenbezogene Himmelsrichtung dressiert (s. auch Kap. VI). Bei derartigen Dressuren mit der künstlichen Sonne ist uns merkwürdigerweise niemals ein Einfluß der Sonnenhöhe aufgefallen. Wir haben das aber noch nicht systematisch geprüft. Es gelingt leicht, Fische mit der künstlichen Sonne zu dressieren und dann an ihre Stelle die natürliche Sonne zu setzen. Wir haben das oft erfolgreich angewandt; das Umgekehrte ist uns nur ausnahmsweise in zwei oder drei Fällen gelungen, obwohl wir es oft probiert haben. Bei Testen mit der künstlichen Sonne bekamen wir in der überwiegenden Mehrzahl der Fälle nur die unerwünschte Azimutwinkelkonstanz.

Für die Spiegelversuche wählten wir Fische aus, bei denen wir uns durch zahlreiche Versuche am Vor- und Nachmittag und zur Mittagszeit an verschiedenen Tagen von ihrer Richtungskonstanz überzeugt hatten (s. Kap. III, Abb. 4). Wir prüften zuerst, ob die Fische nicht durch den Spiegel oder den notwendigen Schattenwerfer, die beide Landmarken darstellen, verstört waren. Dazu prüften wir fünf Tiere an verschiedenen

Tagen, zunächst wie üblich in 4—5 Einzelläufen. Danach stellten wir den etwas geneigten, in seiner Höhe verstellbaren Schattenwerfer zwischen Sonne und Versuchswanne so auf, daß die gesamte Wanne vollständig im Schatten lag. Den vertikalen Spiegel stellten wir der Sonne gegenüber dicht an den Wannenrand. Die Wanne war damit von der anderen Seite vollkommen ausgeleuchtet, das Azimut der Sonne um 180° verstellt. Dann ließen wir die 5—10 min vorher mit der natürlichen Sonne untersuchten Fische wieder 4—5mal schwimmen. Das zusammengefaßte Ergebnis zeigt die Abb. 5. Alle Fische flüchten in die gegenüberliegende

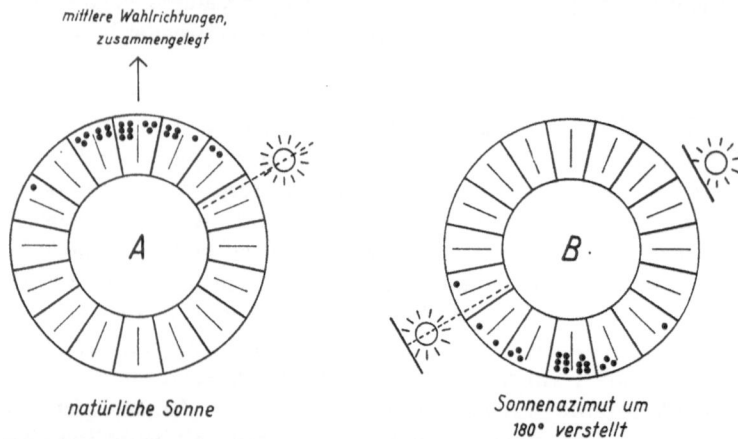

Abb. 5. Der Versuch mit der gespiegelten Sonne. Jeder Punkt bedeutet eine Wahl. In *A* sind die mittleren Wahlrichtungen der fünf Tiere alle nach oben zusammengelegt, oben ist deshalb nicht Norden, und in *B* wurden die etwa 15 min später getroffenen Wahlen mit verstelltem Azimut auf diese mittlere Wahlrichtung bezogen. Der Sonnenstand ist zur Illustration der Bedingungen willkürlich eingezeichnet (Zusammenfassung von Versuchen an 5 Tieren)

Himmelsrichtung. Die Sonne scheint tatsächlich der einzige äußere Bezugspunkt zu sein, die Fische befinden sich jedenfalls nicht in einem Konflikt, etwa zwischen zwei Himmelsrichtungen, die sich aus dem ungespiegelten polarisierten Himmelslicht über ihnen und dem durch den Spiegel veränderten Sonnenazimut ergeben.

Wir haben nur einmal in den weiter unten beschriebenen Versuchen mit verstelltem Sonnenazimut und gleichzeitig verstellter Sonnenhöhe kurz hintereinander bei zwei Fischen bevorzugtes Schwimmen in die alte Himmelsrichtung gefunden. Das trat sowohl bei verstellter Sonnenhöhe, als auch bei allein verstelltem Azimut auf. An diesem Tage war zwar die Sonne gut sichtbar, aber der Himmel außerdem mit konturierten Wolkenmustern bedeckt, die in nordwestlicher Richtung zogen. Unter diesen Wolken waren die Fische vor dem Spiegelversuch geprüft worden. Außerdem wurde der Spiegel durch den Wind bewegt. Wir haben diese Versuche hier nicht mit ausgewertet und führten dann nur noch Versuche bei völlig wolkenlosem Himmel aus.

Uns interessierte in den weiteren Versuchen, wie sich die Fische bei künstlich veränderten Sonnenhöhen verhielten. Hierbei gingen wir im allgemeinen wie in Abb. 6 dargestellt vor. Wir prüften alle Fische morgens bei relativ niedrigem Sonnenstand. Die Sonne wurde durch den gegenüberliegenden und geneigten Spiegel im Durchschnitt 22,5° höher präsentiert. Jeden Fisch ließen wir 4—5mal schwimmen. Danach stellten wir den Spiegel senkrecht, so daß nur das Azimut um 180° verstellt war, und schließlich prüften wir den Fisch nach Abbau der Spiegeleinrichtung mit der Sonne allein. Zwischen den einzelnen Versuchen wurde der Tank

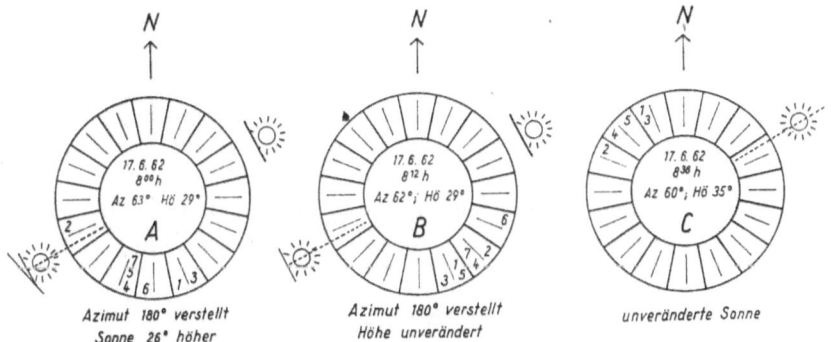

Abb. 6. Beispiel eines Versuchs mit der gespiegelten und auch in der Höhe verstellten Sonne. Richtung nach oben = Norden. In den Diagrammen: Datum, Zeit des Versuchs und Position der Sonne. A = Azimut 180° verstellt, Sonne 26° höher. B = Azimut 180° verstellt, Höhe unverändert. C = unveränderte Sonne. Die Ziffern bedeuten die einzelnen Wahlen in ihrer Reihenfolge

natürlich immer gedreht, so daß eine Orientierung nach Landmarken im Tank ausgeschlossen war. Wir bevorzugten diese Reihenfolge, weil jedes Tier kurz nacheinander 15mal schwimmen mußte. Der wichtige Sonnenhöhenversuch sollte nicht durch Ermüdung beeinträchtigt werden. In einigen Fällen verstellten wir zuerst das Azimut allein und dann die Höhe der Sonne. Wie man in Abb. 6 sieht, flüchten bei morgens künstlich erhöhter Sonne die Fische in einem solchen Winkel zum Azimut der Sonne, wie sie ihn zu späterer Tagesstunde, wenn die Sonne tatsächlich höher ist, bei Richtungskonstanz auch einhalten. Die Streuung ist im allgemeinen größer; in weniger Versuchen ist überhaupt keine Vorzugsrichtung zu erkennen. In der Abb. 7 haben wir alle Spiegelversuche, mit Ausnahme der schon erwähnten zwei Versuche mit Wolken, miteinander verglichen. Die Mittelrichtungen aus den einzelnen Spiegelazimutversuchen wurden übereinander gelegt, oben ist jetzt nicht Norden, sondern die mittlere Richtung, in die die Fische mit 180° verstelltem Sonnenazimut schwimmen! daneben sind alle kurz vorher oder kurz danach gemachten Versuche mit erhöhter Sonne eingezeichnet. Wie man sieht, weichen, wie bei Abb. 6 bei künstlich erhöhter Sonne, die Fische gerichtet ab gegenüber den Versuchen, bei denen die Sonne in ihrer natürlichen Höhe belassen wurde. Diese Abweichung ist statistisch gesichert. (Zur statistischen Auswertung von Kreisverteilungen siehe SCHWASSMANN und BRAEMER, 1961.)

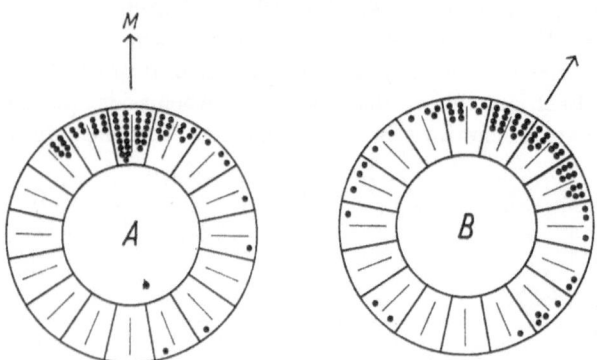

Abb. 7. Zusammenfassung der Spiegelversuche. Jeder Punkt bedeutet eine Wahl. In *A* wurden die mittleren Wahlrichtungen (*M*) unter der gespiegelten, aber in normaler Höhe belassenen Sonne alle nach oben gezeichnet. *A* = Sonnenazimut um 180° verstellt; *B* = Sonnenazimut um 180° verstellt und Sonne 22,5° höher. Unterschied der beiden Richtungen 34°

V. Umstimmungsversuche

Das Ziel der Umstimmungsversuche war es, zu messen, wie die Kompensation der Sonne in der Nacht fortgesetzt wird. Entweder schwimmen in der Mitternacht die Fische in demselben Winkel zum Azimut der Sonne wie mittags; das hieße, in der Nacht findet eine Richtungsumkehr in der Einrechnung der Sonnenbewegung statt; oder aber die Fische flüchten mitternachts in einem Winkel 180° entgegengesetzt zu dem Mittagswinkel, was bedeuten würde, daß in der Nacht in der gleichen Richtung weiterkompensiert wird wie am Tage.

Bei der Interpretation der Versuche müssen wir beachten, daß die Sonnenhöhe einen Einfluß auf die Richtung hat, in die die Fische schwimmen. Dieser Einfluß muß sich geltend machen, weil die Versuchstiere in einem künstlichen Lichtrhythmus, der gegenüber dem natürlichen Tag phasenverschoben ist, gehalten werden. Es sei daran erinnert, daß uns der Einfluß der Sonnenhöhe bei solchen Versuchen überhaupt erst auffiel. Wir müßten in der physiologischen Nacht den Fischen vermutlich „negative" Sonnenhöhen bieten, um die nächtliche Einrechnung der Sonnenwanderung korrekt zu messen.

Wir haben drei Fische Ende Mai um 9 Std und einen Fisch Ende Juli um 7 Std umgestimmt. Der künstliche Tag begann für die erste Gruppe 15⁰⁰ nachmittags und endete 3⁰⁰ morgens. In den späten Nachmittagsstunden überschneidet sich somit der Anfang des Fischtages mit dem Ende des natürlichen Tages, und wir konnten diesen Fischen um 16³⁰, das ist für sie 7^{30}, die natürliche Sonne in der Höhe bieten, die der physiologischen Zeit entsprach. Versuche zu dieser Zeit sind die Kontrollen 1. dafür, ob die Umstimmung stattgefunden hat und 2. ob die dann folgenden Versuche in der subjektiven Nacht der Tiere Nacheffekte haben. Bei dem Versuch im Juli ließen wir den Tag um 23^{00} ananfangen und um 1^{00} enden. Die Kontrollversuche mit der richtigen Sonnenhöhe können somit um 8^{30} gemacht werden, das ist für den Fisch 15^{30}.

Wir wissen aus früheren Versuchen, daß die Umstellung auf den phasenverschobenen Lichtzyklus vier Tage dauert. Für die Versuche in der Fischnacht beleuchteten wir das betreffende Versuchstier zuerst mit gedämpftem Tageslicht und setzten es dann allmählich und ohne direkte Sonnensicht dem hellen Tag aus, bis sie auf visuelle Reize reagierten. Das dauert bei Cichliden etwa eine Stunde. Davon, daß die Fische dann sehen können, kann man sich durch Handbewegungen über dem Eimer überzeugen. Vor dem Ablauf dieser Adaptationszeit sind Cichliden anscheinend geblendet, wenn sie in der Nacht mit Licht aufgestört werden, und es erscheint unsinnig, dann sofort Versuche im hellen Sonnenlicht zu machen, bei denen sie in ein Versteck flüchten müssen. Nach solchen Versuchen in seiner Nacht kommt der Fisch wieder in seinen phasenverschobenen Lichtzyklus zurück.

Zur Kontrolle dafür, ob diese einstündige Beleuchtung während der „Nacht" der Fische einen Einfluß auf die Phasenlage des Orientierungsrhythmus hat, dienten die Versuche mit der „richtigen" Sonnenhöhe am folgenden Morgen resp. am selben Nachmittag.

In der Abb. 8 zeigen wir die Richtungswahlen des 7 Std umgestimmten Fisches. Auf der Abszisse ist die Fischzeit aufgetragen, auf der Ordinate die Winkel, die der Fisch zum Azimut der Sonne eingeschlagen hat. Alle ausgefüllten Punkte sind Einzelwahlen des Fisches im phasenverschobenen Tag. Die nichtausgefüllten Kreise sind die Werte, die von diesem Fisch vorher, bei Versuchen im natürlichen Tag, gewonnen wurden. Die hindurchgelegte Kurve gibt die Sonnenazimutkurve zur Zeit der Versuche an. Wie man sieht, ist der Fisch nach Norden dressiert, denn er schwimmt mittags auf die Sonne zu. Bei dem Pfeil um 15^{30} muß man die ausgefüllten Kreise mit den nichtausgefüllten vergleichen. Sie zeigen, daß der Fisch umgestimmt ist. Die nicht ausgefüllten Einzelwahlpunkte stammen von Versuchen um 15^{30} vor der Umstimmung, die ausgefüllten von Versuchen um 15^{30} Fischzeit, also 8^{30} Ortszeit nach der Umstimmung, wo der Fisch vorher einen Winkel weit links von der Sonne einschlug. Bei den Versuchen in der Mitternacht streuten die Wahlrichtungen. Es ist keine Vorzugsrichtung erkennbar. Diese Versuche sind um 17^{00} abends gemacht. Die Höhe der Sonne entspricht dann der Höhe um 7^{00} morgens, wie man in der Abbildung oben an den Sonnenhöhenkurven sehen kann. Die zeitliche Differenz zwischen Mitternacht, dem physiologischen Zustand des Fisches, und 7⁰⁰ morgens oder 17⁰⁰ abends beträgt in beiden Fällen 7 Std. Es scheint, daß immer, wenn diese Differenz gleich ist, wir die größte Streuung erhalten. Würde der Fisch in dem Winkel zum Azimut der Sonne schwimmen, den er um 7⁰⁰ morgens einschlägt, dann müßte er etwa 70° links von der Sonne schwimmen;

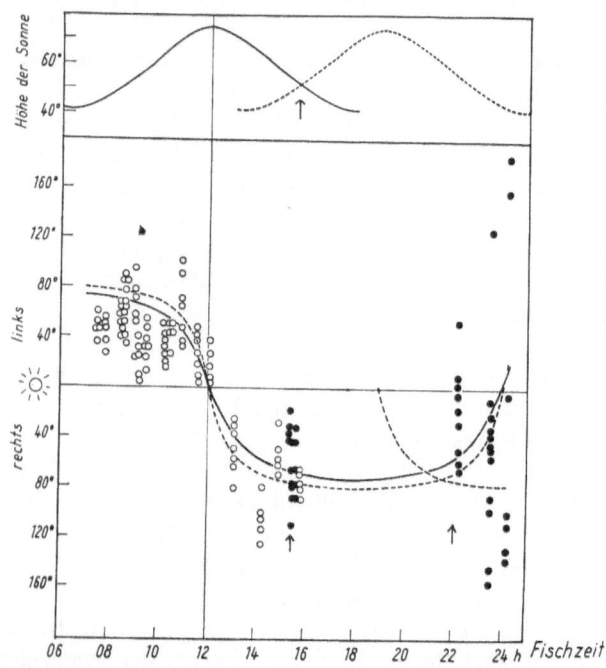

Abb. 8. Richtungswahlen eines dressierten Fisches vor (offene Kreise) und nach (ausgefüllte Kreise) der Umstimmung um 7 Std. Abszisse: Fischzeit, entspricht vor der Umstimmung der Ortszeit. Ausgezogene Kurve: Sonnenazimutkurve zu Beginn der Umstimmung am 7. 8. 1962, gestrichelte Kurve: Ende der Umstimmung am 27. 8. 1962, links, wie der umgestimmte Fisch sie „erwartet", rechts, wie er sie wirklich sieht. Im oberen Teil des Diagramms die Kurve der Höhenänderung der Sonne, ausgezogen vor, gestrichelt nach der Umstimmung. Die Refraktion an der Wasseroberfläche ist berücksichtigt

würde er in dem Winkel zum Azimut der Sonne schwimmen, der 17⁰⁰ entspricht, müßte er etwa 70° rechts von der Sonne schwimmen. Kurz vor Mitternacht erhalten wir dagegen gerichtete Wahlen, siehe 22¹⁵ = 15¹⁵ Ortszeit (Pfeil), die gut mit dem, was man bei einer Richtungsumkehr in der Einrechnung der Sonnenbewegung in der Nacht erwarten müßte, übereinstimmen (s. Azimutkurve). Diese Versuche sind jedoch mit der Sonnenhöhe von 15¹⁵ gemacht, die auch der Sonnenhöhe um 8⁴⁵ entspricht. Der zeitliche Abstand zwischen 22¹⁵, dem physiologischen Zustand des Fisches, und 15¹⁵ beträgt 7 Std, der zwischen 22¹⁵ und 8⁴⁵ 10 Std 30 min. Der vom Fisch eingeschlagene Winkel zur Sonne paßt auch einigermaßen zu dem Winkel, den der Fisch um 15^{15} geschwommen ist. Wir können nicht entscheiden, ob der umgestimmte Fisch um 22^{15} überwiegend in einem Winkel rechts von der Sonne schwimmt, weil er wie in den Spiegelversuchen bei einer bestimmten Sonnenhöhe einen bestimmten Winkel zum Azimut der Sonne einschlägt oder, weil in seiner Nacht ihm der von der physiologischen Uhr gesteuerte Prozeß eine Richtungsumkehr in der Einrechnung der Sonnenazimutbewegung vorschreibt. Am wahrscheinlichsten ist, daß wir ein Gemisch von beiden bekommen. Das gilt für alle unsere Umstimmungsversuche. Deshalb sind wir auf dieses eine Beispiel so ausführlich eingegangen.

Wir finden bei allen umgestimmten Fischen um Mitternacht eine starke Streuung. In den Versuchen kurz vor und kurz nach Mitternacht liegen die Meßpunkte in unmittelbarer Nähe der die Richtung umkehrenden Sonnenazimutkurve und nicht $180°$ dazu entgegengesetzt. Wir können aber vorläufig nicht entscheiden, inwieweit das auf der Mitwirkung der Sonnenhöhe beruht.

Interessant scheint uns noch zu sein, daß wir bei ganz ähnlichen Versuchen in nördlichen Breiten richtige Konfliktsituationen bei den Fischen beobachtet haben, die wie Gleichgewichtsstörungen aussahen und sich gelegentlich in einer Aufspaltung der Richtungswahlen äußerten. Dieses trat immer dann auf, wenn die von der Sonnenhöhe vorgeschriebene Richtung und die von der jeweiligen Nachtzeit vorgeschriebene Richtung einander gegenüberlagen. Diesen Konflikt haben wir hier trotz der höheren Sonne niemals gesehen; vielleicht ist das auch ein Hinweis dafür, daß diese beiden Richtungen hier einander nicht gegenüberliegen, mit anderen Worten, daß in der Nacht auf dem Äquator der Sinn der Verrechnung der Sonnenazimutbewegung der am Tage entgegengesetzt ist.

VI. Dressuren in der Nacht

Uns steht eine zweite Methode für die Untersuchung der Sonnenorientierung in der Nacht zur Verfügung. Wir haben sie schon in nördlichen Breiten erfolgreich angewandt (BRAEMER, 1959, 1960). Es wurden damals 13 grüne Sonnenfische und 2 Lachse in einem geschlossenen Raum in der Nacht dressiert, auf eine $50°$ hohe künstliche Sonne zuzuschwimmen, jedes Individuum zu einer bestimmten Zeit, der erste Fisch um 20^{00}, der zweite um 21^{00}, der dritte um 22^{00} usw. Nachdem die Fische das gelernt hatten, prüften wir am Tage mit der Sonne, in welche Himmelsrichtung diese Fische schwammen. Von den 15 Fischen hatten zehn einen Sonnenkompaß, die anderen fünf waren azimutwinkelkonstant, schwammen also immer auf die Sonne zu. Die vor Mitternacht dressierten richtungskonstanten Fische schwammen nach NW bis N, die mitternachts dressierten Fische schwammen nach Norden und die nach Mitternacht dressierten nach NO. Die Fische fahren demnach auf dem 43. nördlichen Breitengrad in der Nacht fort, im Sommer die im Uhrzeigersinn umlaufende Sonne weiter zu verrechnen[1]. Auf diesen und

---

[1] Eine einzige Ausnahme zu dieser Regel ist uns bisher bekannt (SCHWASSMANN, 1962).

weiteren Ergebnissen von Dressuren mit einer künstlichen Sonne, auch am Tage, gründet sich unsere Ansicht, daß die Höhe der künstlichen Dressursonne keine wesentliche Bedeutung hat für die in den nachfolgenden Versuchen gemessene Einrechnung der Sonnenbewegung.

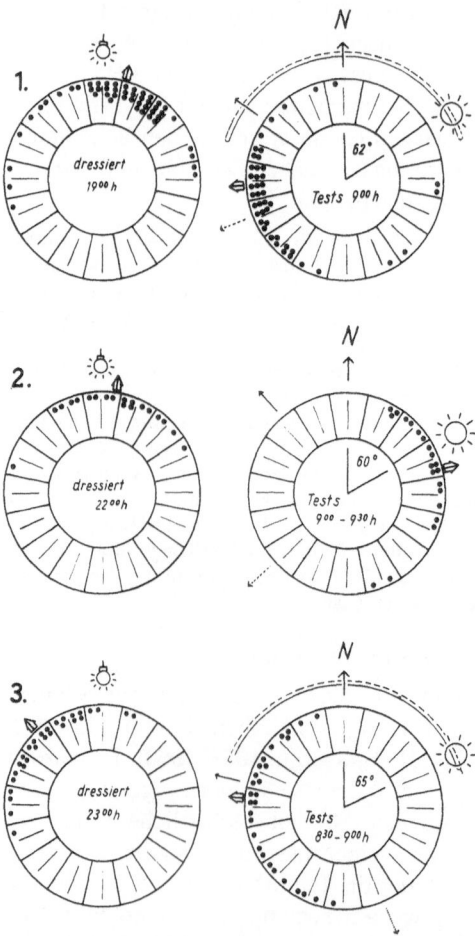

Abb. 9. Ergebnisse der Nachtdressuren vor Mitternacht. Die Fische sind dressiert, auf die künstliche Sonne zuzuschwimmen. Linke Reihe = Tests zur Dressurzeit (Zeit im Diagramm). Dicker Pfeil = Mittelrichtung der Wahlen. Rechte Reihe = Tests unter der natürlichen Sonne. Dicker Pfeil = Mittelrichtung der Wahlen, Pfeil mit N = Nordrichtung, ausgezogener Pfeil = Richtung, in die der Fisch schwimmen müßte, wenn er die nächtliche Richtungsumkehr der Sonnenbahn kompensierte, gestrichelter Pfeil = dasselbe ohne Richtungsumkehr. Eingezeichneter Winkel = Sonnenazimut zur Testzeit. In 1 und 3 ist die Sonnenbahn für Tag und Nacht eingezeichnet. Jeder Punkt bedeutet eine Einzelwahl

Dieselbe Methode wandten wir nun auf dem Breitengrad 1° 27′ S an. Allerdings stand uns hier kein geschlossener Raum zur Verfügung. Die Fische wurden im Freien auf dem gleichen Platz dressiert, auf dem wir alle bereits beschriebenen Versuche ausgeführt haben. Die künstliche Sonne war eine 300 Watt-Glühbirne, die 50° hoch, 3,5 m von der Mitte des Versuchstankes entfernt, an einem schmalen Rohr befestigt war.

Anfang August dressierten wir 5 *Crenicichla saxatilis*, in der *ersten* Nachthälfte auf die künstliche Sonne zuzuschwimmen. Einen um 19⁰⁰, einen um 20⁰⁰, usw. Die künstliche Sonne war im Osten aufgestellt. Eine Stunde vor der Dressur wurden die Fische mit einer 40 Watt-Birne beleuchtet, d. h. also, für den um 19⁰⁰ dressierten Fisch wurde es gar nicht erst dunkel. Jeder Fisch wurde jeweils 10 min dressiert. Von den 5 Fischen starben 2 an Flossenfäule, bevor mit den Messungen begonnen werden konnte. Die Ergebnisse der anderen drei Tiere sind in der Abb. 9 dargestellt. Links sind die Versuche zur Dressurzeit vom 6. bis zum 10. Tag. Wie man sieht, schwimmt der 19⁰⁰ Fisch bei Prüfungen zur Dressur-

zeit etwas rechts von der künstlichen Sonne, der 23^{00} Fisch etwa 40°
links davon. Das muß natürlich bei der Berechnung der sonnenbezogenen
Himmelsrichtung berücksichtigt werden. Der um 22^{00} dressierte Fisch
ist azimutwinkelkonstant, denn er schwimmt auch bei Versuchen zwischen
9^{00} und 9^{30} auf die Sonne zu. Der um 19^{00} dressierte Fisch schwimmt zu
seiner Dressurzeit 12,5° rechts von der ,,Sonne" und in Versuchen am
Morgen zwischen 8^{30} und 9^{00} nach 291°. Er schwimmt also in die Himmelsrichtung, in der sich das Azimut der Sonne kurz nach Sonnenuntergang
befand. Es kann bei ihm jedoch nicht entschieden werden, ob er weiter in
gleicher Richtung rechnet oder ob er die Richtung umkehrt, da beide
Wahlrichtungen zu nahe beieinander liegen. — Auch der um 23^{00} dressierte
Fisch ist auf eine Himmelsrichtung eingestellt, denn er schwimmt in
5 Versuchen zwischen 8^{30} und 9^{00} im Mittel nach 277°. Um 23^{00} befindet
sich das Azimut der Sonne bei 323°, und da der Fisch zu seiner Dressurzeit 40° links von der Sonne schwimmt, ist er auf die Himmelsrichtung
283° dressiert, und das stimmt fast genau mit der Himmelsrichtung
überein, in die er tatsächlich schwimmt. Würde er in der Nacht nicht die
Richtung seiner Verrechnung umkehren, dann müßte er fast genau entgegengesetzt flüchten, nämlich nach 177°.

    In der Zeit vom 23. August bis zum 31. August dressierten wir sechs
Fische in der *zweiten* Nachthälfte. Die künstliche Sonne stand im Osten.
Wir dressierten diese Fische, von der künstlichen Sonne wegzuschwimmen, da alle sechs Fische spontan diese Richtung bevorzugten. In der
Tat stellten sich Dressurerfolge früher ein, und mit den ersten Fischen
konnte schon nach dreimaliger Dressur mit Versuchen am Tage begonnen
werden. Nur der um 5^{00} morgens dressierte Fisch schwamm noch nach
8tägiger Dressur in alle Himmelsrichtungen. Danach gaben wir unsere
Bemühungen mit ihm auf. Es kommt ab und zu vor, daß sich ein Fisch
schlecht oder gar nicht dressieren läßt. Die Richtungswahlen der anderen
5 Fische sind in der Abb. 10 zusammengefaßt dargestellt. Links sind die
Richtungswahlen der Fische zur Dressurzeit und rechts Versuche des
Morgens, am frühen und am späten Nachmittag mit der natürlichen
Sonne wiedergegeben.

    Aus diesen Richtungswahlen geht hervor, daß der um 24^{00}, der um
1^{00} und der um 4^{00} dressierte Fisch sich so verhalten, als ob sie in der
Nacht auf eine in der nördlichen Himmelshälfte stehende Sonne dressiert
worden wären. Der um 2^{00} dressierte Fisch verhält sich azimutwinkelkonstant, denn er schwimmt immer von der Sonne weg. Der um 3^{00}
dressierte Fisch kehrt als einziger in der Nacht nicht die Richtung seiner
Verrechnung um, denn er schwimmt am Tage, am Vor- und am Nachmittag, nach Norden.

    Von insgesamt 11 in der Nacht dressierten Fischen gingen somit 2 ein,
bevor Versuche mit ihnen gemacht werden konnten, einer ließ sich in

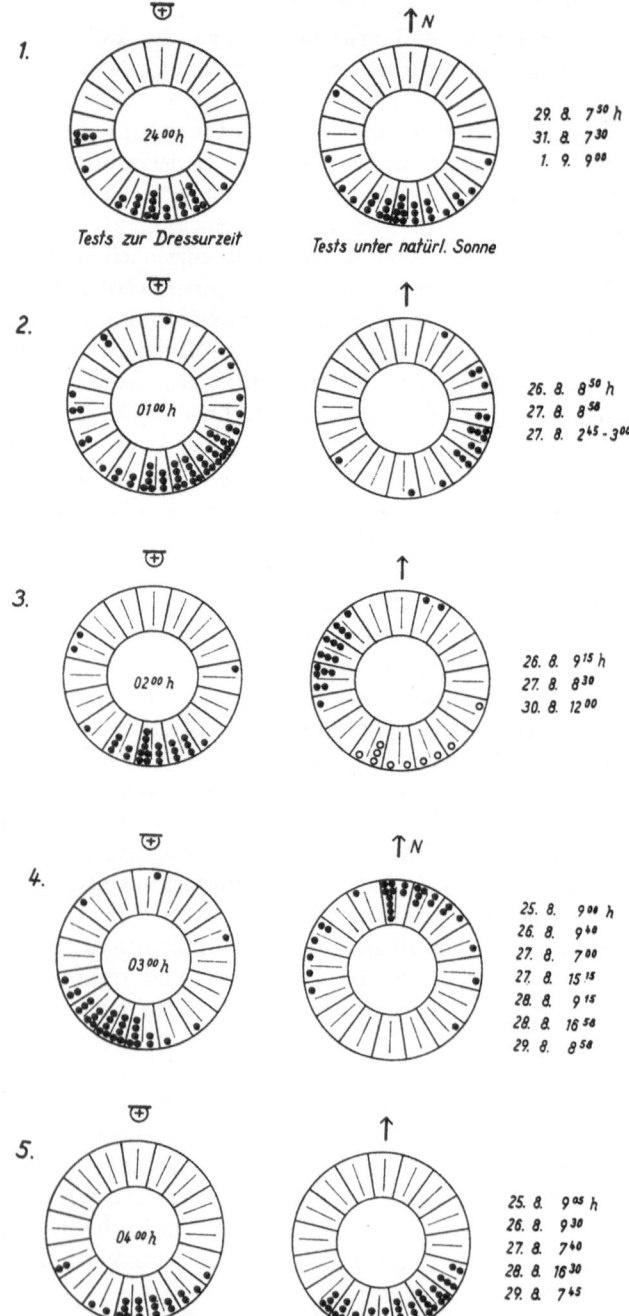

Abb. 10. Ergebnisse der Nachtdressuren nach Mitternacht. Die Fische sind dressiert, von der künstlichen Sonne wegzuschwimmen. Linke Reihe = Tests zur Dressurzeit (Zeit im Diagramm). Rechte Reihe = Tests unter der natürlichen Sonne, Daten und Tageszeiten stehen rechts daneben. Jeder Punkt bedeutet eine Einzelwahl. Nr. 2: Statt 2⁴⁵—3⁰⁰ lies 14⁴⁵—15⁰⁰

acht Nächten nicht dressieren, 2 waren azimutwinkelkonstant, und 6 hielten eine Himmelsrichtung ein. Davon verrechneten 4 mit Sicherheit eine in der Nacht ihre Richtung umkehrende Sonne, von dem 5., dem um 19⁰⁰ dressierten Fisch läßt sich das nicht mit Sicherheit sagen. Der 6. Fisch kehrt in der Nacht nicht die Richtung seiner Verrechnung um. Da auch die Versuche mit Fischen im phasenverschobenen Tag für eine Richtungsumkehr in der Verrechnung der Sonnenbewegung sprechen, ist nicht mehr daran zu zweifeln, daß für *Crenicichla saxatilis* auf dem Äquator in der Mitternacht die Sonne dasselbe Azimut hat wie mittags.

VII. Ergebnis

Unsere Versuche zeigen, daß auf dem Äquator, mindestens von Mai bis Ende August, am Tage die im Norden kulminierende Sonnenbewegung verrechnet wird. Einige Fische kompensieren Mitte April noch anders herum, obschon vom 21. März an die Sonne mittags im Norden steht. Die notwendige Umstellung von der im Süden kulminierenden Sonne auf die im Norden kulminierende dauert bei Fischen offenbar mehrere Wochen. Die Versuche sprechen ferner dafür, daß in den Mittagsstunden die gegenüber den Abend- und Morgenstunden schnellere Azimut-änderung eingerechnet wird (Abb. 3 und 4). In Spiegelversuchen mit 180° verstelltem Azimut schwimmen dressierte Fische in die jeweils gegenüberliegende Himmelsrichtung; bei morgens künstlich erhöhter Höhe des einfallenden Sonnenlichtes finden wir eine statistisch gesicherte gerichtete Abweichung. Die Fische verhalten sich so, als ob sie später mit der höheren natürlichen Sonne geprüft würden. Danach hat die Sonnenhöhe bei Fischen einen Einfluß auf die Einrechnung der kompensierten Sonnenbewegung. Das bestätigt frühere Versuche von uns. Wir möchten hier nur am Rande erwähnen, daß solche „Störungen" durch die Sonnenhöhe in der von MITTELSTAEDT veröffentlichten Theorie noch nicht vorgesehen sind.

Am Äquator lassen sich Fische durch Phasenverschiebungen des Licht-Dunkel-Wechsels ebenso umstimmen wie in nördlichen Breiten. Versuche mit phasenverschobenen Fischen in ihrer subjektiven Nacht sprechen für eine Richtungsumkehr der eingerechneten Sonnenbewegung in der Nacht. Es läßt sich jedoch nicht entscheiden, inwieweit diese Befunde auf dem Einfluß der Sonnenhöhe beruhen. Fische, die in der Nacht dressiert wurden, auf ein künstliches Licht zu- oder von ihm weg-zuschwimmen, zeigen mit einer Ausnahme, daß in der Nacht tatsächlich eine Richtungsumkehr in der Einrechnung der Sonnenbewegung erfolgt. Damit haben wir zusätzlich zu der schon erwähnten breitengrad- und jahreszeitspezifischen Verrechnung der Sonnenbewegung am Tage einen drastischen Unterschied im Verhalten auf dem Äquator gegenüber gleichen Versuchen in nördlichen Breiten gefunden. Dieses Ergebnis ist in Übereinstimmung mit der von MITTELSTAEDT veröffentlichten Theorie. Es erhebt sich die Frage: Woher beziehen die Fische die Information über den Breitengrad, auf dem sie sich befinden?

Wir wissen von früheren Versuchen an zwei Fischarten, daß die Fähigkeit zur Verrechnung der Sonnenbewegung angeboren ist. Nordamerikanische grüne Sonnenfische, die niemals in ihrem Leben die tägliche Sonnenbewegung gesehen hatten, verrechneten auf dem 43. Breitengrad die im Süden kulminierende Sonne, während *Aequidens portalegrensis* H., eine tropische Cichlidenart, eine Aufspaltung in ihren Richtungswahlen aufweist, die zeigt, daß diese Art am Tage gleichzeitig eine rechts- und eine linksherumlaufende Sonne kompensieren kann und dann lernt, sich auf die richtige Richtung festzulegen. Das paßt gut zu der Herkunft dieser Art aus den Tropen, wo in der Tat die Sonne in der einen Jahreshälfte im Norden kulminiert und in der anderen Jahreshälfte im Süden, der Kompaßfisch demnach zweimal im Jahr die Richtung seiner Kompensation ändern muß. Über die Einrechnung der Sonnenbewegung in der Nacht von sonnenunerfahrenen Fischen wissen wir noch nichts. Uns fiel bereits bei Versuchen mit sonnenunerfahrenen Fischen auf, daß die quantitative Verrechnung der Sonnenbewegung bei diesen Fischen, die niemals in ihrem Leben die tägliche Sonnenbewegung sahen, sowohl bei der tropischen Art, als auch bei der Art, in deren Verbreitungsgebiet wir die Versuche gemacht hatten, zu dem Breitengrad paßte, auf dem wir die Versuche durchführten. Da die Fische in der Tageslänge gehalten wurden, die der Jahreszeit und dem Breitengrad entsprach, und die Versuche mit der zur Tageszeit genau passenden Sonnenhöhe gemacht wurden, müssen wir annehmen, daß diese beiden Faktoren die quantitative Einrechnung der Sonnenbewegung so beeinflussen, daß sie für Breitengrad und Jahreszeit richtig wird. Die Fähigkeit zur Verrechnung der Sonnenbewegung ist Fischen angeboren. Jedoch Richtung und quantitative Verrechnung der Sonnenbewegung hängen von der Phase des Licht-Dunkel-Wechsels, von der Länge des Tages und von der Höhe der Sonne ab.

Literatur

BIRUKOW, G.: Lichtkompaßorientierung beim Wasserläufer *Velia currens* F. (Heteroptera) am Tage und zur Nachtzeit I. Z. Tierpsychol. 13, 463—484 (1956).
— Innate types of chronometry in insect orientation. Cold. Spr. Harb. Symp. quant. Biol. 25, 403—412 (1960).
— u. E. BUSCH: Lichtkompaßorientierung beim Wasserläufer am Tage und zur Nachtzeit II. Z. Tierpsychol. 14, 184—203 (1957).
BRAEMER, W.: Versuche zu der im Richtungsgehen der Fische enthaltenen Zeitschätzung. Verh. Dtsch. Zool. Ges. Münster 1959, 276—288.
— A Critical Review of the sun-azimuth hypothesis. Cold Spr. Harb. Symp. quant. Biol. 25, 413—427 (1960).

BRAEMER, W., u. H. O. SCHWASSMANN: In Vorbereitung: Sonnenorientierung bei sonnenlos aufgezogenen Fischen.

HASLER, A. D., R. M. HORRALL, W. J. WISBY and W. BRAEMER: Sun-orientation and homing in fishes. Limn. and Oceanogr. 3, 353—361 (1958).

— and H. O. SCHWASSMANN: Sun orientation of fish at different latitudes. Cold Spr. Harb. Symp. quant. Biol. 25, 429—441 (1960).

HOFFMANN, K.: Die Richtungsorientierung von Staren unter der Mitternachtssonne. Z. vergl. Physiol. 41, 471—480 (1958).

LINDAUER, M.: Dauertänze im Bienenstock und ihre Beziehung zur Sonnenbahn. Naturwissenschaften 41, 506 (1954).

— Sonnenorientierung der Bienen unter der Äquatorsonne und zur Nachtzeit. Naturwissenschaften 44, 1—6 (1957).

MITTELSTAEDT, H.: Control systems of orientation in insects. Ann. Rev. Entomol. 7, 177—198 (1962).

PARDI, L.: Esperience sull'orientamento di *Talitrus saltator* (Montagu) (Crustacea Amphipoda): L'orientamento al sole degli individui a ritmo nictimerale invertito, durante la ,,lore notte". Boll. Ist. Mus. Zool. Univ. Torino 4, 127—134 (1953).

— Innate components in the solar orientation of littoral amphipods. Cold Spr. Harb. Symp. quant. Biol. 25, 395—401 (1960).

SCHMIDT-KÖNIG, K.: Die Sonnenorientierung richtungsdressierter Tauben in ihrer physiologischen Nacht. Naturwissenschaften 48, 110 (1961).

SCHWASSMANN, H. O.: Environmental cues in the orientation rhythm of fish. Cold Spr. Harb. Symp. quant. Biol. 25, 443—449 (1960).

— 1962, in Vorbereitung.

— and W. BRAEMER: The effect of experimentally changed photoperiod on the sun-orientation rhythm of fish. Physiol. Zool. 34, 273—286 (1961).

The Concepts of Home Range and Homing in Stream Fishes

With a Discussion of Sensory Implications*

By GERALD E. GUNNING

Department of Zoology, Tulane University, New Orleans, Louisiana (U.S.A.)

With 4 Figures

Contents

I. Introduction

Home range may be defined as the area over which an animal normally travels. GERKING (1950, 1953) found that longear sunfish *(Lepomis megalotis)*[1], rock bass *(Ambloplites rupestris)*, and green sunfish *(Lepomis cyanellus)* in two Indiana streams moved about very little from one year to the next. These sunfishes were estimated to have remained within a home range of 100—200 linear feet of stream. The home ranges of smallmouth bass *(Micropterus dolomieui)* and spotted bass *(Microp-*

---

* These studies were aided by grants from the U.S. Public Health Service (National Institutes of Health, RG-7125), National Science Foundation (G-10697), and the Sport Fishing Institute, Washington, D. C.
[1] Scientific names of fishes are those recommended by the American Fisheries Society (1960).

terus punctulatus) were believed to range from 200—400 linear feet of stream. In a later paper, GERKING (1959) listed 33 species of fishes which exhibit restricted movement or occupy home ranges.

Homing as used here means the return of a fish to a home range following experimental or natural displacement. Homing behavior of longear sunfish was studied in Richland Creek, Indiana (GUNNING, 1959). Longear sunfish were able to return to a home range following experimental displacement to an unfamiliar area 200 to 350 feet away.

The objectives of this paper are: 1) To compare occupancy of home range in the longear sunfish, bluegill *(Lepomis macrochirus)*, American eel *(Anguilla rostrata)*, and blacktail redhorse *(Moxostoma poecilurum)*. One could surmise that sensory mechanisms are involved in the ability of these species to occupy a restricted area for a considerable period of time, but experimental proof is lacking on this point. 2) To discuss homing in one of these species, the longear sunfish. Experiments to determine the sensory basis for homing in this form are described.

II. Methods

A. Home range

Home range investigations in stream environments are generally conducted by arbitrarily dividing a continuous stream section into several segments (Fig. 1), marking the fish in each segment distinctively, and subsequently sampling the segments, as well as adjacent areas upstream and downstream, to determine how restricted fish movements are (GERKING, 1953; GUNNING, 1959; GUNNING and SHOOP, 1962; GUNNING and SHOOP, in Press). For all species studied by us, the streams were sampled with an electro-fishing device; the details of the sampling procedure are given elsewhere (GUNNING and SHOOP, in Press). Longear sunfish, bluegill, and blacktail redhorse were marked distinctively by removing one, or a combination of two, pectoral or pelvic fins. American eels were marked either by removing one of the pectoral fins or attaching a monel-metal strap tag (fingerling size). Eel tags were attached dorsally a short distance in front of the posterior end of the vertebral column, in such manner that the hold-fasts of the tags penetrated the musculature.

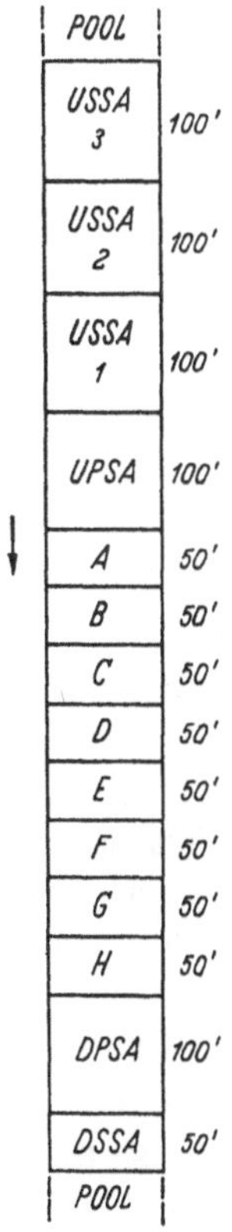

POOL	
USSA 3	100'
USSA 2	100'
USSA 1	100'
UPSA	100'
A	50'
B	50'
C	50'
D	50'
E	50'
F	50'
G	50'
H	50'
DPSA	100'
DSSA	50'
POOL	

Fig. 1. Schematic representation of a study area at Big Creek, Louisiana, near Alexandria. The stream segment is actually S-shaped. USSA, Upstream Secondary Stray Area; UPSA, Upstream Primary Stray Area; DPSA, Downstream Primary Stray Area; DSSA, Downstream Secondary Stray Area

The study areas utilized were as follows:

1) Six study areas were located on Richland Creek, a bedrock stream in Greene County, Indiana. This stream meanders through limestone country where it has cut through the topsoil to the underlying bedrock.

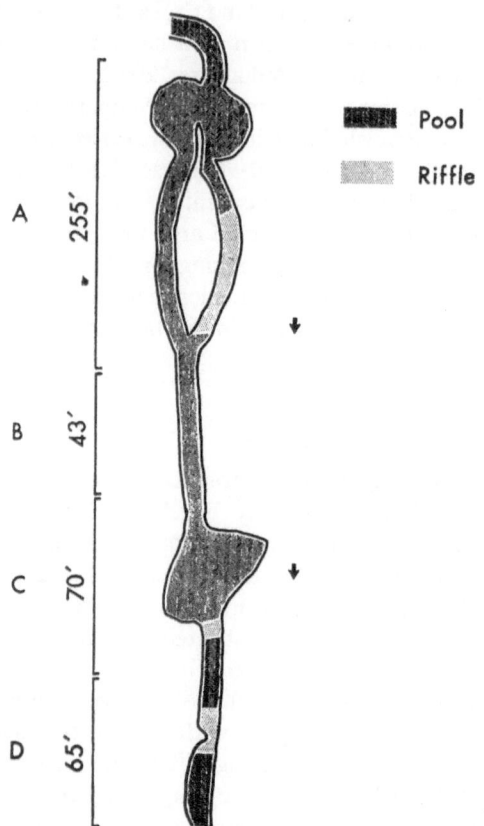

Fig. 2. Diagram of Study Area I, Talisheek Creek, Louisiana. Note that section A is not drawn to scale. Direction of stream flow is indicated by arrows

The stream is fed by springs, hence some flow of water is generally maintained. The width of the stream in the areas studied ranged from 10 to 35 feet; depth ranged from one to four feet. Diagrams and photographs of these study areas are to be found in GUNNING (1959).

2) Two study areas (Figs. 2 and 3) at Talisheek Creek, St. Tammany Parish, Louisiana, were employed (GUNNING and SHOOP, 1962). Talisheek Creek is 10—20 feet wide at the study areas used, with a few expansions of greater width. Water depth ranged from a few inches to five feet.

3) A single study area (Figs. 1, 4) at Big Creek, Grant Parish, Louisiana, was used. Big Creek is 25—35 feet wide at the point of our investigation. Water depth ranged from a few inches to five feet. Numerous logs and overhanging banks afford considerable cover for the fish population.

Fig. 3. Portions of sections C—D, Study Area I, Talisheek Creek, Louisiana. Note the clarity of the water and the abundance of cover for fishes

It should be emphasized that all study areas considered above are located in the headwaters of the streams where water depth is generally not prohibitive to successful collection of fishes.

B. Homing and sensory experiments

All work of this nature was done on the longear sunfish. The first objective was to determine whether or not longear sunfish would return to a home range after being displaced 200 to 350 feet away from it. Samples of untreated longear sunfish were removed from a segment of stream and transported either upstream or downstream from this area. The assumption was made that the fish were transported outside of their home areas. GERKING (1953) found that the size of the home range of the longear sunfish was no greater than 100 to 200 feet.

Should the longear sunfish home as expected, it was decided to concentrate first on the organs of smell and vision as possible sensory mechanisms enabling the fish to accomplish homing responses. Three

Fig. 4. Study area at Big Creek, Louisiana, facing upstream. The tree touching the water on the right side of the photograph marks the lowermost limit of the downstream secondary stray area (Fig. 1)

groups of fish were to be used in these experiments: 1) control or untreated individuals, 2) specimens with the olfactory mechanism impaired, and 3) fish with impaired vision. The relative abilities of these different groups to return to the home range following experimental displacement were studied. All fish were fin-clipped for future identification. Details of the sensory-impairment procedures are found in GUNNING (1959).

III. Results

A. Home range

1. Longear sunfish. A short-term experiment was conducted at Study Area II, Richland Creek, Indiana, to corroborate the home range concept for the longear sunfish as envisioned by GERKING (1953). Study Area II consists of sections A (105 ft), B (274 ft), C (125 ft), and D (84 ft), proceeding from upstream to downstream (GUNNING, 1959). On July 17, 1956, a sample of 34 longear sunfish was captured from section B of Study Area II, marked, and released in the middle of the same section. The great majority of recaptures taken on later dates were in section B (Table 1). Only two longear sunfish strayed; both were found in section A,

Table 1. *Movements of 34 control longear sunfish following capture on July 17, 1956, marking, and return to the stream area from which they were removed* (from GUNNING, 1959)

Date	No. fish recaptured	No. of home fish	No. of stray fish
7/23/56	16	14	2
7/30/56	15*	15	0

* Since all recaptures taken on July 23 were returned to the section from which they were captured, they were available for subsequent recapture on July 30, 1956. All recaptures are released for subsequent recapture at a later date unless otherwise stated in the experiments which follow.

upstream. The recapture rate was high in both samples, 44 and 47 per cent, and it is doubtful that many fish strayed beyond the boundaries of the study area. Adjacent areas were sampled downstream for a period of three summers, but no strays from this experiment were found (GUNNING, 1959).

A long-term experiment was conducted at Study Area III, Richland Creek, Indiana, to add further support for the home range concept (GUNNING, 1959). Study Area III consists of sections A (200 ft) and B (150 ft). Sixty-six fish were marked in Study Area III during 1956 and 1957, and sampled in 1958 in order to study the extent to which longear sunfish remain in the same pool from year to year. All individuals were removed from section A. The history of ten longear sunfish that remained in their home area from year to year is given in Table 2. Only two (not included in Table 2) strayed into section B; these were marked in August, 1956, and were recaptured during August, 1958. Adjacent areas of stream both upstream and downstream were sampled on numerous occasions, but no other strays were found.

Further studies on the various aspects of occupancy of home range by longear sunfish have been conducted at Talisheek Creek, Louisiana. Conclusions from this work included the following (GUNNING and SHOOP, in Press): 1) The size of the home range for longear sunfish in this population does not exceed 70 feet of stream for most individuals; 2) Intensive collecting between two study areas approximately 0.4 mile

Table 2. *Year-to-year occupancy of section A, Study Area III, by ten longear sunfish* (from GUNNING, 1959)

No. of fish	Originally marked in August 1956[1]	Subsequent Dates Collected in III-A		
		July 1957	July 1958	August 1958
2	Marked LP	—	—	Recaptured
2	—	Marked RV	Recaptured	—
2	—	Marked RV	—	Recaptured
1	Marked LP	Recaptured and marked RV	Recaptured	—
1	Marked LP	Recaptured and marked RV	—	Recaptured
1	—	Marked RV Recaptured and marked LV	Recaptured	—
1	—	Marked LV	—	Recaptured

[1] LP signifies removal of the left pectoral fin. RP designates right pectoral, RV right ventral, and LV left ventral.

apart demonstrated that straying outside the study areas does occur. However, in my opinion straying should be considered as predictable biological variation; such straying does not destroy the usefulness of the home range concept as long as it is limited in comparison to the total number of fishes involved; 3) The longer a given longear sunfish is at large subsequent to fin-clipping, the greater is the tendency for this fish to stray outside the limits of a restricted segment; and 4) Older longear sunfish appear to have larger home ranges than younger individuals.

2. Bluegill. An experiment was set up at Study Area I, Talisheek Creek, Louisiana (Figs. 2 and 3), in order to test the home range concept for the bluegill (GUNNING and SHOOP, in Press). A total of 84 bluegill was marked in Study Area I between February 20, 1960, and July 15,

Table 3. *Days at large between fin-clipping and subsequent recapture for bluegill of Study Area I, Talisheek Creek, Louisiana* (from GUNNING and SHOOP, in press)

Stream Section	No. of Fish and % of total	Days at Large	
		Mean	Range
A	6 (9%)	122	95-147
B	28 (40%)	73	17-235
C	30 (42%)	45	7-188
D	6 (9%)	40	7-95

1960. Collecting in Study Area I yielded 70 recaptures (Table 3). Thirty (42 per cent) remained within the marking section (C); the remaining 40 strayed into sections A, B, and D. Almost as many strayed into section B as remained in section C. A tendency to stray in the upstream direction is evident.

Study Area II, Talisheek Creek, consists of an upstream section (A; **135** ft), a middle section (B; **125** ft), and a downstream section (C; **150** ft). Sixty-eight bluegill were marked in section B, Study Area II, between July 21, 1960, and September 7, 1960. A total of 44 recaptures was taken from Study Area II; one was taken from section A, thirty-nine were taken from section B, and four were taken from section C (GUNNING and SHOOP, in Press). These fish were at large 7 to 134 days between marking and recapture.

On the basis of these experiments we estimate the home range of bluegill to be no more than 125 linear feet of stream, since over 80 per cent of the individuals of both study areas remained within this limit. Other details and conclusions may be found in the above-mentioned paper.

Table 4. *Movements of American eels in Talisheek Creek, Louisiana* (from GUNNING and SHOOP, 1962)

Eel Number	Date Marked[1]	Date Recaptured	Total Length in mm	Distance Travelled (In Feet)
1	9/ 7/60	—	390	—
		9/30/60	—	0—50
		10/28/60	—	0—50
		11/18/60	—	0—50
2	11/18/60	—	310	—
		12/ 9/ 60	310	0—50
3	9/ 7/60	—	255	—
		9/30/60	—	51—100
4	9/ 7/60	—	355	—
		9/30/60	—	51—100
5	10/28/60	—	280	—
		12/ 9/60	285	151—200
6	9/ 7/60	—	560	—
		10/28/60	—	51—100
7	9/ 7/60	—	305	—
		10/28/60	—	51—100
8	9/30/60	—	390	—
		10/28/60	—	101—150
9	7/21/60	—	305	—
		7/28/60	—	51—100
10	7/21/60	—	360	—
		7/28/60	—	51—100
11	8/ 9/60	—	—	—
		9/ 7/60	321	101—150
		11/ 4/60	—	101—150
12	10/ 7/60	—	567	—
		11/ 4/60	—	51—100
		12/ 2/60	—	101—150
13	8/10/60	—	360	—
		10/ 7/60	—	151—200
14	7/21/60	—	360	—
		7/26/61	498	101—150
15	7/21/60	—	360	—
		7/26/61	685	101—150

[1] All eels were identified by metal tags except numbers 9, 10, 14, and 15; these were fin-clipped.

3. American eel. The movements of 15 American eels in Talisheek Creek, Louisiana, are shown in Table 4. Numerical designations are given to facilitate discussion. Multiple recaptures (eels 1, 11, and 12) resulted in a total of 19 recorded movements for the 15 eels. The first stream segment in which a given eel was first captured is taken as an assumed home area. All subsequent statements concerning extent of movement are thus based on this assumed home area. Hence, an eel marked and subsequently recaptured in the expanded pool region at the upper end of section A, Study Area I (Fig. 2), would be recorded in our field notes as having moved 0 to 50 feet. Extent of movement of the 19 recaptures may be summarized as follows: 1) four eels moved 0—50 feet, 2) seven eels moved 51—100 feet, 3) six eels moved 101—150 feet, and 4) two eels moved 151—200 feet.

Four eels were marked in the study area at Big Creek (Fig. 1). No other unmarked eels were taken in the 950-foot segment of stream during the period it was studied; a single eel was taken in a basket trap in the

Table 5. *Recapture data for three American eels marked in Big Creek, Louisiana* (from GUNNING and SHOOP, 1962)

Eel Number	Date Marked[1]	Stream Segment Marked In	Date Recaptured	Stream Segment Recaptured In	Total Length in mm
1	7/ 7/60	Section A	—	—	—
			7/ 8/60	USSA No. 2	—
			10/21/60	USSA No. 3	660
			11/25/60	USSA No. 3	—
			3/25/61	UPSA	678
			4/22/61	USSA No. 1	685
			5/ 6/61	UPSA	—
2	4/14/60	Section D	—	—	760
			6/ 7/60	D	—
			6/ 8/60	A	—
			7/ 7/60	F	—
			8/ 2/60	D	—
			5/30/61	C	—
3	9/ 1/60	USSA No. 3	—	—	915
			11/25/60	USSA No. 3	—

[1] Eel number 3 was marked using a monel-metal strap tag; the other two were fin-clipped.

large pool below the study area, however (Figs. 1, 4). One of the four eels (610 mm total length) was tagged with a monel-metal strap tag on November 25, 1960; it was never taken again. A second eel (Table 5, eel 3) was tagged in the same manner on September 1, 1960. It was subsequently recaptured in the same stream segment on November 25, 1960, but was not taken again. The remaining two eels (Table 5, eels 1—2) were fin-clipped; these were taken repeatedly (GUNNING and SHOOP, 1962). Eel number 1 was recaptured six times between July 7, 1960, and

May 6, 1961. During this 10-month period, the eel probably remained within a home range consisting of 450 linear feet of stream (Table 5; Fig. 1). Eel number 2 was recaptured five times between April 14, 1960, and May 30, 1961. During this 13-month period, the eel presumably remained within a home range consisting of 300 linear feet of stream (Table 5; Fig. 1).

4. Blacktail redhorse. In order to test the home range concept for the blacktail redhorse, individuals were marked by fin-clipping at Big Creek, Louisiana (Figs. 1, 4), between June 7, 1960, and August 30, 1960 (seven collecting trips). Sixty-seven blacktail redhorse were marked. During the period June 8, 1960 to March 30, 1962, a total of 61 marked blacktail redhorse was taken (Table 6); multiple recaptures were possible, that is the same fish could be taken more than once. Of the 61 blacktail redhorse recaptures taken, 45 (74 per cent) remained within 100 linear feet of stream. A few individuals travelled greater distances up to 450 feet, however (Table 6).

Table 6. *Occupancy of home range by 61 blacktail redhorse recaptures*

Number of Fish	Distance Moved in Feet
23	0—50
22	51—100
6	101—150
1	151—200
1	201—250
2	251—300
1	301—350
4	351—400
1	401—450
Total 61	

Further light is shed upon occupancy of home range by blacktail redhorse when one considers another aspect of the problem. Table 7 shows the minimum number of days at large for 19 blacktail redhorse recaptures that were marked between June 7, 1960 and August 2, 1960. Some individuals thus remain in a limited area for a considerable period of time, or return to it should they leave periodically.

Occupancy of home range in the blacktail redhorse must be described as seasonal for large individuals. Blacktail redhorse spawn in Big Creek, Louisiana, during the latter part of April and the month of May. At this time, and through the month of November, movement of these fishes is restricted to a home range of approximately 100 feet for most individuals. Larger blacktail redhorse (one foot total length and over) become

Table 7. *Minimum number of days at large for 19 blacktail redhorse recaptures that were marked RV (removal of right ventral fin) between June 7, 1960, and August 2, 1960.[1] All fish were marked in Section A of Big Creek (Fig. 1)*

Number of Fish	Minimum Days at Large
4	25
4	28
2	30
5	79
1	113
1	354
1	374
1	606

[1] For purposes of calculation it is assumed that all fish were fin-clipped on August 2, 1960. Note that multiple recaptures are possible.

progressively harder to collect in the study area during the latter part of November. From December through early April one can collect only juvenile blacktail redhorse with total lengths of eight inches or less; large blacktail redhorse can not be found in the study area at this time although collecting has been intense during this period for the past two years. It is our belief, although not verified, that the blacktail redhorse drop downstream to deeper water during the winter. During the middle part of April, blacktail redhorse can be collected again in the study area as they arrive prior to the spawning period.

Consider how misleading the above results would have been had collecting not been possible on a year-around basis. STEFANICH (1952) studied the population and movement of fish in Prickley Pear Creek, Montana. He found the western white sucker *(Catostomus commersoni suckli)* to be most numerous during the spring; the longnose sucker *(Catostomus catostomus)* was abundant in the spring and early summer, but decreased very markedly in the fall. STEFANICH records sucker movements up to several miles in length, with rates approximating one mile per day. He found that the movement of suckers captured outside of the regular sections in which they had been tagged was generally downstream.

B. Homing

Displacement experiments were designed to determine whether or not longear sunfish would return to their home range following experimental transfer to an unfamiliar area. Two such experiments, taken from

Table 8. *Homing performance of control longear sunfish following experimental displacement upstream or downstream from their home ranges* (from GUNNING, 1959)

Experi-ment	Direction Displaced	Date	No. Fish Marked	No. Fish Recaptured	No. Fish Homing	No. Fish Not Homing	Distance Travelled by Fish Homing
A	Down-stream	7/ 9/57	36	—	—	—	
		7/10/57	—	19	19	0	200 ft
B	Up-stream	7/30/57	46	—	—	—	
		7/31/57	—	13	10	3	265 ft

GUNNING (1959) are given in Table 8. All of the longear sunfish recaptures displaced downstream returned to their home range within 24 hours. Of those displaced upstream, 77 per cent returned to their home range within 24 hours. Additional experiments are to be found in GUNNING (1959). The fact that longear sunfish will return to a home range following experimental displacement allowed us to set up experiments to shed light on the possible sensory mechanisms being used by the fish to enable such returns.

C. The sensory basis for homing in the longear sunfish

Displacement experiments wherein the homing ability of control groups of longear sunfish was compared with groups in which sensory structures were impaired allowed interpretation of a sensory basis for homing. For example, in order to determine the role of vision in homing, 114 longear sunfish with the lenses removed from the eyes were displaced from their home ranges; 41 of these were recaptured. Thirty-one of the 41 recaptures (70 per cent) returned to their home ranges (GUNNING, 1959). The control experiments involved a total of 109 fish; 35 of these were recaptured. Twenty-seven of the 35 recaptures (77 per cent) returned to their home ranges. The similarity of the return of control and treated groups is striking. A chi-square test of independence of the two groups shows that it is extremely unlikely that they represent separate populations (chi-square $= 0.047$ with 1 d.f., $p = 0.81$) (GUNNING, 1959). It was concluded on the basis of these and additional data from other types of experiments that visual recognition of the environment is not necessary for homing to occur in the longear sunfish population studied.

In order to determine the role of olfaction in the homing ability of the longear sunfish, the return of control fishes was compared with fishes which had the olfactory mechanism impaired by heat cautery. Thirty-one of the 136 fish in the olfactory experiments were recaptured. Seventeen (55 per cent) homed, while 14 (45 per cent) remained in the area to which they were displaced (GUNNING, 1959). Thirty-two recaptures were obtained from the 112 controls that were released. Twenty-nine (91 per cent) returned to their home ranges, and three (9 per cent) stayed in the area to which they were transported. A chi-square test of independence of these data indicates that the difference between the heat cautery group and control group is significant (chi-square $= 8.38$ with 1 d.f., $p = <0.005$). It was concluded that a random distribution of recaptures resulted from random movement following loss of the olfactory sense, and that the olfactory sense thus mediates the return of fish that home (GUNNING, 1959).

Data from experiments in which sensory structures are impaired present direct evidence that a sensory structure is or is not involved in homing. The olfactory mechanism was believed to be involved in this instance. Hence, fish that move upstream toward their home ranges should have an advantage over those moving downstream, since odoriferous substances would be carried downstream by the current. A study of the data from untreated fish should shed light on this point. Fifty of the 139 control longear sunfish moving upstream toward their home ranges were recaptured; 48 of these returned to their home ranges, while two failed to return (GUNNING, 1959). Fifty-eight of the 133 control longear sunfish moving downstream toward their home ranges were recaptured. Forty-three travelled back to their home ranges, but 15 remained in the area to which they were transferred. A chi-square test of independence indicates that there is a significant difference between the homing abilities of the two groups (chi-square = 8.47, with 1 d.f., $p = <0.005$). The above results are, therefore, in accord with what would be expected if the longear sunfish were relying on the sense of smell to return to the home range.

The odoriferous substance, or substances, to which the fish are believed to be orienting have not been identified. This would, of course, be an interesting line of investigation.

Discussion

Stream fishes live in a continuous system, except during periods of low water when pools may be isolated. A study of occupancy of home range in the longear sunfish and bluegill led us to the conclusion that these species do not travel long distances up and down the length of a stream in which they live. Each pool may be envisioned as having its own fish population with limited exchange between pools. Occupancy of home range in these two species seems to differ only in terms of size, the bluegill having a larger home range than the longear sunfish. In comparison, the blacktail redhorse occupies a restricted area, or home range, for part of a calendar year, but subsequently leaves to occupy an area elsewhere in the stream system. The American eel presumably travels from a spawning area in the Sargasso Sea to inland streams of North America. VLADYKOV (1957) recorded eel movements of 200 miles extent in the St. Lawrence River and other waters of Quebec. SMITH and SAUNDERS (1955) described runs of the American eel from various lakes in New Brunswick, Canada, which represented the beginning of their return trip to the ocean. The movements of eels, and other fishes as well, must be considered with respect to the specific environment they occupy and the phase of their life history being completed. Hence, our results indicate that once an eel has selected a headwater stream, this individual can be expected to occupy a restricted segment or home range. Following maturity in fresh water, the home range would ultimately be abandoned when the return trip to the Sargasso Sea is begun. Hence, the characteristics of occupancy of home range in the four species of fishes studied vary considerably.

In my opinion, a knowledge of the size of home range for a particular species of fish, as well as the length of time the fish can be expected to occupy such a home range, makes a consideration of homing more meaningful. It is important to the investigator to know whether a fish accomplishing a homing response is returning to a particular stream system in general, a certain tributary within a given stream system, or a restricted segment of several hundred feet in extent within a given tributary. This would be difficult to determine in some instances. However, in my opinion, studies involving homing, and more particularly the sensory bases for homing, should not be attempted until the required basic information concerning fish movements is available. The home range concept is useful in that the investigator knows what the fish in question is orienting to. Sensory experiments conducted with this information as a basis should be meaningful.

Acknowledgments. The author is indebted to Messrs. C. ROBERT SHOOP, JOSEPH E. HANEGAN, JOE BLACK, JOHN BRIDGMAN, JOE ZAMBERNARD, and JAMES LANGHAMMER for assistance in carrying out field experiments. Mr. LEON TRICE aided in preparation of the illustrations.

References

American Fisheries Society, Committee on Names of Fishes. A list of common and scientific names of fishes from the United States and Canada. Special Publication No. 2, 102 p. (1960).

GERKING, S. D.: Stability of a stream fish population. J. Wildl. Mgt. **14**, 193—202 (1950).

— Evidence for the concepts of home range and territory in stream fishes. Ecology **34**, 347—365 (1953).

— The restricted movement of fish populations. Biol. Rev. **34**, 221-242 (1959).

GUNNING, G. E.: The sensory basis for homing in the longear sunfish, *Lepomis megalotis megalotis* (Rafinesque). Invest. Indiana Lakes and Streams **5**, 103—130 (1959).

— and C. R. SHOOP: Restricted movements of the American eel, *Anguilla rostrata* (LE SUEUR), in freshwater streams, with comments on growth rate. Tulane Stud. Zool. **9** (5), 265—272 (1962).

— — Occupancy of home range by longear sunfish, *Lepomis megalotis megalotis* (Rafinesque), and bluegill, *Lepomis macrochirus macrochirus* Rafinesque. Anim. Behav. (in press).

SMITH, M. W., and J. W. SAUNDERS: The American eel in certain fresh waters of the Maritime Provinces of Canada. J. Fish. Res. Bd. Canada **12** (2), 238—269 (1955).

STEFANICH, F. A.: The population and movement of fish in Prickley Pear Creek, Montana. Trans. Amer. Fish. Soc. **81**, 260—274 (1952).

VLADYKOV, V. D.: Fish tags and tagging in Quebec waters. Trans. Amer. Fish. Soc. **86**, 345—349 (1957).

Die Sonnenkompaßorientierung der Eidechsen

Von Georg Birukow, Klaus Fischer und Horst Böttcher

I. Zoologisches Institut der Universität Göttingen

Mit 11 Abbildungen

Inhalt

I. Einleitung

Das Vermögen, sich kompaßgerecht nach dem Stand der Sonne zu orientieren und ihre tageszeitliche Wanderung einzurechnen, ist seit den grundlegenden Entdeckungen von K. v. Frisch (1950, 1951) an der Biene auch bei zahlreichen anderen Vertretern der Arthropoden und Wirbeltiere nachgewiesen worden. Soweit wir heute wissen, setzt das Richtungsfinden mittels eines Himmelsgestirnes mindestens zwei Grundleistungen des tierischen Organismus voraus: erstens die Fähigkeit, „Zeit zu messen" (physiologische oder innere Uhr; vgl. Bünning, 1958; Aschoff, 1960); zweitens den zeitgerechten Gebrauch eines Orientierungsmechanismus, der die tageszeitlichen Azimutänderungen des Gestirnes „rechnerisch" mit Hilfe der inneren Uhr kompensiert. Während das erste Grundvermögen wohl eine sehr allgemeine Eigenschaft von Lebewesen ist, wie es der tagesrhythmische, annähernd mit der Erdumdrehung synchronisierte Ablauf zahlreicher Lebensprozesse bei Tieren und Pflanzen zeigt, sichert das zweite Vermögen speziellere Lebensbedürfnisse des tierischen Organismus; die Fähigkeit, Zeit zu messen, wird dabei in den Dienst der Orientierungsfunktion gestellt.

Sowohl die Funktionsweise der inneren Uhr als auch des mit ihr gekoppelten Taxismechanismus sind heute Gegenstand eingehender experimenteller Arbeit und neuerdings auch regeltheoretischer Ansätze (WEVER, 1960; MITTELSTAEDT, 1960, 1961). Für beides ist zunächst eine möglichst breite vergleichende Basis zu fordern, die jedoch zur Zeit besonders bei der Sonnenkompaßorientierung von Wirbeltieren noch nicht gegeben ist. Vielleicht könnte hier das Auffinden eines besonders günstigen Versuchsobjektes die Weiterarbeit entscheidend fördern.

In den letzten Jahren konnten wir erstaunlich gute Orientierungs-leistungen bei der Sonnenkompaßorientierung von Eidechsen feststellen (FISCHER und BIRUKOW, 1960; FISCHER, 1961), die manche Einblicke in die oben angedeuteten Zusammenhänge zu verheißen scheinen. Dieser Beitrag soll über den derzeitigen Stand der Untersuchungen informieren und ihren weiterhin geplanten Verlauf andeuten.

II. Die bisherigen Befunde

1. Die Richtungsdressur in der Versuchsarena

Smaragdeidechsen (*Lacerta viridis* LAUR.), Ruinen- und Mauer-eidechsen (*L. sicula* und *L. muralis* LAUR.) lassen sich leicht auf Himmels-richtungen dressieren, wenn man dabei ihr natürliches Wärmebedürfnis ausnützt. Ihre Vorliebe, warme, von der Sonne beschienene Plätze auf-zusuchen, läßt sie auch im Laboratorium unter einer ,,künstlichen Sonne'', die ein Scheinwerfer von 6 V/25 W nachahmt, die Himmels-richtung eines warmen Platzes inmitten einer kühlen Umgebung erlernen. Als Versuchsapparatur dient eine Rundarena (Abb. 1a), deren Boden aus drei konzentrisch gegeneinander beweglichen Teilen besteht: einer Mittelplatte von 50 cm Durchmesser, einem 10 cm breiten Zwischenring und einem 5 cm breiten Außenring, der außen einen Abschirmrand trägt. Im Zwischenring ist, für das Versuchstier unsichtbar, eine elektrische Heizplatte von 10×10 cm eingelassen; der Boden der Mittelplatte und des übrigen Teiles des Zwischenringes ist hohl und durch fließendes Wasser kühlbar. Alle drei Ringe sind gegeneinander verschiebbar, außerdem läßt sich auch die ganze Apparatur (Abb. 1b) samt der daran befestigten oder daneben stehenden künstlichen Sonne auf Rollen und Schienen beliebig im Raum orientieren. Höhen- und Seitenwinkel des Scheinwerfers sind beliebig einstellbar, bei den meisten Versuchen betrug die Entfernung Schweinwerfer—Arenamittelpunkt 2,20 m.

Setzte man die Tiere im kalten Teil der Arena aus, so fanden sie nach einer Eingewöhnungszeit von 2—3 Tagen sehr rasch die warme Stelle und merkten sich ihre Lage relativ zur ,,künstlichen Sonne'', deren Wärme-strahlung bei diesen Versuchen zu vernachlässigen war. Dabei erlernten die Eidechsen zuerst den *absoluten* Winkel zur Lichtquelle zu *einer*

Dressurzeit und behielten ihn auch zu beliebigen anderen Uhrzeiten während des Tages bei. Anders als die Vögel (vgl. KRAMER, 1947; v. SAINT PAUL, 1958; HOFFMANN, 1960), die meisten Fische (BRAEMER, 1959) und erst recht als die Bienen (v. FRISCH, 1951; LINDAUER, 1957 u. a.) erlernten die Eidechsen also stets zuerst die *winkeltreue* Orientierung, die natürlich noch kein Richtungsfinden unter Einschätzung der

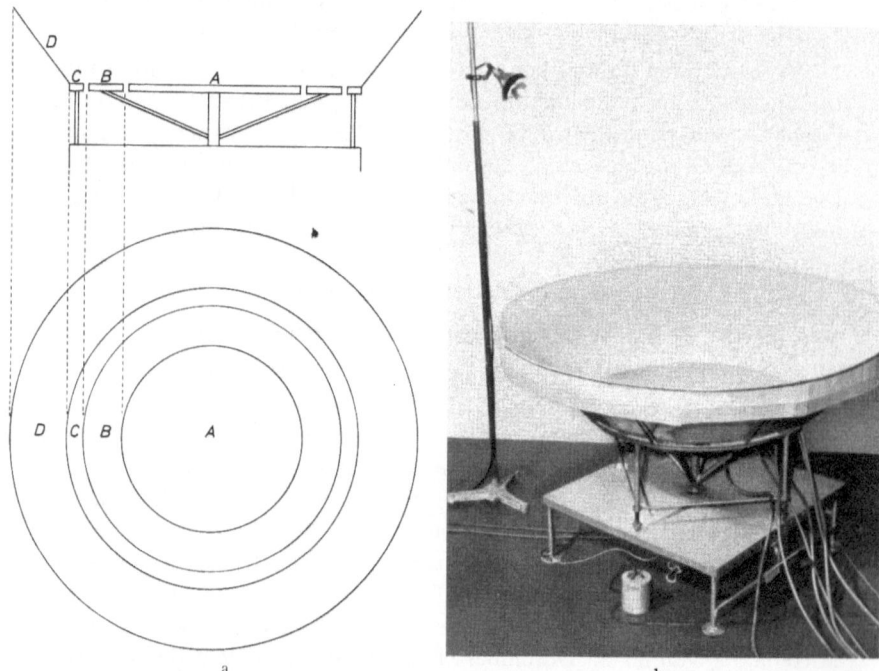

a   b

Abb. 1a u. b. a) Schema der Dressurarena in Querschnitt und Aufsicht: *A* Zentralplatte, *B* Zwischenring mit unsichtbar eingebauter Heizplatte, *C* Außenring mit Abschirmrand *D*.
b) Gesamtansicht der Versuchsanordnung

tageszeitlichen Sonnenwanderung (kompaßtreue Orientierung) erlaubt. Diese lernten sie erst gleichsam mit Nachhilfe, wenn man ihnen auch zu anderen Uhrzeiten als der Dressurzeit weitere tageszeitlich bestimmte Winkel bot, die für die betreffende Himmelsrichtung galten. Die Frage, auf welche Art und Weise die Eidechsen dabei in einer räumlich begrenzten Arena den Richtungswinkel relativ zu einer sehr nahen nicht parallelstrahligen Lichtquelle bestimmten, und wie viele tageszeitlich gegebene Winkel sie kennenlernen mußten, um sich während des *ganzen* Tages kompaßtreu zu orientieren, soll vorerst offen bleiben. Nach dieser „Nachdressur" mußten die Eidechsen in unbelohnten kritischen Wahlen zeigen, was sie gelernt hatten; dabei blieb die künstliche Sonne am Ort; die Heizplatte war entweder ausgeschaltet oder stand in einer anderen als der Dressurrichtung. Als Richtungswahl galt bei den Testversuchen das „Suchen am Ort", wo die Eidechsen die warme Stelle erwarteten; hier prüften sie lebhaft züngelnd den Boden, blickten wiederholt nach der Lichtquelle oder ließen sich in typischer Sonnstellung mit gespreizten Rippen auf der völlig kalten Stelle nieder. Als Richtpunkt für die Bestimmung des gewählten Winkels galt stets der Kopf des Versuchstieres; den Richtungswinkel konnte man so auf ±10° genau bestimmen. Wie Abb. 2 als Beispiel zeigt, wählten in den unbelohnten Tests die nach Norden dressierten Tiere Winkel, die recht genau zur Göttinger Sonnenazimutkurve für diese Himmelsrichtung paßten; die nach Osten und nach Westen dressierten Tiere verhielten sich entsprechend.

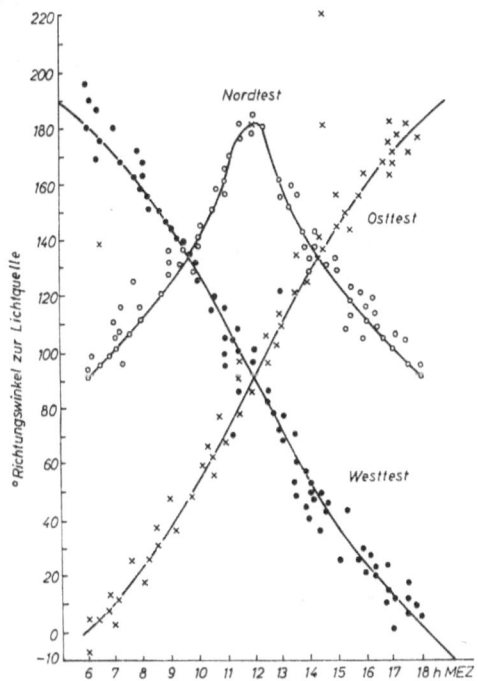

Abb. 2. Unbelohnte Testwahlen von drei Versuchstieren nach Dressur auf Nord (O‑O), Ost (×‑×) und West (●‑●). Ordinate: Richtungswinkel zur künstlichen Sonne. Abszisse: Uhrzeiten. Die Sonnenazimutkurven sind jeweils auf die Dressurrichtung bezogen

Wie sich schon in den ersten Versuchen zeigte (FISCHER, 1961), rechneten die Eidechsen zur Nachtzeit eindeutig die Weiterwanderung der stillstehenden künstlichen Sonne im Uhrzeigersinne ein. Sie verhielten sich also ähnlich wie die unter der Mitternachtssonne richtungsdressierten Stare (HOFFMANN, 1958), obwohl sie in unseren Breiten die Mitternachtssonne nie in ihrem Leben gesehen haben dürften. Der Nachtorientierungsmechanismus der Eidechsen ist auch dem der Bienen (LINDAUER, 1957) und mancher anderen Arthropoden vergleichbar, während einige Insekten, aber auch die Tauben (SCHMIDT-KÖNIG, 1961) zur Nachtzeit sich so orientieren, als liefe die Sonne in der umgekehrten Richtung, d. h. von Westen über Süden nach Osten zurück. Die Nachtazimutkurve der Sonne brauchten jedenfalls unsere Tiere nicht mehr zu erlernen, wenn sie die Tagesazimutkurve bereits kannten. Es schien uns aber, als ob sie auch diese keineswegs als Ganzes erlernen mußten, denn der Übergang von der winkeltreuen zur kompaßtreuen Orientierung gelang ihnen verhältnismäßig leicht. Wir vermuteten daher, daß sie die ganze Sonnenbahn für die betreffende Himmelsrichtung auch aus ihren

Teilstücken gleichsam rekonstruieren konnten, und prüften daher als nächstes, wie viele Punkte der Sonnenazimutkurve dazu notwendig waren. Eindeutige Ergebnisse waren jedoch nur von solchen Tieren zu erwarten, die in ihrem Leben noch keinerlei Erfahrungen bei der Sonnen-kompaßorientierung gemacht haben konnten. Das war bei den im Frei-land gefangenen Versuchstieren keineswegs sicher; die biologische Be-deutung der Sonnenkompaßorientierung von Eidechsen ist noch nicht geklärt.

2. Die Orientierung sonnenlos aufgezogener Eidechsen

Die Prüfung sonnenlos, jedoch im normalen Tag-Nacht-Wechsel auf-gezogener Versuchstiere erlaubt erste Aussagen darüber, welche Anteile der Sonnenkompaßorientierung bei ihnen angeboren und welche er-worben sein könnten. So ist z. B. nach LINDAUERS (1960) Untersuchungen

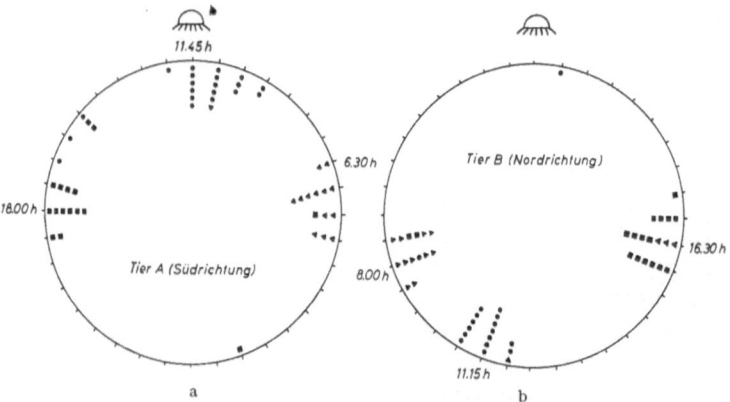

Abb. 3 a u. b. Unbelohnte Testwahlen von zwei auf Süd bzw. Nord dressierten, sonnenlos aufgezogenen Smaragdeidechsen zu den drei am Rande bezeichneten Dressurzeiten

den sonnenlos aufgezogenen Bienen das Grundvermögen der Sonnen-kompaßorientierung *angeboren*; die Richtung der Sonnenwanderung müssen sie aber auf der nördlichen und auf der südlichen Halbkugel der Erde *erlernen*. Bei Fischen aber (BRAEMER, 1959; HASLER et al., 1958; SCHWASSMANN, 1960) scheinen starke angeborene Komponenten auch bei der Verrechnung der Wanderungsrichtung mitzuspielen.

Es war uns gelungen, eine ausreichende Anzahl von Smaragd-eidechsen aus in Gefangenschaft abgelegten Eiern im Brutschrank bei 30° C zu erbrüten und die Jungtiere in der Thermokammer aufzuziehen. Sie lebten ab ovo in einem Langtag von 16 Std hell und 8 Std dunkel, ohne bis zum Eintritt der Geschlechtsreife jemals die Sonne zu sehen; zwei der erfahrungslosen Weibchen legten Eier ab, so daß die Aufzucht auch in der zweiten Generation möglich gewesen wäre — leider mußten wir die Zuchten aus äußeren Gründen vorerst abbrechen. Zwei der erwachsenen Eidechsen wurden in einem 12/12stündigen Tag auf Kompaßrichtungen dressiert, diesmal jedoch so, daß die Tiere den Lichtkompaß nach der feststehenden künstlichen Sonne nur zu drei feststehenden Uhrzeiten kennenlernten; die Heizplatte stand also dreimal während eines Dressurtages unter drei für die betreffende Himmelsrichtung gültigen Azimuten zur künstlichen Sonne. Jede Dressur währte etwa 15 min. Das eine Tier (A) lernte um 6^{30}, 11^{45} und 18 Uhr die Südrichtung; das andere um 8^{00}, 11^{15} und 16^{30} die Nordrichtung. Nach 6 Tagen folgten unbelohnte Testwahlen, die wir zunächst zu den drei Dressurzeiten vornahmen, um die winkeltreue Orientierung zu prüfen. Wie Abb. 3 zeigt, beherrschten die beiden Tiere in den nächsten zwei Wochen die Dressurwinkel zu den betreffenden Uhrzeiten recht gut: das „Südtier" (a) wählte morgens, mittags und abends die zu diesen Tageszeiten andressierten Winkel ebenso sicher wie das „Nordtier" (b). Darauf prüften wir die beiden Tiere im kritischen Versuch zu anderen als den Dressurzeiten: Wie Abb. 4 belegt, verfielen zwar beide Tiere gelegentlich wieder in die winkeltreue Orientierung (was auch erfahrene Tiere manchmal tun); die drei erlernten Richtungswinkel wurden auch zu anderen als den Dressurzeiten etwas bevorzugt. Im übrigen

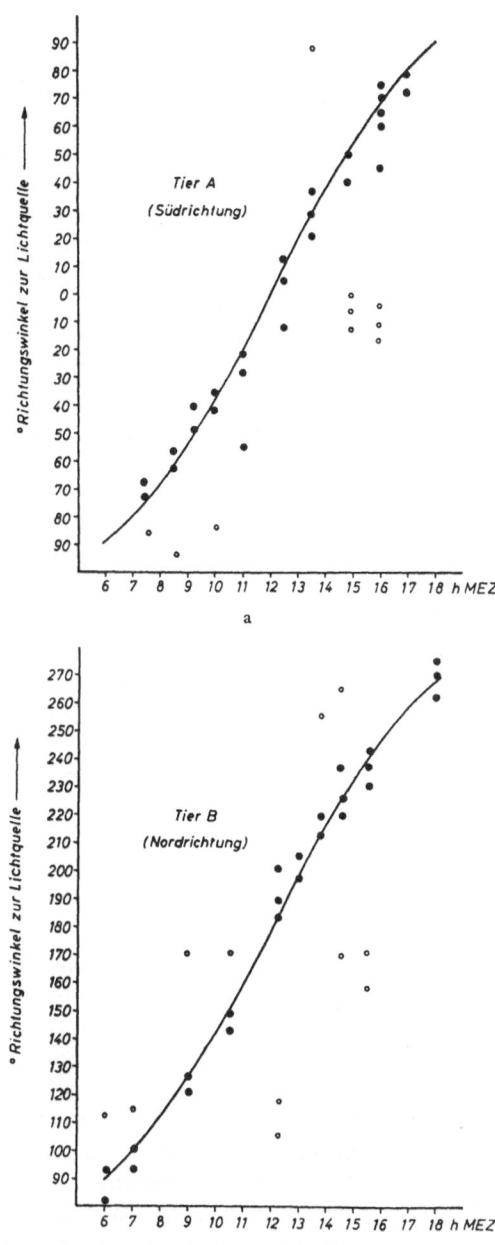

Abb. 4a u. b. Testwahlen während 12 Std von zwei auf Süden bzw. Norden dressierten, erfahrungslosen Smaragdeidechsen, die zuvor auf drei Punkte der Sonnenazimutkurve (vgl. Abb. 3) dressiert worden waren. Leere Kreise: winkeltreue, volle Kreise: kompaßtreue Wahlen

verhielten sich aber beide Tiere durchaus kompaßtreu, wie der Vergleich ihrer Richtungswahlen mit den entsprechenden Sonnenazimutkurven lehrt. Ebenso sicher rechneten sie auch die Bewegung der künstlichen Sonne zur Nachtzeit ein, genauso wie die erfahrenen, erwachsenen Tiere.

Nach diesen Befunden können also die Eidechsen aus drei diskreten Punkten der Sonnenbahn die vollständige Sonnenazimutkurve für die betreffende Himmelsrichtung bestimmen. Soweit wir bisher wissen, genügen nur zwei Punkte hierfür noch nicht; doch wäre in weiteren Versuchen noch zu klären, wieweit z. B. auch der Abstand der Punkte voneinander und die Dressurdauer von Einfluß sind. Die sonnenlos aufgezogenen Eidechsen zeigen uns jedenfalls, daß sie zum Einrechnen der Sonnenbahn keine Erfahrung über die tageszeitliche *Bewegung* der Sonne benötigen; das Vermögen dürfte demnach angeboren sein. Man darf aber daraus noch nicht folgern, daß auch die *Richtung* der tageszeitlichen Sonnenbewegung ebenfalls angeborenermaßen richtig eingeschätzt wird. Den Drehsinn der Azimutänderungen rechneten zwar auch die erfahrungslosen Tiere auf Anhieb richtig ein, d. h. so wie es die Umstände auf der nördlichen Halbkugel erfordern — auf der südlichen hätten sie mit umgekehrtem Drehsinn rechnen müssen. Doch könnten die Eidechsen ja schon aus den drei tageszeitlich bestimmten Richtungswinkeln eine Information über den erforderlichen ,,Verrechnungsmodus'' gewonnen haben, da die drei Dressurwinkel auf der ,,richtigen'' Sonnenbahn lagen. Daher prüften wir in weiteren Versuchen, ob die erfahrungslosen, sonnenlos aufgezogenen Tiere sich ebenso gut auch auf eine ,,inverse'', d. h. gegen den Uhrzeiger, so wie auf der Südhalbkugel, laufende Sonnenbahn dressieren lassen würden.

3. Die Orientierung der Smaragdeidechsen nach Dressur auf eine ,,inverse'' Sonnenbahn

Falls *nur* die Erfahrung über den erforderlichen Verrechnungsmodus entscheidet und keinerlei angeborene Bevorzugungen des einen oder des anderen Drehsinnes mitspielen, müßte die Dressur auf eine verkehrt umlaufende Sonne den Eidechsen keine Schwierigkeiten bereiten. Um sie den Tieren noch mehr zu erleichtern, dressierten wir diesmal wieder auf beliebige Punkte einer inversen Sonnenbahn; normalerweise lernen die Eidechsen nach dieser Methode schneller und leichter (vgl. Fischer, 1961). Zwei sonnenlos aufgezogene Tiere lernten jetzt die Nordrichtung relativ zu einer künstlichen Sonne, die für sie von Westen über Süden nach Osten zu wandern schien. Das Ergebnis zeigt Abb. 5: Obwohl die Orientierung jetzt ungenauer ist, ordnen sich die unbelohnten Testwahlen um zwei sich kreuzende Kurven an, deren eine (von 270° über 180° nach 90°) für die Azimute der verkehrten Sonnenbahn gilt, während die andere, in umgekehrter Richtung, die richtigen Azimutänderungen für die Nordrichtung wiedergibt. Einige Wahlen streuen allerdings im Bereich zwischen den beiden Kurven. Beide Tiere entschieden sich demnach etwa gleichhäufig *entweder* für die eine *oder* für die andere Verrechnungsweise.

Dieses Ergebnis läßt sich wie folgt deuten: Nach der Dressur auf eine inverse Sonnenbahn verhalten sich die Tiere im Testversuch offenbar ambivalent: sie können *beide* Richtungen der Sonnenwanderung gleich gut einschätzen, und zwar nicht nur die „falsche", die wir ihnen beizubringen versuchten, sondern ebenso gut auch die „richtige", die sie gar nicht lernen sollten und auch nie zuvor in ihrem Leben erfahren haben konnten. Man kann das Verhalten der Eidechsen mit dem der Bienen auf der südlichen und auf der nördlichen Halbkugel (LINDAUER, 1960) vergleichen; in beiden Fällen ist der Verrechnungsmodus offenbar weitgehend erlernbar. Die Eidechsen scheinen indessen die in unseren Breiten richtige Verrechnungsweise leichter als die falsche zu erlernen, denn bei der Dressur auf die richtige Sonnenbahn irrten sie sich auch unter erschwerten Dressurbedingungen (Darbietung von nur drei Punkten der Sonnenbahn) niemals zugunsten der falschen. Nach

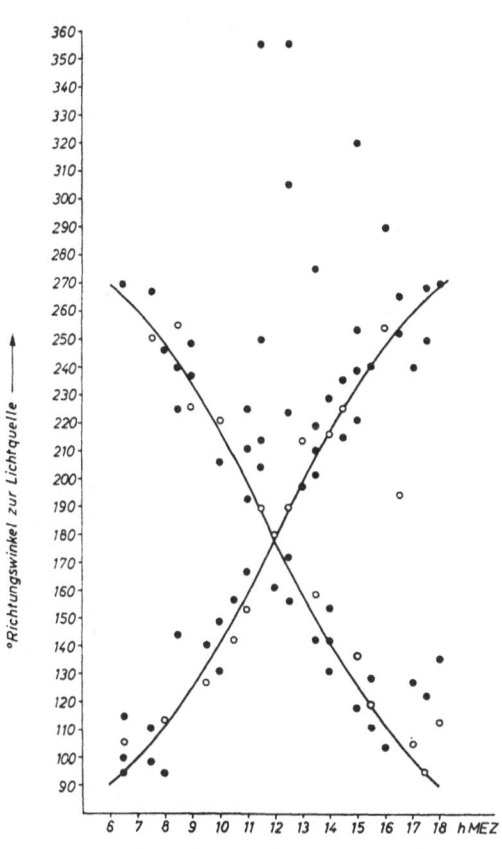

Abb. 5. Unbelohnte Testwahlen von zwei Smaragdeidechsen (○ und ●) nach Dressur auf eine inverse Sonnenbahn. Weitere Erklärungen im Text

der Dressur auf die falsche Sonnenbahn dagegen bevorzugten sie gleichoft auch die richtige. Bisher fanden sich keine Anzeichen dafür, daß eine längere und intensivere Dressur auf die falsche Sonnenbahn das Ergebnis wesentlich verbessert hätte.

4. Der Einfluß der Höhe der richtenden Lichtquelle

Den Einfluß der Sonnenhöhe auf das Richtungsfinden der Smaragdeidechsen hatte schon FISCHER (1961) mit negativem Ergebnis geprüft.

Stimmt man die Versuchstiere auf einen phasenverschobenen Tag um,
dann kann man ihnen auch unter einer natürlichen Sonne zu einer bestimmten Tageszeit Sonnenhöhen bieten, die sie nach ihrer „subjektiven"
Zeit nicht erwarten. Tiere, die auf einen um 3 Std verspäteten oder ver-
frühten Tag umgestimmt worden waren, hielten jedoch zu Zeiten der
größten Differenz zwischen der erwarteten und wirklichen Sonnenhöhe
stets die dressurgemäßen Richtungswinkel ein. Auch in Laboratoriums-
versuchen traten keine Abweichungen von der Dressurrichtung ein,
wenn man ihnen die Lichtquelle morgens möglichst hoch und mittags
möglichst tief bot. Demnach richten sich die Eidechsen nur nach dem
Sonnenazimut; nichts spricht dafür, daß auch die tageszeitlich verschiedene Sonnenhöhe eine vielleicht nur zusätzliche Information über
die einzuschlagende Richtung liefert. Das bestätigte sich auch in neueren
Versuchen an Ruinen- und Mauereidechsen, die wir in der Arena stets
bei einer konstanten Höhe der Lichtquelle von 45° (gemessen vom
Zentrum der Arena aus) dressierten, denen wir aber im Testversuch eine
andere Höhe boten. In Vorversuchen hatte sich gezeigt, daß die Eidechsen unter einer „künstlichen Äquatorsonne" (vgl. LINDAUER, 1957),
d. h. wenn die Lampe senkrecht über dem Zentrum der Arena leuchtete,
völlig desorientiert waren. Nach einer Richtungsdressur, bei der die
richtende Lichtquelle stets 45° hoch stand, erhöhten wir den Scheinwerfer in den Tests soweit, bis die Tiere zu versagen begannen. Eine
Ruinen- und eine Mauereidechse vertrugen eine Höhe von 79° gerade
noch, versagten aber bei 85°; zwei Smaragdeidechsen aber orientierten
sich bei einem Höhenwinkel von 85° noch einigermaßen gut. Ein Winkelabstand der künstlichen Sonne von rund 11 bzw. 5° vom Zenit reicht
demnach für die Azimutbestimmung gerade noch aus; dem Richtungsfinden nach der extrem hochstehenden Lichtquelle gehen jedoch fast
stets besonders heftige „Peilbewegungen" (vgl. oben S. 219) voraus.
Diese Leistung würde in unseren Breiten den Eidechsen auch im Hoch-
sommer zur Zeit des höchsten Sonnenstandes eine sichere Azimutbestimmung erlauben; in Göttingen erreicht die Sonne ihren höchsten
Stand am 21. 6. bei 72° ($\varphi = 51{,}32°$). Ob die Unterschiede zwischen den
Ruinen- und Mauereidechsen einerseits und den Smaragdeidechsen
andererseits artspezifisch sind und mit der geographischen Verbreitung
der Arten zusammenhängen, läßt sich z. Z. noch nicht mit Sicherheit
sagen.

Die Höhe der richtenden Lichtquelle spielt demnach bei der Sonnenkompaßorientierung von Eidechsen nur soweit mit, als sie die Azimutbestimmung auf Höhenwinkel begrenzt, die unterhalb eines zenitnahen,
kritischen Punktes liegen. Die Fische (vgl. Beitrag BRAEMER) scheinen
sich anders zu verhalten.

5. Die Orientierung nach zwei gleichzeitig gebotenen Lichtquellen

Die sehr zuverlässige Sonnenkompaßorientierung der Eidechsen erlaubte uns zu prüfen, wie ein Tier sich in einer sehr ungewöhnlichen Situation verhalten würde, nämlich wenn nach der Dressur auf *eine* Lichtquelle im Testversuch gleichzeitig deren *zwei* aufleuchteten. Bei der tageszeitlichen Sonnenkompaßorientierung ist ein solcher „Zweilichter-Versuch" bisher unseres Wissens nur beim Wasserläufer *Velia* unternommen worden (Birukow, 1957); in dieser Konfliktsituation gaben die Wasserläufer ihre Lichtkompaßorientierung auf und stellten sich tropotaktisch in die Resultierende zwischen den beiden Lichtquellen ein; ihre Orientierung war also vom tropotaktischen Erregungsgleichgewicht auf beiden Receptoren bestimmt. Die Eidechsen (*Lacerta sicula* und *L. viridis*) verhielten sich jedoch anders: Abb. 6 veranschaulicht das Ergebnis der Versuche an drei Smaragd- und einer Ruineneidechse: Bei der Dressur stand die richtende Lichtquelle 45° hoch; im Test mußten die Eidechsen ihren Kurs nach zwei gleichhellen Lampen bestimmen, die einander gegenüberstanden — die Dressurrichtung war Nord. Die Richtungswahlen (Abb. 6) zeigen, daß die Tiere zwar oft desorientiert waren, jedoch bei weitem nicht so stark, wie wir

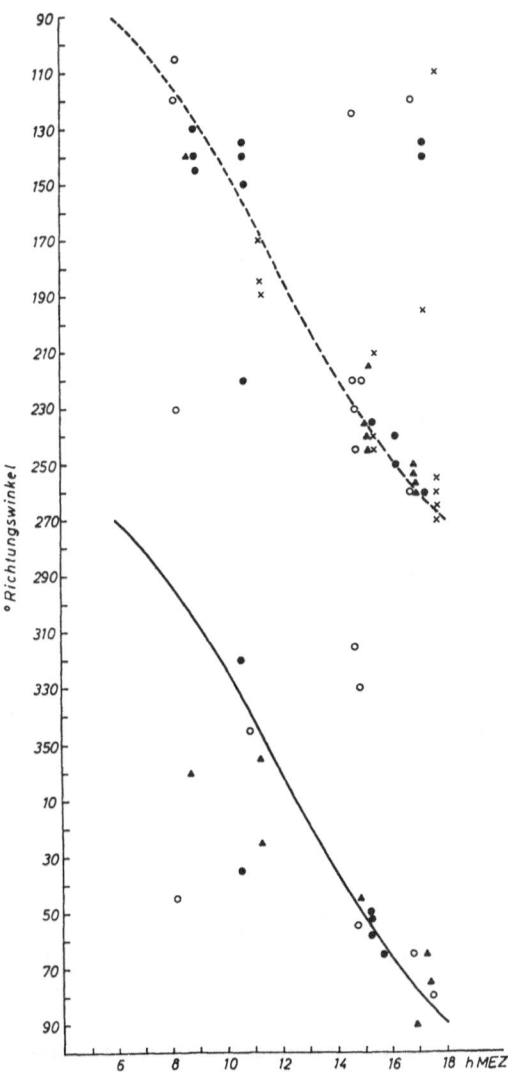

Abb. 6. Testwahlen einer Ruineneidechse (○) und dreier Smaragdeidechsen (▲, ●, ×) im „Zweilampenversuch". Winkelabstand der beiden Lichtquellen 180°; Höhe 45°

es erwartet hatten. Die Testpunkte liegen nämlich auf den Azimutkurven *beider* Lampen; ein und dasselbe Tier konnte also zu einer bestimmten

Zeit den Kurs entweder nach der einen Lampe bestimmen — dann ließ es die andere unbeachtet — oder umgekehrt. In einem anderen Versuch boten wir einer Smaragd- und einer Ruineneidechse die beiden Lampen unter 90° Winkelabstand: auch unter diesen erschwerten Bedingungen versagten sie keinesfalls (Abb. 7), doch streuten jetzt die Werte natürlich stärker.

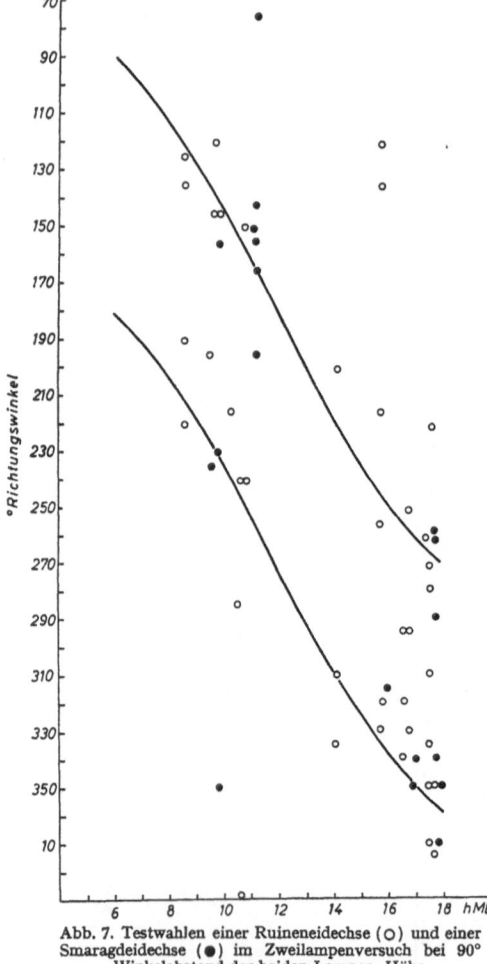

Abb. 7. Testwahlen einer Ruineneidechse (○) und einer Smaragdeidechse (●) im Zweilampenversuch bei 90° Winkelabstand der beiden Lampen. Höhe beider Lichtquellen 45°

In der Konfliktsituation des Zweilichtversuches treffen also die Eidechsen recht eindeutige *Alternativentscheidungen*, die an ähnliche Verhaltensweisen bei der Menotaxis von Käfern (BIRUKOW, 1954) erinnern. Doch kann man die Entscheidung in Grenzen beeinflussen, indem man im kritischen Test die Höhe der *einen* Lampe verändert (Abb. 8): Stand die eine der beiden „Sonnen", so wie bei der Dressur, unter 45° Höhe, die andere aber unter 74°, d. h. in der Nähe des kritischen Punktes, so wählten die Tiere den Kurs fast ausschließlich nach der niedrigeren Lampe und vernachlässigten die höhere. Daraus kann man folgern, daß die Eidechsen die Höhe der richtenden Lichtquelle durchaus wahrnehmen und beurteilen können, obwohl der Höhenwinkel nach allen bisherigen Befunden keine für die Azimutbestimmung notwendige Information enthält.

6. Das Wegfinden und die Azimutbestimmung der Eidechsen in der Arena

Die Leistungen unserer Eidechsen sind um so erstaunlicher, als sie den Lichtkompaß nach einer endlichen, radiärstrahligen Lichtquelle innerhalb einer eng umgrenzten Dressurfläche bestimmen mußten. Wie alle bisherigen Versuche und Befunde zeigen, richten sie sich jedoch stets *nur* nach der Lichtquelle; irgendwelche falschen Orientierungsschlüssel oder Orientierungshilfen sind mit Sicherheit auszuschließen, da in jedem Versuch sowohl die Lage der drei Ringe zueinander (vgl. Abb. 1) als auch die Stellung der Versuchsapparatur im Raume wechselte.

Die „künstliche Sonne" war der einzige Bezugspunkt, nach dem die Tiere den zur gegebenen Zeit gültigen Richtungswinkel einstellen konnten. Wie alle Orientierungskurven zeigen, ließ trotzdem die Genauigkeit der Orientierung, auch die der erfahrungslosen Tiere, normalerweise nichts zu wünschen übrig.

Besonders erstaunlich schien uns, daß die Versuchstiere in unbelohnten Testversuchen den richtigen Ort auf dem Zwischenring auf sehr verschiedenen Wegen erreichen konnten: Manchmal liefen sie vom Zentrum der Arena geradlinig dorthin, wo sie das warme Plätzchen vermuteten; meist erreichten sie aber den Dressurort auf einem mehr oder weniger komplizierten Umweg, indem sie z. B. auf dem Außenring entlangliefen, dabei gelegentlich auch umkehrten, einen anderen Weg einschlugen, dann auf dem Zwischenring weiterliefen usw. Es sah so aus, als ob sie während der Dressur eine sehr genaue Kenntnis aller räumlichen Beziehungen in der Arena erworben hätten und dabei auch die unvermeidlichen parallaktischen Verschiebungen mitberücksichtigten.

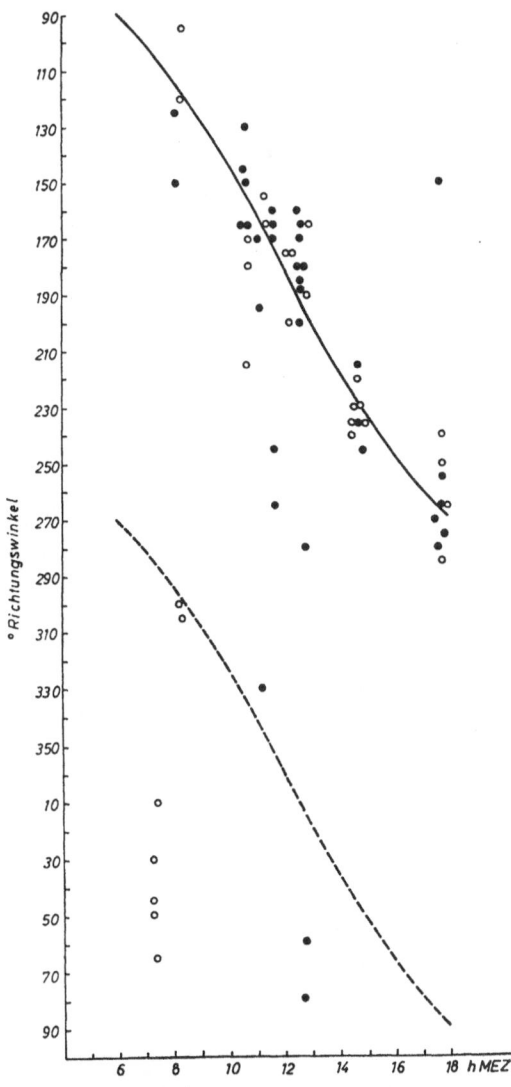

Abb. 8. Testwahlen einer Ruineneidechse (○) und einer Smaragdeidechse (●) im Zweilampenversuch, wenn die eine Lichtquelle (——) 45°, die andere (------) 74° hoch stand

Um zu prüfen, welche Orientierungshilfen die Eidechsen dabei benutzten, ließen wir sie zunächst im Testversuch einen anderen Weg als bei der

Dressur laufen. Während in den bisherigen Versuchen stets vom Zentrum der Arena aus dressiert und ebenso getestet wurde, dressierten wir jetzt zwar ebenfalls vom Zentrum aus, ließen aber die Tiere im Testversuch

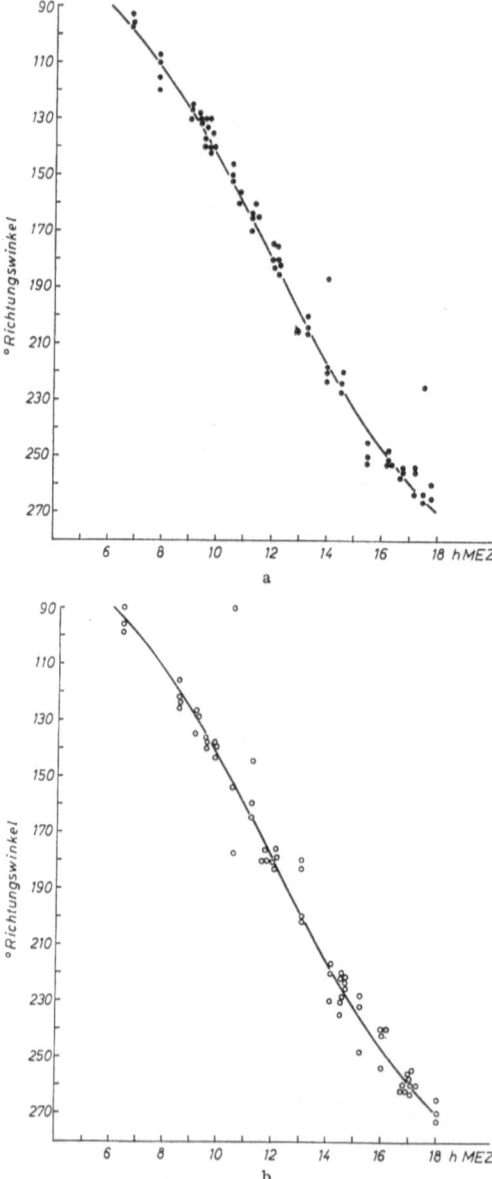

von einem beliebigen Punkt der Arena starten. Wie Abb. 9b zeigt, orientierten sich die Eidechsen keinesfalls schlechter, als wenn die Startorte bei der Dressur und beim Test übereinstimmten (Abb. 9a). Demnach können die richtungsdressierten Eidechsen den Dressurort von jedem beliebigen Punkt der Arena finden; eine Kenntnis des Weges ist jedenfalls nicht erforderlich.

Danach prüften wir, ob der künstliche Horizont, den die äußere Umrandung der Arena bildete, irgendwie mitspielte; es war zumindest denkbar, daß die Tiere die richtende Lichtquelle von verschiedenen Stellen der Arena aus in ungleichen Höhen über dem Horizont sahen und vielleicht so den Dressurpunkt orteten. Erhöhten wir jedoch den Horizont im Testversuch auf das Doppelte der Dressurhöhe, so störte das die Orientierung der Eidechsen keineswegs. Schließlich gingen wir dazu über, die Arena linear zu vergrößern und auch gestaltlich zu verformen, indem wir z. B. im Test die runde Arena durch eine flächengleiche quadratische oder rechteckige ersetzten.

Einen wichtigen Hinweis auf das Ortungsverfahren brachten bereits Testwahlen in einer vergrößerten Arena nach der

Abb. 9a u. b. a) Testwahlen von Ruineneidechsen bei Dressur und Test vom Zentrum der Arena aus, b) bei Dressur vom Zentrum und Test von beliebigen Punkten der Arena aus

Dressur in einer kleineren. Abb. 10 veranschaulicht als Beispiel das Verhalten einer Ruineneidechse. Dressiert wurde in diesem Fall auf einen konstanten Winkel von 90° relativ zur Lichtquelle in der kleineren Arena von 80 cm Durchmesser (innerer Kreis in Abb. 10); die Lichtquelle stand in

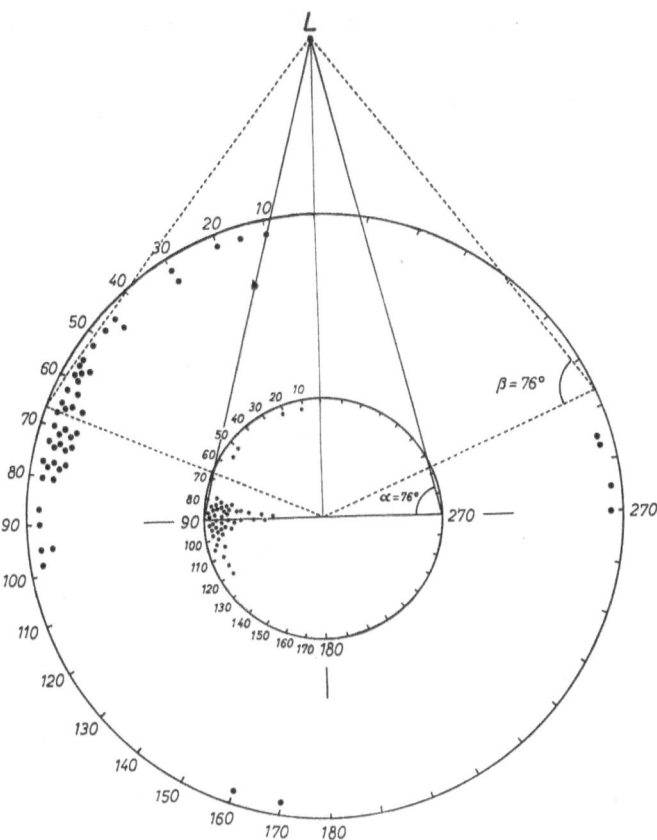

Abb. 10. Unbelohnte Richtungswahlen einer Ruineneidechse in der großen ($r = 1$ m) und in der kleinen ($r = 0,40$ m) Arena nach der winkeltreuen Dressur in der kleineren Arena. L = Standort der Lampe. Weitere Erklärungen im Text

der Projektion in 1,6 m Entfernung vom Arenamittelpunkt. Bei gleichem Lampenabstand testeten wir danach abwechselnd in der kleinen und in einer auf 2 m Durchmesser vergrößerten Arena (äußerer Kreis in Abb. 10). In der kleinen Arena liegen die Testwahlen dicht um den 90°-Punkt, in der großen aber um 70°. Die Winkel α und β, die der Radius des Dressurortes in beiden Fällen mit der Richtung zur Lichtquelle einschließt, sind jedoch beide Male etwa gleich groß. Demnach übertrug die Eidechse den in der kleinen Arena erlernten Winkel *größenrichtig* in das Bezugssystem der vergrößerten Arena, wobei sie die Richtung zur Lichtquelle offenbar auf den Radius des Dressurortes bezog. Das lebhafte „Peilen" kurz vor dem Niederlegen am Dressurort dient wohl dazu, diese beiden Koordinaten zu ermitteln. Eine solche „Zwei-Koordinaten-Ortung" setzt genaue Kenntnis aller räumlichen Daten in der kreisförmigen Arena voraus, die das Versuchstier vor und während der Dressur erwirbt.

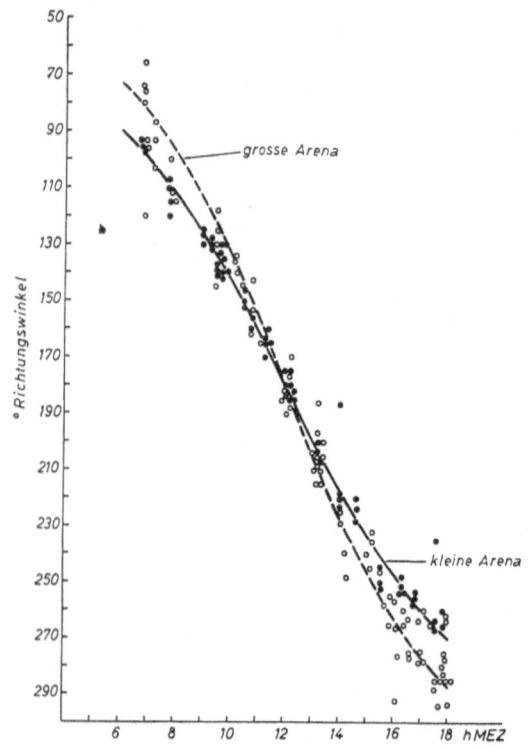

Abb. 11. Kompaßtreue Testwahlen einer Ruineneidechse in der großen (○) und in der kleinen (●) Arena

Erst nach 2—3 Tagen sind die Tiere im allgemeinen soweit mit den räumlichen Gegebenheiten vertraut, daß man mit Richtungsdressuren überhaupt beginnen kann; vorher sind sie unruhig und versuchen aus der Arena auszubrechen. Das Kennenlernen der Arena schließt wohl auch ein „Abstasten" des Arenarandes und seiner Krümmung ein; dabei wirken neben propriozeptorischen wahrscheinlich auch labyrinthäre Reize mit. Es ist daher nicht erstaunlich, daß alle bisher geprüften Tiere beim Test in einer flächengleichen quadratischen oder rechteckigen Arena nach vorheriger Dressur in einer runden vollständig versagten; in einer ovalen Arena konnten sie sich noch fehlerhaft orientieren.

Ob die Eidechsen neben den beiden Koordinaten noch weitere Orientierungshilfen gebrauchen, indem sie z. B. auch die Entfernung zur Lichtquelle abschätzen, wissen wir z. Z. noch nicht. Gegen die Entfernungsschätzung spricht allerdings, daß die Höhe der Lichtquelle keinen nachweisbaren Einfluß auf die Azimutbestimmung hat. Trifft aber die oben vorgeschlagene Deutung zu, dann muß auch bei der kompaßtreuen, tageszeitlichen Orientierung die Differenz zwischen den Richtungswinkeln in beiden Arenen je nach der Uhrzeit und nach der Himmelsrichtung gesetzmäßig variieren. Bei einem nach Norden dressierten Tier müßte z. B. der Unterschied um die Mittagszeit am kleinsten, morgens und abends aber am größten sein, da am Mittag die Winkel α und β 180° betragen und auf dem Lampenvertikal, d. h. dem Radius, der zur Lichtquelle weist, liegen. Dies belegt Abb. 11 für eine nach N dressierte Ruineneidechse: die Differenz der Richtungswinkel ist bei 90° und 270° am größten, bei 180° aber gleich null.

III. Schlußfolgerungen, Ausblicke und Zusammenfassung

Die physiologische Analyse der bei der Sonnenkompaßorientierung beteiligten Taxismechanismen und „inneren Uhren" wird um so erfolgreicher sein, je mehr Vergleichsmaterial aus den verschiedensten Gruppen der Wirbeltiere und Wirbellosen zur Verfügung steht. Die Eidechsen, deren Lernvermögen bekanntlich recht gut ist (BERGER, 1924; EHRENHARDT, 1937), bieten hier den Vorteil leichter Züchtbarkeit, so daß man auch die angeborenen und erworbenen Anteile ihrer Sonnenkompaßorientierung sowie ihre Entwicklungsgeschichte mit Erfolg untersuchen kann. Vermöge ihrer außerordentlich genauen und zuverlässigen Orientierung am Tage und zur Nachtzeit, wobei sie (FISCHER, 1961) auch die tages- und jahreszeitlichen Schwankungen der Azimutwinkelgeschwindigkeit der Sonne erstaunlich genau beachten, sind sie besonders gut für Laboratoriumsexperimente geeignet. Die weitere experimentelle Arbeit erstrebt sowohl eine nähere Analyse des Mechanismus der „inneren Uhr" als auch der Funktionsweise der steuernden Taxismechanismen. Zur Zeit prüfen wir den Einfluß verschiedener Narkotica, Pharmaka und der Temperatur auf den Gang der inneren Uhr, der sich ja am Verlauf der Azimutkurven richtungsdressierter Tiere sehr genau ablesen läßt. Eine weitere Frage betrifft das Zusammenspiel der Sinnesorgane, die den Eidechsen noch vor der Dressur räumliche Daten der Arena vermitteln und so den Dressurerfolg sichern. Schließlich gestatten unsere Versuchstiere auch die Verknüpfung von Orientierungsexperimenten mit elektrophysiologischen Hirnreizversuchen und Hirnausschaltungen; wir hoffen, dadurch Einblicke in das zentralnervöse Funktionsgefüge der Lichtkompaßorientierung zu erhalten.

Vorerst können aber folgende Befunde als gesichert gelten:

1. Eidechsen (*Lacerta viridis* Laur., *Lacerta sicula*, *Lacerta muralis* Laur.) können unter einer „künstlichen Sonne" die Himmelsrichtung eines warmen Platzes bestimmen, dessen Lage sie bei der Richtungsdressur erlernt hatten. Alle bisher untersuchten Arten erlernen zuerst die „winkeltreue" Orientierung zu *einer* Dressurzeit und orientieren sich erst dann „kompaßtreu", wenn man auf weitere, tageszeitlich bestimmte Winkel für die betreffende Himmelsrichtung nachdressiert. Die Kenntnis von nur drei Punkten der Sonnenbahn genügt ihnen, um daraus die *ganze* Azimutkurve für die betreffende Dressurrichtung zu errechnen.

2. Zur Nachtzeit verrechnen die Eidechsen ohne weitere Nachhilfe eine scheinbare Sonnenwanderung im Uhrzeigersinn, wobei sie die Tagesazimutkurve der Sonne von West über Nord nach Ost fortsetzen; diese Verrechnungsweise erhält sich sowohl in Bedingungen des 12/12-stündigen Normaltages als auch im Langtag von 16/8 Std. Im Kurztag (8/16 Std) sind die Eidechsen am Tage und zur Nachtzeit stets desorientiert.

3. Im phasenverschobenen, vor- oder zurückgestellten Tag weichen die richtungsdressierten Tiere voraussagbar von der erlernten Richtung ab; ihre innere Uhr ist umstimmbar. Nach Umstimmung auf einen Langtag rechnen die Eidechsen auch jahreszeitliche Unterschiede in der Azimutwinkelgeschwindigkeit der Sonne ein.

4. Auch erfahrungslose, ohne Sonnen- oder Himmelssicht aus dem Ei aufgezogene Smaragdeidechsen können unter der stillstehenden künstlichen Sonne die richtige Sonnenwanderung einrechnen, wenn sie auf drei diskrete Punkte der Sonnenbahn dressiert werden. Nach einer Dressur auf die inverse, gegen den Uhrzeigersinn laufende Sonnenbahn, rechnen sie in unbelohnten Testwahlen zu je 50% *sowohl* die erlernte inverse, *als auch* die ihnen unbekannte richtige Sonnenwanderung ein. Beide Befunde machen es wahrscheinlich, daß das Vermögen, die tageszeitliche Wanderung der Sonne einzuschätzen den Eidechsen angeboren ist. Die Richtung der Sonnenwanderung ist aber erlernbar; dabei lernen unsere einheimischen Eidechsen die in unseren Breiten richtige Wanderungsrichtung besser als die falsche.

5. Ein Einfluß der Höhe der richtenden Lichtquelle auf das Richtungsfinden ließ sich bisher nicht erbringen, obwohl die Eidechsen anderen Versuchen zufolge die Höhe der Lichtquelle durchaus wahrnehmen können.

6. Die richtungsdressierten Eidechsen sind desorientiert, wenn die Lichtquelle näher als 5° (*Lacerta viridis* Laur.) bzw. 11° (*Lacerta sicula* und *L. muralis* Laur.) zum Zenit steht.

7. Werden den Eidechsen im Testversuch simultan zwei richtende Lichtquellen geboten, deren Winkelabstand 180° oder 90° beträgt, so

orientieren sie sich entweder nach der einen oder nach der anderen. Steht die eine Lichtquelle beträchtlich höher als die andere, so bevorzugen die Eidechsen die niedrigere.

8. Die Ortung des dressurgemäßen Platzes in der Arena, an dem die Tiere die Heizplatte „erwarten", nehmen sie wahrscheinlich mittels zweier Koordinaten vor, indem sie von einem Punkt des Arenarandes aus die Richtung zur Lichtquelle ermitteln und diese auf den zugehörigen Radius beziehen. In einer vergrößerten Rundarena sind daher die Tiere sofort orientiert; dagegen werden gestaltliche Verformungen der Dressurarena nicht vertragen. Dem eigentlichen Richtungslernen geht das Kennenlernen der räumlichen Beziehungen innerhalb der Arena voraus.

Literatur

Aschoff, J.: Exogenous and endogenous components in circadian rhythms. Cold Spr. Harb. Symp. quant. Biol. 25, 11 (1960).

Berger, K.: Experimentelle Studien über Schallperzeption bei Reptilien. Z. vergl. Physiol. 1, 517 (1924).

Birukow, G.: (1) Photo-Geomenotaxis bei Geotrupes silvaticus Panz. und ihre zentralnervöse Koordination. Z. vergl. Physiol. 36, 176 (1954).

— (2) Lichtkompaßorientierung beim Wasserläufer Velia currens F. am Tage und zur Nachtzeit. Z. Tierpsychol. 13, 463 (1956).

— (3) Innate types of chronometry in insect orientation. Cold Spr. Harb. Symp. quant. Biol. 25, 403 (1960).

Braemer, W.: (1) Versuche zu der im Richtungsgehen der Fische enthaltenen Zeitschätzung. Verh. D. Zool. Ges. Münster/Westf., 276 (1959).

— (2) A critical review of the sun-azimuth hypothesis. Cold Spr. Harb. Symp. quant. Biol. 25, 413 (1960).

Bünning, E.: Die Physiologische Uhr. Berlin, Göttingen, Heidelberg: Springer Verlag 1958.

Ehrenhardt, H.: Formensehen und Sehschärfebestimmung bei Eidechsen. Z. vergl. Physiol. 24, 248 (1937).

Fischer, K.: (1) Experimentelle Beeinflussung der inneren Uhr bei der Sonnenkompaßorientierung und der Laufaktivität von Lacerta viridis Laur. Naturwissenschaften 47, 287 (1960).

— (2) Untersuchungen zur Sonnenkompaßorientierung und Laufaktivität von Smaragdeidechsen (Lacerta viridis Laur.). Z. Tierpsychol. 18, 450 (1961).

— u. G. Birukow: Dressur von Smaragdeidechsen auf Kompaßrichtungen. Naturwissenschaften 47, 93 (1960).

Frisch, K. von: (1) Die Sonne als Kompaß im Leben der Bienen. Experientia (Basel) 6, 210 (1950).

— (2) Orientierungsvermögen und Sprache der Bienen. Naturwissenschaften 38, 105 (1951).

Gould, E.: Orientation in Box Turtles Terrapene c. carolina (Linnaeus). Biol. Bull. 112, 336 (1957).

Hasler, A. D., R. M. Horrall, W. J. Wisby and W. Braemer: Sun orientation and homing in fishes. Limnol. and Oceanogr. 3, 353 (1958).

— and H. O. Schwassmann: Sun orientation of fish at different latitudes. Cold Spr. Harb. Symp. quant. Biol. 25, 429 (1960).

HOFFMANN, K.: (1) Versuche zu der im Richtungsfinden der Vögel enthaltenen Zeitschätzung. Z. Tierpsychol. 11, 453 (1954).
— (2) Die Richtungsorientierung von Staren unter der Mitternachtssonne. Z. vergl. Physiol. 41, 471 (1958).
— (3) Experimental manipulation of the orientational clock in birds. Cold Spr. Harb. Symp. quant. Biol. 25, 379 (1960).
KRAMER, G.: Experiments on bird orientation and their interpretation. Ibis 94, 196 (1957).
LINDAUER, M.: (1) Sonnenorientierung der Bienen unter der Äquatorsonne und zur Nachtzeit. Naturwissenschaften 44, 1 (1957).
— (2) Angeborene und erlernte Komponenten in der Sonnenorientierung der Bienen. Z. vergl. Physiol. 42, 43 (1959).
— (3) Time-compensated sun-orientation in bees. Cold Spr. Harb. Symp. quant. Biol. 25, 371 (1960).
MITTELSTAEDT, H.: Probleme der Kursregelung bei frei beweglichen Tieren. Aufnahme und Verarbeitung von Nachrichten durch Organismen. 138. Stuttgart: S. Hirzel Verlag 1961.
— (2) Control systems of orientation in insects. Ann. Rev. Entomol. 7, 177 (1962).
PARDI, L.: Innate components in the solar orientation of littoral Amphipods. Cold Spr. Harb. Symp. quant. Biol. 25, 395 (1960).
SCHMIDT-KOENIG, K.: Die Sonnenorientierung richtungsdressierter Tauben in ihrer physiologischen Nacht. Naturwissenschaften 48, 110 (1961).
ST. PAUL, U. v.: Neue experimentelle Ergebnisse über die Fernorientierung der Tiere. Naturwissenschaften 45, 123 (1958).
SCHWASSMANN, H. O.: Environmental cues in the orientational rhythm of fish. Cold Spr. Harb. Symp. quant. Biol. 25, 443 (1960).
WEVER, R.: Possibilities of phase-control, demonstrated by an electronic model. Cold Spr. Harb. Symp. quant. Biol. 25, 197 (1960).

Psychophysical Limits of Celestial
Navigation Hypotheses

By Helmut E. Adler

The American Museum of Natural History, New York 24, New York
and Yeshiva University, New York 33, New York (U.S.A.)

With 15 Figures

Contents

I. Psychophysics

The term psychophysics was first used by G. T. Fechner in 1860 to describe methods of measuring sensory functions. Fechner himself thought he had solved philosophy's classical mind-body problem, but what he did in fact was to show that the only psychophysical measurement possible was a quantification of the relationship between a response and a physical stimulus, expressed in physical units.

Psychophysical methods can be applied to animals by bringing the animal's behavior under the control of some physical stimulus or sets of stimuli and measuring the appropriate response. Analysis of the response pattern establishes the animal's perception of its environment and allows quantitative formulation of the limiting values of the stimulus which evoke a specifiable probability of responding. Used appropriately, they can tell us not only what kind of stimuli an animal can detect, but also the extent to which it can discriminate one degree of stimulation from another and, thus, how well it can utilize information provided by environmental cues.

Two problems arise in this connection, and refinement in method has focused on achieving greater precision in their solution. One is the specification of the stimulus, which must be measured accurately and controlled in such a way that one can be sure the animal is responding to it, and not to some irrelevant cue. Contrary to popular opinion, for example, it appears that the red cloth waved in front of the bull does not provoke his attack because of its red color, but because of its movement. The second problem is to find a simple and dependable response, which also has some definite characteristic which ties it to the specific behavior under investigation. These two requirements are more or less in conflict with one another. Responses which are naturally evoked by the stimulus, whether physiological or behavioral in nature, are simple and dependable, but often one cannot be sure whether they are expressions of the specific behavior which is being studied. If measuring visual acuity, for example, we can record the optomotor response to a revolving striped drum, but we cannot be sure the animal actually perceives the stimulus, i.e., whether it could utilize cues provided by such fine visual detail in its behavior.

All measurements made reduce to the obtaining of two fundamental quantities, the absolute and the relative threshold. The absolute threshold may be defined as the least amount of energy which will still result in a specified change in behavior. The relative threshold is the least amount of change in stimulus energy to which the organism is able to react. The actual numerical values of these thresholds will depend on many factors, such as the operations by which they were obtained and the conditions under which the observations were made, but the results can be used to establish in the first place whether a given type of stimulation is capable of arousing an organism to make an appropriate response and, once this capacity is established, to trace the changes in the sensitivity of the animal's receptor processes as the stimulus is varied systematically. Thus we can establish, using the appropriate procedures, that cats need a minimum amount of light in order to see, and we will note that this minimum varies from cat to cat, and depends on the time that the cat has been in the dark (BRIDGMAN and SMITH, 1942).

Methods

Each kind of measurement involves the solution of problems connected with the nature of the stimulus and the capacities of the animal. A great number of approaches have been used successfully, but we can divide the methods in general use into four general types (as modified from WARDEN, JENKINS and WARNER, 1935):

1. *Preference method:* In its simplest form, an animal is placed in a position to choose the preferred stimulus among two or more choices offered. In a more complex variant, a graduated series of stimulus values

is presented to the organism. The animal indicates its reaction to the stimulus by moving along the stimulus gradient to a preferred position. Animals are often introduced in groups and their aggregation at a preferred portion of the gradient noted, although they may be tested individually.

Recently this method was used, for example, to establish the presence of odor preference to water from various streams in salmon fingerlings (HASLER and LARSEN, 1955).

2. *Stimulus-test method:* Any kind of adequate stimulation may be employed and the resulting behavior observed. The presence or absence and direction of a response is noted. This method differs from the foregoing in that no choice is offered to the animal. It is similar to it in that no training is required. A single stimulus may be used or two or more stimuli may be presented, and the behavior analyzed in terms of the resultant of forces represented by the degree of effectiveness of the stimuli. An example of this method, using two stimuli, is represented by BRAEMER's (1957) measurements of optical functions in fish along lines developed by v. HOLST (1950), balancing the effect of gravity and of lateral illumination to effect a measurable inclination, which could be interpreted as a quantitative effect of light stimulation.

3. *Discrimination response method:* An animal is trained to respond differentially to two or more stimuli presented simultaneously, usually by making a locomotor response. We may take v. FRISCH's (1915) technique as an example, where bees were trained to fly to a colored square of paper among a checkerboard arrangement of gray papers in order to measure their capacity to discriminate hues.

4. *Conditioned response method:* In this method, animals are trained to respond differentially to two or more stimuli, usually presented in succession. The response used as an indicator may not be under the control of the organism, in which case a Pavlovian conditioning model is followed. Thus, for example, McCLEARY and BERNSTEIN (1959) employed heart rate in their study of color vision of the goldfish, pairing electric shock and the positive stimulus, while the neutral stimulus was never so paired. After training, a change in heart rate to the positive stimulus, presented alone, indicated discrimination. On the other hand, if a response under the control of the organism is employed, an operant conditioning model is followed. If the stimuli are presented simultaneously, this method may then become indistinguishable from the discrimination response method, which can be classified as a special case of the conditioned response method. More typically, however, rate of responding is recorded, and response rate in the presence of the positive stimulus compared with the rate in the presence of a neutral or negative stimulus. A further development of this method consists in putting intensity of

stimulation under control of the animal, following the procedure first outlined by VON BÉKÉSY (1947) and CRAIK and VERNON (1941) for human threshold determination. This method, also known as the 'tracking' method, was adapted to animal psychophysics by RATLIFFE and BLOUGH (1954). Its essential feature is a feedback system in which the animal's correct choices lower the stimulus intensity, while failure to discriminate increases stimulus intensity. By appropriate scheduling of reinforcement, a stable performance level is reached in which the animal's responses control the stimulus, while the stimulus in turn controls the animal's behavior, resulting in a state of equilibrium in which the stimulus intensity is kept oscillating about its threshold value. This method has now been employed to measure several kinds of visual thresholds in birds (ADLER and DALLAND, 1959; BLOUGH, 1956, 1957). For an example, see the section on spectral sensitivity below.

II. Limits of a celestial navigation hypothesis

We can use psychophysical methods to establish the limits imposed on animals by their sensory capacities as far as theories of celestial navigation are concerned. These capacities relate to their ability to utilize a grid or position lines made up, among other possibilities, of the sun's altitude, its extrapolated apparent movement (MATTHEWS, 1951), its apparent rate of change of altitude (PENNYCUICK, 1960) and an accurate determination of the difference between the local time and a fixed reference time. If we substitute any fixed star or constellation or the moon for the sun, the same arguments hold for navigation at night (SAUER, 1957).

a) The margin of error

Animals must rely on their own senses. A margin of error of appropriate dimensions should therefore be allowed in any theory. The case is not too bad if the sun is used only for orientation. A 15 minute error in time under idealized conditions would bring an animal travelling in a straight line for 100 miles within $6^1/_2$ miles of its goal. Even this error decreases rapidly with shorter travel distance, if it is assumed that the sun's position can be accurately determined, its movement is uniform, and that there is no lateral drift.

On the other hand, when fixing position in our latitudes (i.e., 40° N), each 4 minutes error in time misplaces the animal by 53 miles east or west, and each degree of misjudgment of the sun's altitude causes a net displacement of 69 miles. These values are taken from idealized curves for the movement of the sun. The azimuth of the sun (Fig. 1) actually changes in a more complex manner depending on latitude, time of day, and season. Fishes at least (BRAEMER, 1960), do not seem to compensate.

The sun's apparent path (Fig. 2) also is not a circle, but changes in a sinusoidal manner.

Our laboratory has been engaged since 1955 in a series of studies designed to examine critically the basis for an accurate visual estimate of movement and position of celestial objects and an accurate time sense

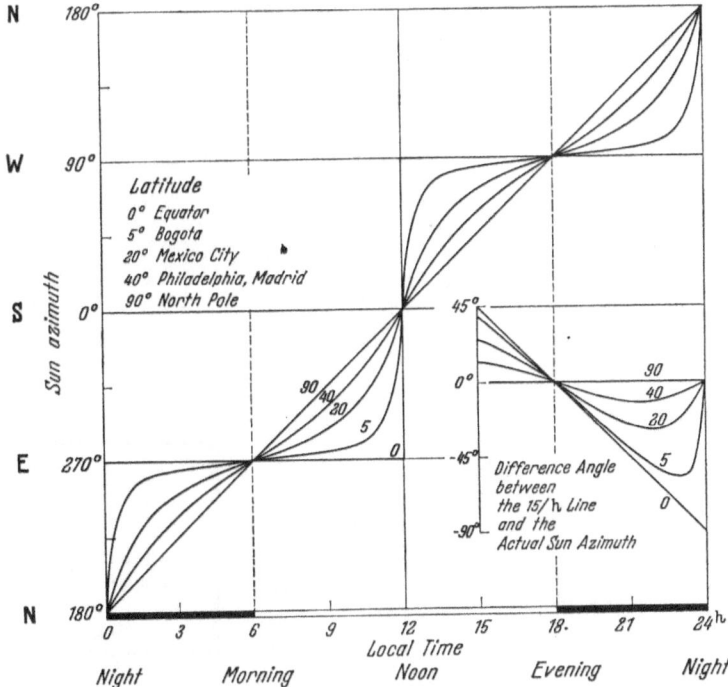

Fig. 1. Sun azimuth as a function of local time, at the equinox, at five different latitudes. Insert at lower right: Difference between azimuth position of the sun at five latitudes and the straight line corresponding to an idealized azimuth movement of 15°/hour (from BRAEMER, 1960)

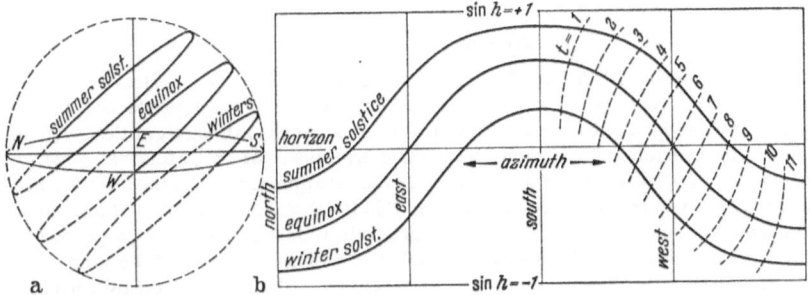

Fig. 2. The sun's apparent motion as a function of azimuth at the equinox and at the summer and winter solstices, at a latitude of approximately 51°. Left: Idealized perspective diagram. Right: Actual appearance of sun's path to an observer on the earth. t = lines of equal time (from KRAMER, 1957)

which are the sensory factors in migration. Starlings *(Sturnus vulgaris)* and robins *(Turdus migratorius)* have been studied so far to establish the spectral sensitivity curve (ADLER and DALLAND, 1959), to explore the more complex visual functions (ADLER, 1959, 1960), and to investigate the accuracy of time judgments (ADLER and GIANUTSOS, 1962).

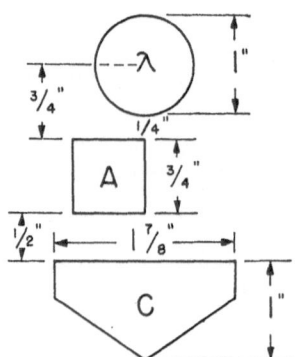

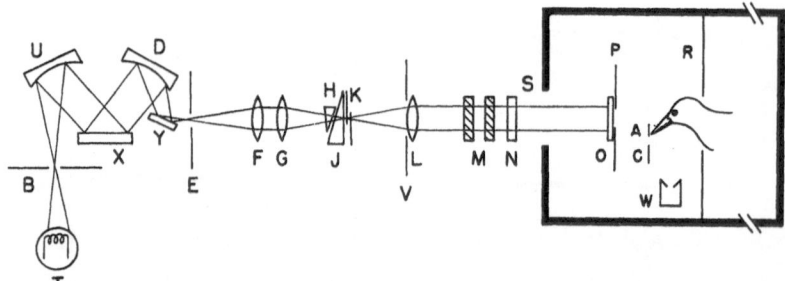

Fig. 3. Schematic diagram of apparatus (bottom) and scale diagram of stimulus patch (λ) and pecking keys A and C (top). T 150 w. Xenon arc light source; B, U, X, D, Y, E light path through Farrand ultra-violet visible grating monochromator (No. 103420); F, G, L Lenses; H, J Kodak circular neutral density wedge and balancer; K Shutter; V Field stop; M Neutral density filters; N Cut-off filter; S Box aperture; O Flashed opal glass; P 1 in. dia. stop; R Partition and opening to fix bird's head position; W Food tray (from ADLER and DALLAND, 1959)

b) Spectral sensitivity

Apparatus and Procedure. The apparatus used to obtain the spectral sensitivity curves (Fig. 3) consisted of a high-pressure xenon arc lamp which produced a monochromatic one-inch dia. spot of light of the desired wavelength via an optical system. The birds controlled their own stimulation by means of an automatic program controlled by a relay system. By a series of gradual approximations the birds learned to peck key A in the presence of light, key C in its absence. Pecks on A showed that the light was seen and caused the luminance to be reduced by means of a wedge (J). Pecks on C indicated the light was not visible to the bird and increased the luminance. At scheduled random intervals pecks on A closed a shutter (K) and by correctly pecking key C the bird caused the food tray (W) to rise, providing it with access to food and reinforcing the whole chain of behavior (for details see ADLER and DALLAND, 1959).

Results. Complete results have been obtained of relative threshold energy for light at 20 wavelengths on three starlings. Records have also been analyzed for one robin. A record made of the position of the wedge (Fig. 4) traces the threshold of the bird. Preadapting the bird to light gave us a record of cone sensitivity (Fig. 4, upper curve at 520 mµ). At the end of two hours in the dark, we interpreted the record as being

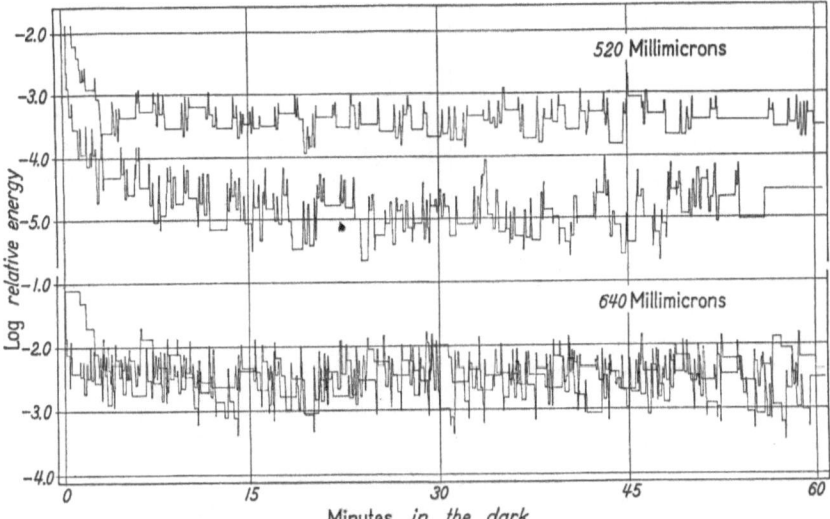

Fig. 4. Superimposed tracings of four samples of threshold curves obtained from one starling during four separate one hour sessions. Curves record wedge position, calibrated in log relative energy, and thus present a continuous tracking of the bird's threshold. Top curve was taken after preadaptation to white light, while next curve down was preceded by dark adaptation. Both curves were obtained at 520 mµ. Two superimposed records at the bottom represent the same two conditions at 640 mµ (from ADLER and DALLAND, 1959)

that of the rods (Fig. 4, lower curve at 520 mµ). Both curves are superimposed at 640 mµ since rods and cones are about equally sensitive to red light.

Compared to the human eye (Fig. 5), the starling detects blue and violet light better than we do with the normal eye, but not better than the lenseless or aphakic eye. Man also has generally more sensitive red vision, except that the starling may see slightly better in the far red at night. The robin has a similar cone curve, but seems to be astonishingly insensitive in the red. (We are still trying to confirm this finding.) Compared to the pigeon (Fig. 6; BLOUGH, 1957), the starling's cone curve is shifted to the short end of the spectrum, perhaps due to the difference in the filtering action of the colored oil droplets found in the respective retinas.

c) Dark adaptation

Turning now to dark adaptation, we used the same apparatus modified to bring white light from a ribbon-filament lamp into the system. Thresholds in the dark were followed as a function of duration of preadaptation. Two starlings (ADLER, 1959) and two robins (ADLER, 1960) have been tested. A sample tracing (Fig. 7)

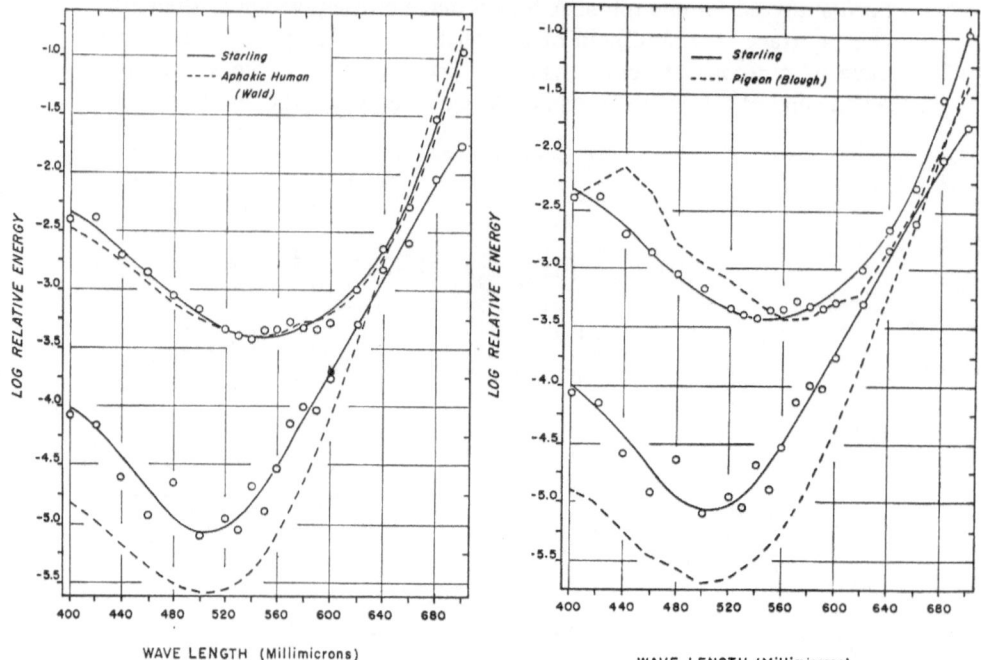

Fig. 5. Comparison of log relative spectral thresholds, averaged for three starlings, with those of the aphakic (lenseless) human eye, as taken from WALD, 1945 (from ADLER and DALLAND, 1959)

Fig. 6. Comparison of log spectral thresholds, averaged for three starlings, with those of the pigeon, as taken from BLOUGH (1957) (from ADLER and DALLAND, 1959)

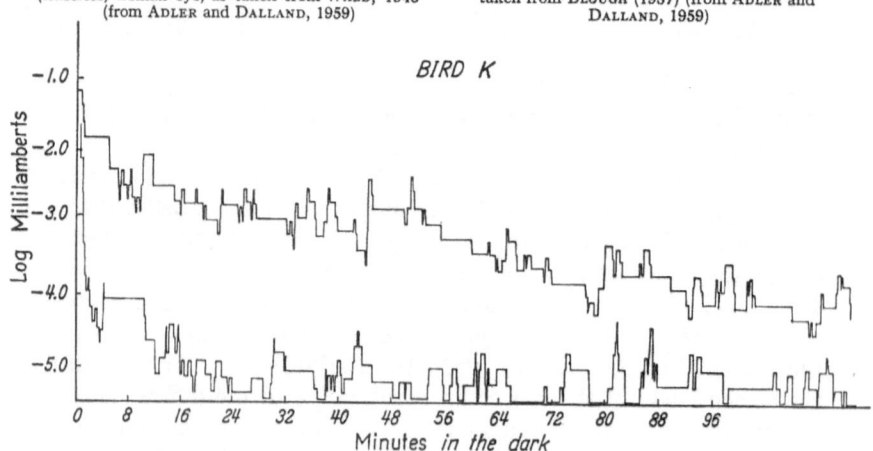

Fig. 7. Superimposed tracings of two samples of threshold curves obtained from one starling during separate 120 min. sessions. The top curve records wedge position, calibrated in log millilambert, after 60 min. of dark adaptation and 10 min. of preadaptation to white light. It is displaced upward by 0.5 log units for legibility. The lower curve was taken from a continuous tracking of the threshold for 120 min. in the dark after 60 min. of dark adaptation and 0 min. of preadaptation. The sharp drop during the first few minutes is an artifact

illustrates the main features of this experiment. Birds were confined in the dark for 60 min., then exposed for the scheduled duration to an adapting light. The lower curve represents the threshold for 120 min. after no light adaptation. The upper curve, displaced upward by 0.5 log mL shows a curve for 10 min. preadaptation. Horizontal portions of the curves represent pauses of the bird for feeding or imposed rest periods.

Results. Fig. 8 shows the averaged dark adaptation curves for two robins. Each point is derived from the two minute average of all threshold

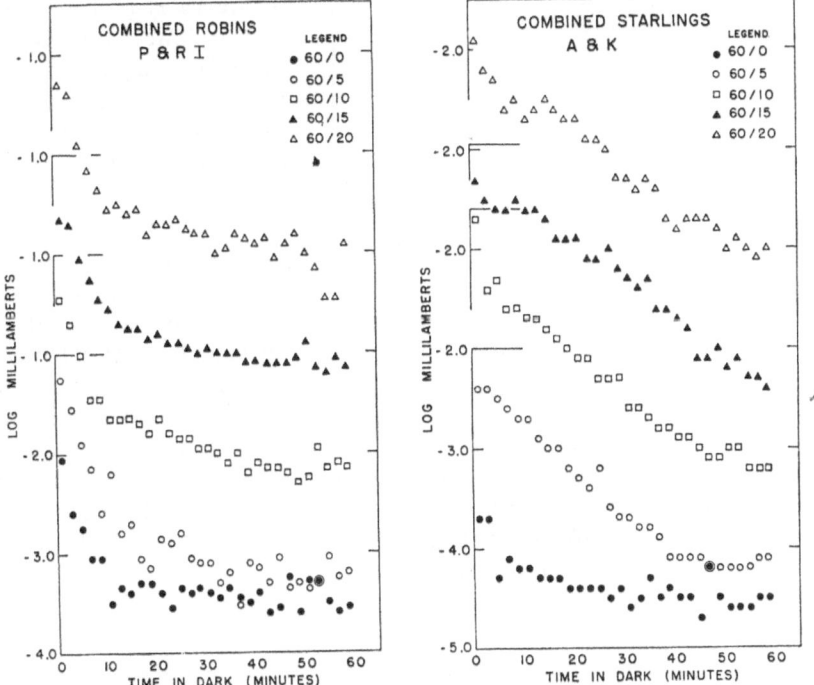

Fig. 8. Averaged dark adaptation curves for two robins at five durations of preadaptation to white light as shown in the legend at upper right. The curves above 5 min preadaptation are displaced upward by one log unit each

Fig. 9. Averaged dark adaptation curves for two starlings at five durations of preadaptation to white light as shown in the legend at upper right. The curves above 5 min. preadaptation are displaced upward by one log unit each

points falling into this interval over all trials. The curves above 5 min. preadaptation are displaced upward by one log unit for legibility. Dark adaptation of the robin is a slow and rather shallow function. The sharp drop in the first few minutes should be discounted as it includes the rapid driving down of the wedge by the bird when presented with a supra-threshold stimulus. Final thresholds were of the order of -3.8 log mL; much higher than human thresholds of -5.3 log mL obtained in the same apparatus.

The comparable starling curves (Fig. 9) are somewhat steeper, especially during the later or rod portion. Comparison of starling (ADLER, 1959), robin (ADLER, 1960), pigeon (BLOUGH, 1956) and human (SLOAN, 1928) (Fig. 10) curves show similar slopes for cone adaptation, but slower adaptation and higher final thresholds for birds. This difference is probably due to the slower processes of pigment migration and photomechanical changes found in the bird retina, as compared to the human eye.

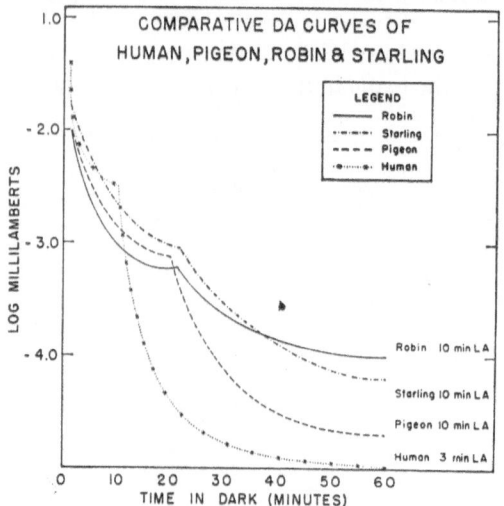

Fig. 10. A comparison of dark adaptation curves of robin (ADLER, 1960), starling (ADLER, 1959), pigeon (BLOUGH, 1956) and man (SLOAN, 1928)

Particular interest is attached to the absolute threshold for light. Although this threshold is affected by many factors, such as wavelength, duration, size, and location of the stimulus, I have tried to put some comparable data together (Table 1). Cat and owl, specialized nocturnal animals, can still see in what seems to us total darkness. On the other hand, robins, starlings and pigeons must find the world at night much darker than it seems to us.

Table 1. *Some comparative absolute thresholds*[1]

	Absolute threshold log mL	Source
Cat	— 7.13	⎫
Human	— 6.36	⎬ BRIDGMAN and SMITH (1942)
Owl	— 6.10	⎭
Human	— 5.43	⎱ DICE (1945)
Pigeon	— 4.70	BLOUGH (1956)
Robin P + R I	— 3.80	⎫
Robin P + R II	— 4.30	⎪
Starling I	— 4.29	⎬ ADLER (1959, 1960)
Starling II	— 4.40	⎪
Starling K	— 5.02	⎪
Human	— 5.32	⎭

[1] Note: Different operations were employed in optaining thresholds by the several authors cited, except that BLOUGH und ADLER used the same tracking technique.

d) Visual acuity

Our own work on visual acuity has not yet borne fruit. I shall therefore rely on the work of others. Contrary to expectations and often repeated statements about the exceptional visual acuity of birds, DONNER (1951) found the visual acuity of seven typical passerine species is poorer than that of humans. One of the best, the blackbird *(Turdus merula)*, for example, was determined to have a 1'20" minimal separable angle and the European robin *(Erithacus rubecula)* one of 2'38". GRUNDLACH's (1933) value of 0'23" for the pigeon, often quoted in the literature of migration, falls far out of this range and must be considered quite atypical. HAMILTON and GOLDSTEIN (1933) and CHARD (1939) obtained at best a value of 2'42" for this bird, and 3'12" would be a more realistic value. Human visual acuity, for comparison, falls at about $^1/_2'-1'$ and agrees with that of the chimpanzee and rhesus monkey (SPENCE, 1934; WEINSTEIN and GRETHER, 1940). Although PUMPHREY (1961) in a recent review attempts to explain the superior seeing ability of birds by their better powers of accomodation and wider field of sharp vision, we cannot expect a bird to see anything that we cannot also notice.

e) Perception of movement and radial localization

A fallacy often committed is to assume that sensory threshold values can be directly applied to estimate perceptual phenomena. This point of view is naive and ignores the fact that a multiplicity of variables besides the threshold play a role in perception. Perceptual functions may follow the same psychophysical laws, but always at a level of stimulation much above threshold, or they may behave in an independent way, when the role of the threshold has become negligible compared to the influence of other variables (GRAHAM, 1951). It is therefore imperative to study the factors relevant to each perceptual variable involved in navigation.

The perception of movement is important to both MATTHEWS' (1951) and PENNYCUICK's (1960) theories. BROWN and CONKLIN (1954) determined the threshold for perceived movement, in humans, of a circular spot of light subtending 38'. (Sun and moon subtend about 30'.) The slowest movement his subjects could still discriminate was 9' per second. We may compare this value to the average speed of the sun of 0.25' per second. It is not reasonable, then to assume that birds perceive the motion of the sun directly.

LEIBOWITZ, MYERS and GRANT (1955) examined the ability of subjects to localize the radial position of a spot of light subtending 38'. The average error made was 4.64° (4° 38'). If animals can do no better in localizing the sun's position, these findings imply a very large error in sun navigation.

f) Internal clock

Previous tests of the role of this clock in migration have "reset" the clock by subjecting animals to a six or twelve hour shift of the external synchronizer (Hoffmann, 1960; Schmidt-Koenig, 1960). The shift in oriented movement was taken as an indicator of the effect of the time variable. But since the choice of direction is determined by other variables as well as time, this procedure cannot tell us about the accuracy of the bird's "clock".

In our experiments time judgment was studied in terms of the bird's feeding rhythm. Two conditions were employed. Under one condition starlings were free to develop their own rhythm, under the other a rhythm was imposed on the birds by restricting access to food to four daily periods. [This method is similar to that used by Renner (1960) on bees.]

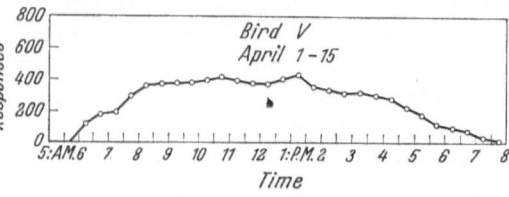

Fig. 11. Sample of distribution of key pecking of one starling during active hours for 15 days. Ordinate represents mean of responses per half hour interval. Food was available at all times

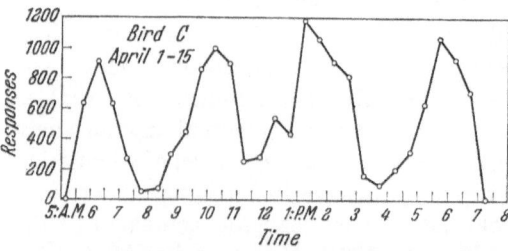

Fig. 12. Sample of distribution of key pecking of one starling during active hours for 15 days. Ordinate represents mean of responses per half hour interval. Food was available for one hour, starting 6 A. M., 10 A. M., 2 P.M., and 6 P.M. Note that peak of activity precedes reinforcement period in anticipation of the two afternoon feeding periods

Three starlings have been tested under conditions of free access to food; data are also available on three starlings under conditions of restricted access to food.

Method. The birds were conditioned to peck keys for food reinforcement at a fixed ratio of twenty. The number of pecks was recorded on counters and also printed out on tape. The bird's output was totaled every half hour for the birds with unrestricted access to food. Birds with restricted access to food received a one-hour access to food starting at 6 A.M., 10 A.M., 2 P.M. and 6 P.M. daily. All scheduling and recording was automatic. The birds had to be disturbed only about once a week at different times of day for cleaning and refilling of food and water reservoirs.

Results. Under unrestricted access to food, birds developed a characteristic feeding rhythm showing one or more daily maxima of activity (Fig. 11 shows a 15 day sample for one bird.) Gradual shift of peak activity could be noted when records were taken over periods as long as a year. Records taken under conditions of four one-hour feeding periods showed four distinct peaks just before and during the reinforcement period (Fig. 12).

Through the courtesy of Dr. SAMUEL J. M. ENGLAND of the Columbus Psychiatric Institute (Columbus, Ohio, USA), computer time was made available and an autocorrelation analysis was performed on the same data. This analysis separates periodic components from random fluctuations. The procedure is to take the wave form, delay it by a time τ and compare it with the original wave form. The autocorrelation function for bird V on unrestricted access to food shows only a 24 hr. circadian period (Fig. 13). The same analysis on bird C, which had four daily reinforcement periods, shows a four hour rhythm superimposed on the basic rhythm of about 24 hrs. (Fig. 14). There is some evidence of the development of beats as a function of the interference between the 4 hr. and the 24 hr. periods (see HALBERG, 1960).

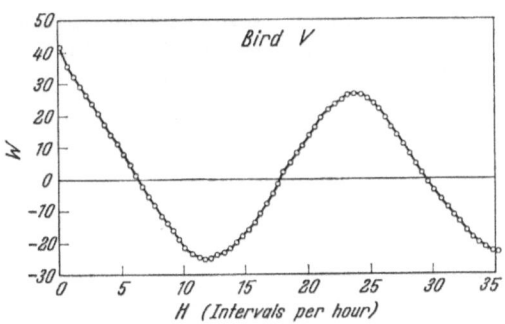

Fig. 13. Autocorrelogram of 15 day sample of same data as Fig. 11. Decimals are omitted for autocorrelation coefficient W on ordinate. Lags were computed by half hour intervals

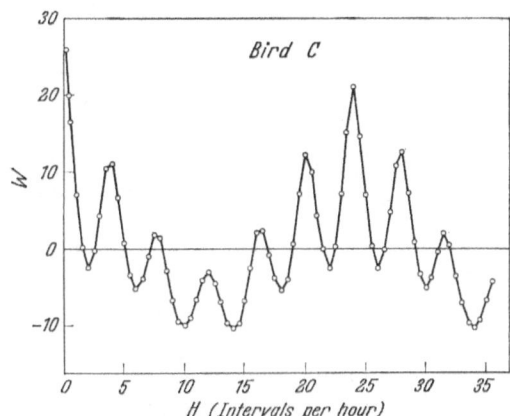

Fig. 14. Autocorrelogram of 15 day sample of same data as Fig. 12. Decimals are omitted for autocorrelation coefficient W on ordinate. Lags were computed by half hour intervals

The starling's pecking depends on daylight and onset of daily activity was closely bound up with the intensity of illumination. During the shorter hours of daylight, as found in the winter season, they may miss the first reinforcement period entirely. Fig. 15 shows onset of key pecking averaged by months for one bird during the period March 1961 to March 1962. Natural daylight conditions prevailed. We note a gradual change in average starting time which typically follows the time course of onset of daylight (ASCHOFF and WEVER, 1962). There is an abrupt change during the months of January and February when onset of pecking falls after the first reinforcement period.

In general we find that starlings will develop a 24 hr. feeding rhythm under natural conditions. This rhythm is synchronized by the day/night

cycle. If we superimpose a 4 hr. rhythm, the birds are locked in phase
with the reinforcement schedule. The degree of accuracy of this 4 hr.
periodicity indicates one aspect of the bird's time judgment. From the
degree of anticipation we estimate an error of 15—20 minutes on the
average. There was no punishment for anticipating, so we do not know
if this accuracy represents the best these birds can do.

If, as an illustrative example, we want to apply this order of magnitude to the error which could result in celestial navigation, we would
find that at our latitude (40° N) a 20 minute error in time judgment

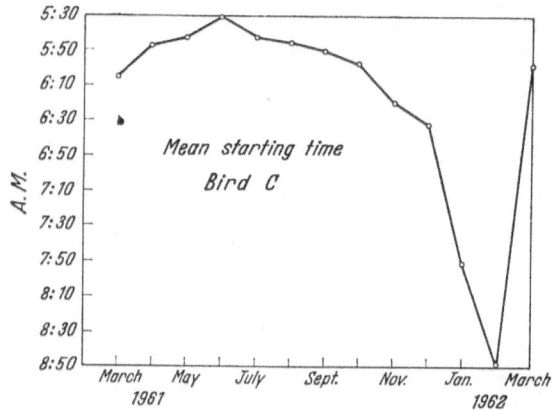

Fig. 15. Mean time of onset of key pecking, averaged per month, throughout one year, for one bird kept under
natural daylight conditions under restricted access to food. The first feeding period commenced at 6 A. M.,
the second at 10 A.M

would result in a dispacement error of 265 miles east or west. Latitude
judgment would not need to be in error. If this degree of accuracy is the
best the birds can manage, it would throw serious doubt on the ability
of the birds to use a complete system of celestial navigation. It would
not necessarily preclude the possibility of orientation by the sun. We
could think of correction during flight, for example.

III. Evaluation

Unless the psychophysical data obtained in the laboratory bear very
little relationship to the actual sense capacities of birds, and experience
has shown that these thresholds in effect represent the best that an
animal can do rather than its average level of functioning, these data
indicate that we must take great care in formulating theories of navigation
in such a way as to take into account the limitations imposed by the
functioning of the sense organs. Too many theories in the past have
either been based on nonexistent or unproved capacities, or have taken
as their starting point purely physiological and anatomical considerations.

Experimental evidence so far does not support any theory that is based on extraordinary sensory capacities, as compared to the human senses. In fact the same arguments apply here, as have been used so successfully to counter the proponents of navigation based on magnetic or electrical forces or forces based on the rotation of the earth (MATTHEWS, 1955). These arguments do not apply with the same force to the non-vertebrates, however, where different capacities, such as sensitivity to the plane of polarization of light (AUTRUM and STUMPF, 1950; WATERMAN, 1960) have evolved.

It is possible to fit a theory of orientation along the lines proposed by KRAMER (1950), making use of the "sun-compass", within the limits imposed by our present knowledge of the psychophysical functions concerned. However, since birds do navigate and home, we should also look further for sensory cues which would fit within the limits imposed by their sensory thresholds or we must look for a presently unrecognized cue.

IV. Summary

Psychophysical methods of measuring sensory functions are based on observing an organism's behavior when presented with a specifiable stimulus situation. The animal's response pattern, under appropriate conditions, establishes its threshold and thereby defines the minimum degree of stimulation it can detect and the least amount of change in the environment it can discriminate. Precautions must be taken, that the stimulus characteristics to which the animal is responding are the characteristics which the experimenter had planned to investigate. The response pattern used as an indicator should be on the same level as the behavior under investigation, i.e. a physiological response, such as might be obtained from the optic nerve, is not sufficient to establish the capacity of an organism to see or discriminate the light stimuli concerned.

Four methods of obtaining psychophysical data from animals may be distinguished: These are (1) the preference method, (2) the stimulus-test method, (3) the discrimination response method, and (4) the conditioned response method. A recent development of the last method is the tracking technique in which the animal controls its own stimulation in such a way that the stimulus intensity stabilizes around the animal's threshold value.

Psychophysical thresholds serve to establish limits of sensory capacity which may be used to evaluate the validity of theories of animal orientation and navigation, although it should be recognized that perfect performance is not required and a certain margin of error is permissible.

Celestial navigation and orientation by celestial cues depend on accurate visual discriminations and time judgments. Data from experiments on starlings and robins are cited, which indicate that their basic visual functions, such as spectral sensitivity and dark adaptation, do not surpass the range of human visual capacities. Visual acuity of passerine birds also does not appear to surpass the range of human thresholds. Since no avian data were available, perception of movement and radial localization were discussed in the light of thresholds obtained on human subjects. Since human beings must supplement their senses with instruments in order to determine their position in celestial navigation, there is no foundation for the assumption that birds can do better with the unaided eye. Evidence from experiments on time judgment indicates that the order of accuracy of the internal clock in starlings also is not sufficient to satisfy the requirements of present hypotheses of celestial navigation.

Conclusions based on these findings must necessarily be tentative since only a few species of birds were tested and results are dependent on the methods employed. Nevertheless, it can be pointed out that present theories of a complete system of navigation seem to make excessive demands on the sense capacities of animals. On the other hand it appears probable that celestial cues may be utilized in direction finding, where the possibility of continuous correction of course exists and where landmarks and visual search may be used to compensate for errors introduced by sensory limitations.

Further research is needed to clarify many unsolved problems, particularly with respect to thresholds for perception of movement and localization of targets, and the accuracy of time judgment. Only when these capacities are established, can an adequate theory of celestial navigation be developed.

Acknowledgments. The author's work was supported by grants from the National Science Foundation and the Ralph Friedman Foundation to the Department of Animal Behavior, American Museum of Natural History.

Drs. JOHN DALLAND and IRWIN SIEGEL and Messrs. DAVID ANDERSON, WILLIAM AYRES, IVAN BAROFSKY, TONY GAHAN, JOHN GIANUTSOS and ARTHUR SNAPPER aided in the collection of data and construction of apparatus. Dr. THOMAS D. NICHOLSON gave helpful advice on navigational problems. LEONORE L. ADLER assisted in many ways.

References

ADLER, H. E.: Dark adaptation in the starling as a function of light adaptation. Amer. Psychol. **14**, 413 (1959) (Abstract).
— Dark adaptation in the robin *(Turdus migratorius)*. Amer. Psychol. **15**, 501 (1960) (Abstract).
— and J. I. DALLAND: Spectral thresholds in the starling *(Sturnus vulgaris)*. J. comp. physiol. Psychol. **52**, 438—445 (1959).
— and J. GIANUTSOS: The accuracy of the 24-hour clock of the starling. Amer. Psychol. **17**, 394 (1962) (Abstract).
ASCHOFF, J., u. R. WEVER: Beginn und Ende der täglichen Aktivität freilebender Vögel. J. Ornithol. **103**, 2—27 (1962).

AUTRUM, H., u. H. STUMPF: Das Bienenauge als Analysator für polarisiertes Licht. Z. Naturforsch. 5, 116—122 (1950).

BÉKÈSY, G. v.: A new audiometer. Acta Otolaryngol. 35, 411—422 (1947).

BLOUGH, D. S.: Dark adaptation in the pigeon. J. comp. physiol. Psychol. 49, 425—430 (1956).

— Spectral sensitivity in the pigeon. J. opt. Soc. Amer. 47, 827—833 (1957).

BRAEMER, W.: Verhaltensphysiologische Untersuchungen am optischen Apparat bei Fischen. Z. vergl. Physiol. 39, 374—398 (1957).

— A critical review of the sun-azimuth hypothesis. Cold Spr. Harb. Symp. quant. Biol. 25, 413—427 (1960).

BRIDGMAN, C. S., and K. U. SMITH: The absolute threshold of vision in cat and man with observations on its relation to the optic cortex. Amer. J. Physiol. 136, 463—466 (1942).

BROWN, R. H., and J. E. CONKLIN: The lower threshold of visible movement as a function of exposure time. Amer. J. Psychol. 67, 104—110 (1954).

CHARD, R. D.: Visual acuity in the pigeon. J. exp. Psychol. 24, 588—608 (1939).

CRAIK, K. J. W., and M. D. VERNON: The nature of dark adaptation: I. Evidence as to the locus of the process. Brit. J. Psychol. 32, 62—81 (1941).

DICE, L. R.: Minimum intensities of illumination under which owls can find dead prey by sight. Amer. Naturalist 79, 385—416 (1945).

DONNER, K. O.: The visual acuity of some passerine birds. Acta Zool. Fenn. 66, 1—40 (1951).

FECHNER, G. T.: Elemente der Psychophysik. Leipzig: Breitkopf und Härtel 1860.

FRISCH, K. v.: Der Farbensinn und Formensinn der Bienen. Zool. Jb. Abt. Zool. Physiol. 35, 1—182 (1915).

GRAHAM, C. H.: Visual perception in S. S. STEVENS (ed.), Handbk. of Experimental Psychology, pp. 868—920. New York: Wiley 1951.

GRUNDLACH, R. H.: The visual acuity of homing pigeons. J. comp. physiol. Psychol. 16, 327—342 (1933).

HALBERG, F.: Temporal coordination of physiologic function. Cold Spr. Harb. Symp. quant. Biol. 25, 289—308 (1960).

HAMILTON, W. F., and J. L. GOLDSTEIN: Visual acuity and accomodation in the pigeon. J. comp. physiol. Psychol. 15, 193—197 (1938).

HASLER, A. D., and J. A. LARSEN: The homing salmon. Sci. Amer. 193, 72—76 (1955).

HOFFMANN, K.: Experimental manipulation of the orientational clock in birds. Cold Spr. Harb. Symp. quant. Biol. 25, 379—387 (1960).

HOLST, E. v.: Die Arbeitsweise des Statolithenapparates bei Fischen. Z. vergl. Physiol. 32, 60—120 (1950).

KRAMER, G.: Orientierte Zugaktivität gekäfigter Singvögel. Naturwissenschaften 37, 188 (1950).

— Experiments on bird orientation and their interpretation. Ibis 99, 196—227 (1957).

LEIBOWITZ, H. W., N. A. MYERS and D. A. GRANT: Radial localization of a single stimulus as a function of luminance and duration of exposure. J. opt. Soc. Amer. 45, 76—78 (1955).

MATTHEWS, G. V. T.: The sensory basis of bird navigation. J. Inst. Nav. 4, 260—275 (1951).

— Bird Navigation. Cambridge, England: Univ. Press 1955.

McCLEARY, R. A., and J. J. BERNSTEIN: A unique method for control of brightness cues in study of color vision in fish. Physiol. Zool. 32, 285—292 (1959).

PENNYCUICK, C. J.: The physical basis of astro-navigation in birds: theoretical considerations. J. exp. Biol. 37, 573—593 (1960).

PUMPHREY, R. J.: Sensory organs: vision. Ch. XV *in* A. J. MARSHALL (ed.), Biology and comparative physiology of birds. Vol. II. New York: Academic Press 1961.

RATLIFFE, F., and D. S. BLOUGH: Behavioral studies of visual processes in the pigeon. USN, ONR. Tech. Rep. Proj. NR 140-072.

RENNER, M.: The contribution of the honeybee to the study of time-sense and astronomical orientation. Cold Spr. Harb. Symp. quant. Biol. 25, 361—367 (1960).

SAUER, F.: Die Sternenorientierung nächtlich ziehender Grasmücken *(Sylvia atricapilla, borin* und *curruca).* Z. Tierpsychol. 14, 29—70 (1957).

SCHMIDT-KOENIG, K.: Internal clocks and homing. Cold. Spr. Harb. Symp. quant. Biol. 25, 389—393 (1960).

SLOAN, L. L.: The effect of intensity of light, state of adaptation of the eye, and size of photometric field on the visibility curve. Psychol. Monogr. 38, 1 (1928).

SPENCE, K. W.: Visual acuity and its relation to brightness in chimpanzee and man. J. comp. Psychol. 18, 333—361 (1934).

WALD, G.: Human vision and the spectrum. Science 101, 653—658 (1945).

WARDEN, C. J., T. N. JENKINS and L. H. WARNER: Comparative Psychology. Vol. I. Principles and methods. New York: Ronald 1935.

WATERMAN, T. H.: Light sensitivity and vision *in* T. H. WATERMAN (ed.), Physiology of Crustacea. New York: Academic Press 1960.

WEINSTEIN, B., and W. F. GRETHER: A comparison of visual acuity in the rhesus monkey and man. J. comp. Psychol. 30, 187—195 (1940).

Bikomponenten-Theorie der Orientierung[*]

Von H. Mittelstaedt

Max Planck-Institut für Verhaltensphysiologie, Seewiesen (Obb.)

Mit 2 Abbildungen

Der erste Teil behandelt die Sachverhalte und Überlegungen, die zur Aufstellung der Bikomponenten-Theorie geführt haben (Mittelstaedt, 1961), der zweite gibt Hinweise für ihre Gültigkeit in einigen Einzelfällen.

Ausgangspunkt ist eine Theorie (1) der statischen Orientierung, die von v. Holst und Mittelstaedt (1950) aufgestellt, von Mittelstaedt (1951) regelungstheoretisch formuliert und von Birukow (1954) und Jander (1957) auf die Lichtorientierung (Photomenotaxis) der Insekten übertragen wurde. Nach ihr wird die Orientierung eines Organismus, d. h. die Erhaltung einer bestimmten Winkelbeziehung einer oder mehrerer Achsen des Organismus zur Richtung eines orientierenden Reizes, durch einen stetigen Regelvorgang bewirkt: Eine Abweichung (x) von dieser „Grundstellung" (um die eine Achse, die wir jeweils betrachten) wird zunächst in eine dem Sinus der Abweichung proportionale „afferente Variable" überführt, die dann eine „Drehtendenz" (r) hervorruft, die ihrerseits die Geschwindigkeit (u) bestimmt, mit der die Effectoren den Organismus im Sinne einer Verringerung der Abweichung drehen. Abweichungen von der Grundstellung können einmal durch mechanische Einwirkung aus der Umwelt entstehen, zum anderen aber auch durch zentralnervöse Aktion, und zwar dadurch, daß eine zentralnervöse Führungsgröße, das „Drehkommando" (w), zur afferenten Variablen hinzuaddiert wird. Solange das Drehkommando einwirkt, hält der Regelkreis dann eine andere, „kommandierte", Winkelbeziehung ein. Denn die Drehtendenz (r) hängt von der Abweichung (x) und dem Drehkommando (w) nach

$$r = w + \sin x \qquad (1)$$

ab, und die Regelung sorgt dafür, daß r im eingeschwungenen, orientierten Zustand Null wird. (Hier und im folgenden wird nur dieser, den

---

[*] Kurzfassung eines Vortrages auf dem Symposium über Orientierung der Tiere in Garmisch-Partenkirchen, 21. September 1962.

orientierten Zustand bestimmende Teil des Regelungsprozesses mathe-
matisch formuliert. Man beachte, daß jeweils alle in den Gleichungen
vorkommenden Größen für *Wirkungsgrößen* stehen, so daß sich das
Wirkungsgefüge unmittelbar aus der Gleichung ablesen läßt.)

Es wird gezeigt, daß ein solcher Mechanismus mit wachsenden
kommandierten Abweichungen von der Grundstellung zunehmend
schlechter und ab ± 90° überhaupt nicht mehr funktioniert (infolge

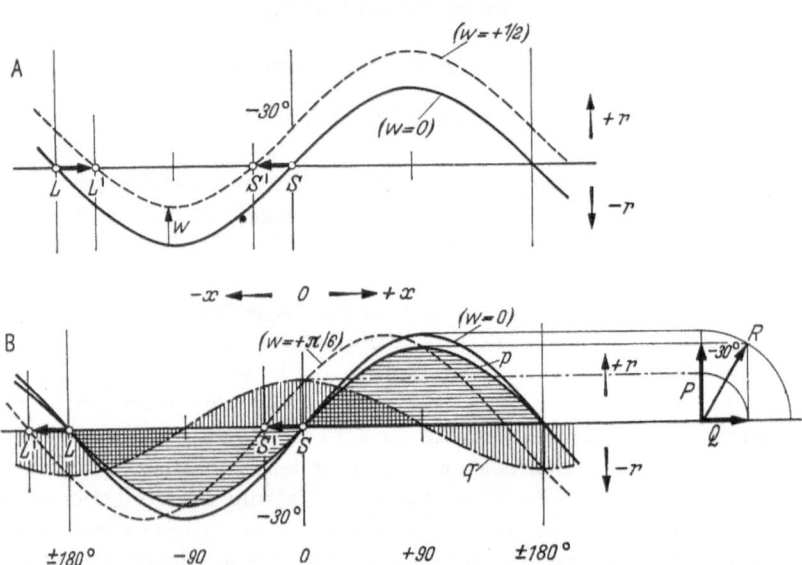

Abb. 1. Die „Drehtendenz" r ($+$ = nach rechts, — = nach links, willkürliche Einheiten; r max = 1) wird als
Funktion der Abweichung x von der Grundstellung dargestellt ($+$ = Lichtquelle rechts von der Körperachse),
und zwar für ein Insekt, welches die Tendenz hat, die Längsachse zur Lichtquelle zu richten (Führungsgröße
$w = 0$) und für ein Insekt, welches eine Abweichung x von $-30°$ einhalten will. A zeigt, wie dies bei der
Theorie (1) einfacher Überlagerung der Führungsgröße mit einer sinusförmigen afferenten Variablen geschieht,
B nach der Theorie (3) der reziproken Modulation zweier Komponenten. S, L Stabilitäts- und Labilitäts-
stellung für $w = 0$. S', L' Stabilitäts- und Labilitätsstellung für $w = +^1/_2$ in A und $w = + \pi/6$ in B. Die Kom-
ponenten der Abweichung nach der reziproken Modulation (s. Abb. 2), die Produkte p (dünne ausgezogene
Kurve) und q (strich-punktierte Kurve) sind einzeln und als Summe ($p + q = r$; gestrichelte Kurve) gezeigt.
Rechts neben der Darstellung in rechtwinkligen Koordinaten sind die Größen als Vektoren (P, Q, R) in
Polarkoordinaten dargestellt. Man beachte, daß die Stabilitäts- und Labilitätsstellungen sich in A entgegen-
gerichtet verschieben, in B hingegen in gleicher Richtung (aus MITTELSTAEDT, 1962)

„Vertikalverschiebung" der Drehtendenzen, siehe Abb. 1A). Diese
Theorie (1) kann deshalb u. a. nicht bei solchen Organismen zutreffen, die
sich nach einer erzwungenen Abweichung stets über den kleineren
Winkel in die kommandierte Stellung zurückdrehen. Zwei grundsätzlich verschiedene Lösungen dieses Problems werden vorgeschlagen:

(2) Die afferente Variable ist der Abweichung selbst proportional
(und nicht deren Sinus), und die — notwendige — Sinustransformation
findet erst *nach* der Addition des Drehkommandos statt, also:

$$r = \sin(x + w) . \tag{2}$$

(3) Es wird außer der dem Sinus proportionalen eine zweite, dem Cosinus der Abweichung proportionale, afferente Variable gebildet. Das Drehkommando beeinflußt stets beide Variable und zwar dergestalt, daß es die eine graduell unterdrückt, wenn es die andere fördert und umgekehrt („Reziproke Modulation"). Dieser Prozeß verläuft ideal, wenn das Drehkommando ebenfalls in eine Sinus- und eine Cosinuskomponente aufgespalten und so mit den beiden afferenten Komponenten „über Kreuz" multipliziert wird. Anschließend werden beide Produkte summiert als Drehkommando den Effectoren zugeleitet. Nach dieser „Bikomponenten"-Theorie (3) hängt die Drehtendenz (r) also von der Abweichung (x) aus der Grundstellung und dem Drehkommando (w) folgendermaßen ab:

$$r = K_s \sin x \cdot \cos w + K_c \cos x \cdot \sin w \qquad (3)$$

(dabei sind K_s und K_c Koeffizienten, die sich experimentell bestimmen lassen). Den Signalflußplan eines solchen Wirkungsgefüges zeigt Abb. 2.

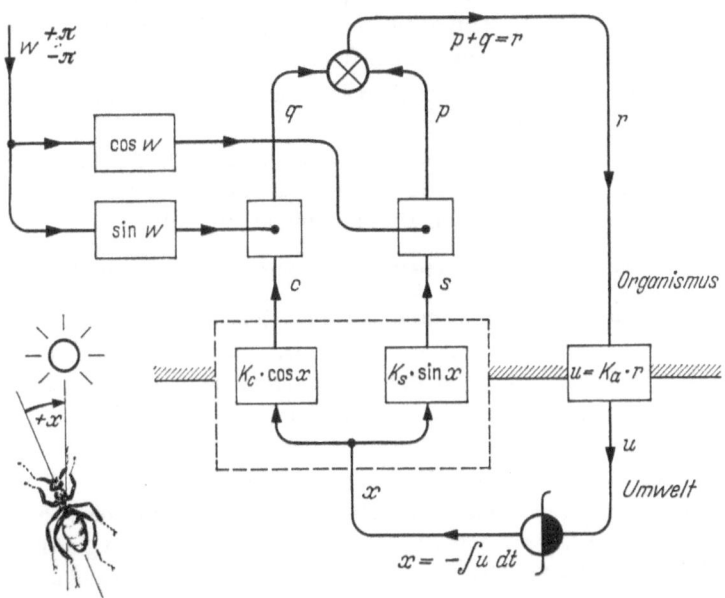

Abb. 2. Signalflußplan der Bikomponentenmodulation. x Abweichung des Tieres von der Grundstellung zur Reizquelle; w Führungsgröße (Drehkommando); c, s Cosinus- und Sinuskomponenten des Lichtwinkels; q, p dieselben, nach reziproker Modulation durch w; r Drehtendenz; u Drehgeschwindigkeit des Insekts relativ zum Untergrund (plus: Rechtsdrehung). Der schwarze Halbkreis symbolisiert die physikalische Beziehung zwischen u und x (Integration und Vorzeichenwechsel) die Quadrate mit dem Punkt symbolisieren die Multiplikation der Eingangsgröße mit der zum Punkt verbundenen Größe (z. B. $q = c \times \sin w$); über die Koeffizienten s. Text (aus MITTELSTAEDT, 1962)

Abb. 1B gibt ein Beispiel für die Drehtendenz (r) in Abhängigkeit von der Abweichung (x) bei einem Drehkommando $w = \pi/6$ („Horizontalverschiebung" statt „Vertikalverschiebung" der Drehtendenzkurve).

Die wichtigsten Hinweise für die Gültigkeit der Theorie (3) im konkreten Fall sind folgende:

I. Auf dem Gebiet der statischen Orientierung:

a) Es existieren Receptoren, die ihrer Anordnung und ihrer Kennlinie nach (Löwenstein und Roberts, 1950) als Cosinusgeber geeignet sind, nämlich Sinneshaare in den Lagenae bzw. Sacculi.

b) Die Ergebnisse von v. Holst und L. Schoen (1950) nach Ausschaltung der Lagenae bzw. Utriculi bei Fischen lassen sich im Sinne der Bikomponenten-Theorie deuten.

c) Das Ergebnis der Ausschaltung der Utriculi beim „Schrägsteher" *Poecilobrycon eques* (Braemer und Braemer, 1957), das nach (1) und (2) unverständlich ist, entspricht der Erwartung nach (3).

d) Die Unabhängigkeit der Rumpflagereflexe bei der Taube von der Kopflage bleibt bei allen Rumpflagen und bei Variation der mechanischen Feldstärke (bis 2 g) erhalten (Mittelstaedt, 1962, unveröff.).

Krebse und Wasserkäfer-Larven (Schöne, 1955, 1962) zeigten indessen das der Abb. 1A entsprechende, auf einfache additive Sollwertverstellung, Theorie (1), hinweisende Verhalten.

II. Auf dem Gebiet der optischen Richtungsorientierung, der Transpositionen und des Sonnenkompasses:

a) Wenn man annimmt, daß die Sinus- und Cosinus-Komponenten von Licht- und Schwerefeld-Abweichung (β bzw. α) gleichnamig summiert und dann von demselben Drehkommando moduliert werden, also:

$$r = (F_s \sin\alpha + L_s \sin\beta)\cos w + (F_c \cos\alpha + L_c \cos\beta)\sin w \qquad (3\,\mathrm{b})$$

und weiter, daß die Koeffizienten von der Reizintensität (L_s u. L_c von der Stärke der Lichtquelle, F_s u. F_c von der Stärke des Schwerefeldes) nach einer Funktion abhängen, die — wie immer sie im übrigen aussehen mag — Null ist, wenn die betreffende Reizintensität Null bzw. unterschwellig ist, erhält man das Ergebnis der Transpositionsversuche an Bienen und Ameisen (v. Frisch, 1948—1962; Vowles, 1954).

b) Unter derselben Annahme erhält man die Ergebnisse von v. Frisch (1948, 1962) bei allein optischer (β_1), allein statischer (α_1) und gleichzeitig statischer und optischer Orientierung (α_2, β_2), nämlich die Beziehung:

$$\beta_1 = \alpha_1 = \frac{\alpha_2 + \beta_2}{2} \quad (\text{„Restmißweisung" gleich Null}$$

angenommen).

Exakt liefert die Theorie (3) dieses Ergebnis allerdings nur unter der Voraussetzung, daß die jeweils von Null verschiedenen der 4 Koeffizienten stets gleich groß sind ($F_s = F_c = L_s = L_c$ bei F, $L > 0$). Dazu paßt gut die Tatsache, daß die gemessenen Abweichungen bei der Biene unter allen diesen Bedingungen oberhalb der Reizschwelle von der Intensität der Lichtquelle unabhängig sind (Metataxis, vgl. JANDER, 1963). Zu Theorie (1) in der Form:

$$r = F \sin\alpha + L \sin\beta + w \qquad (1b)$$

steht das Ergebnis von v. FRISCH in eindeutigem Widerspruch, mit Theorie (2) verträgt es sich nur unter bisher nicht verifizierten Annahmen, nämlich:

$$r = L \sin(\beta + w) + F \sin(\alpha + w) . \qquad (2b)$$

c) Die Fälle, in denen *kommandierte* Abweichungen intensitäts-unabhängig, *reizbedingte* Abweichungen (z. B. bei Divergenz von Schwere-feld- und Lichtrichtung) aber intensitätsabhängig sind, widersprechen Theorie (1), lassen sich aber, wenn L und F in (2b) intensitätsproportional sind, mit Theorie (2) vereinen. Aus Theorie (3) ergeben sie sich, wenn beide Koeffizienten in gleicher Weise von der Reizintensität abhängen ($K_s/K_c =$ konstant). So folgen die Ergebnisse an *Mysis* (JANDER, 1962), die in diese Kategorie gehören, aus der Bikomponenten-Theorie (3) unter der ohnehin plausiblen Annahme:

$$\frac{F_s}{F_c} = \frac{L_s}{L_c} = 1 .$$

d) Die Fälle, in denen sowohl kommandierte wie reizbedingte Ab-weichungen intensitätsabhängig sind (Protaxis, vgl. JANDER, 1963), lassen sich am einfachsten mit Theorie (1) erklären. Zu Theorie (2) stehen sie in Widerspruch. Mit der Bikomponenten-Theorie (3) sind sie nur dann verträglich, wenn man annimmt, daß bei diesen Tieren die Reizintensität beide Koeffizienten K_s und K_c reziprok, und zwar zu-gunsten der Sinus- *oder* der Cosinuskomponente verändert. Es bleibt abzuwarten, ob sich eine solche Annahme experimentell rechtfertigen läßt.

e) Die Bikomponententheorie liefert eine besonders naheliegende Erklärung für den Mechanismus des „Sonnenkompasses" durch die Ein-führung (MITTELSTAEDT, 1962) der vierfachen Modulation durch Cosinus-und Sinuskomponenten der „Inneren Uhr". Die erweiterte Theorie erklärt mehrere scheinbar widersprüchliche Ergebnisse, u. a. warum die nächtliche Kompensation der (nie gesehenen) Sonnenwanderung in gemäßigten Breiten im selben Drehsinn wie tagsüber, in den Tropen aber im Gegensinn erfolgt (W. BRAEMER, 1962, briefliche Mitteilung und dieser Bericht).

Auf dem Gebiet der optischen Orientierung sind einige Phäno-mene beobachtet worden, die mit keiner der drei hier besprochenen Theorien verträglich scheinen. Zur Zeit ist nicht zu übersehen, ob sie sich durch Zusatzannahmen einordnen lassen werden oder ob sie auf ein grundsätzlich verschiedenes Wirkungsgefüge hinweisen.

Literatur

BIRUKOW, G.: Photo-Geomenotaxis bei *Geotrupes silvaticus* PANZ. und ihre zentral-nervöse Koordination. Z. vergl. Physiol. **36**, 176 (1954).

BRAEMER, W., u. H. BRAEMER: Zur Gleichgewichtsorientierung schrägstehender Fische. Z. vergl. Physiol. **40**, 529—542 (1958).

FRISCH, K. VON: Gelöste und ungelöste Rätsel der Bienensprache. Naturwissenschaften **35**, 12—23; 38—43 (1948).

— Die Polarisation des Himmelslichtes als orientierender Faktor bei den Tänzen der Bienen. Experientia (Basel) **5**, 142—148 (1949).

— Die Sonne als Kompaß im Leben der Bienen. Experientia (Basel) **6**, 210—221 (1950).

— Orientierungsvermögen und Sprache der Bienen. Naturwissenschaften **38**, 105 bis 112 (1951).

— Sprache und Orientierung der Bienen. Dr. Albert Wander-Gedenkvorlesung. 3. Bd. Berne, Switzerland: Verlag H. Huber 1961.

— Über die durch Licht bedingte „Mißweisung" bei den Tänzen im Bienenstock. Experientia (Basel) **18**, 49—53 (1962).

HOLST, E. VON: Die Arbeitsweise des Statolithenapparates bei Fischen. Z. vergl. Physiol. **32**, 60—120 (1950).

— u. H. MITTELSTAEDT: Das Reafferenzprinzip (Wechselwirkungen zwischen Zentralnervensystem und Peripherie). Naturwissenschaften **37**, 464—476 (1950).

JANDER, R.: Die optische Richtungsorientierung der roten Waldameise *(Formica rufa L.).* Z. vergl. Physiol. **40**, 162—238 (1957).

— The swimming plane of crustacean *Mysidium gracile* (Dana). Biol. Bull. **122**, 380—390 (1962).

— Insect orientation. Ann. Rev. Entomol. **8**, 95—114 (1963).

LÖWENSTEIN, O., u. T. D. M. ROBERTS: Gleichgewichtsfunktion der Otolithen. J. Physiol. (Lond.) **110**, 392—415 (1949).

MITTELSTAEDT, H.: Zur Analyse physiologischer Regelungssysteme. Verh. Dtsch. Zool. Ges. Wilhelmshaven 1951, 150—157 (1951).

— Probleme der Kursregelung bei freibeweglichen Tieren. Aufnahme und Verarbeitung von Nachrichten durch Organismen, 138—148. Stuttgart: Hirzel-Verlag 1961.

— Control systems of orientation in insects. Ann. Rev. Entomol. **7**, 177—198 (1962).

SCHOEN, L., u. E. VON HOLST: Das Zusammenspiel von Lagena und Utriculus bei der Lageorientierung der Knochenfische. Z. vergl. Physiol. **32**, 552—571 (1950).

SCHÖNE, H.: Kurssteuerung mittels der Statocysten (Messungen an Krebsen). Z. vergl. Physiol. **39**, 235—240 (1957).

— Optisch gesteuerte Lageänderungen (Versuche an Dytiscidenlarven zur Vertikalorientierung). Z. vergl. Physiol. **45**, 590—604 (1962).

VOWLES, D. M.: The orientation of ants. I. The substitution of stimuli. J. exp. Biol. **31**, 341 (1954).

— The orientation of ants. II. Orientation to light, gravity and polarized light. J. exp. Biol. **31**, 356 (1954).

Innate and Learned Components in the Astronomical Orientation of Wolf Spiders[*, **]

By F. Papi and P. Tongiorgi

Istituto di Biologia generale, Università di Pisa (Italia)

With 15 Figures

Contents

I. Introduction

After the discovery of the astronomical orientation of the sandhopper *Talitrus saltator* (Pardi and Papi, 1952, 1953; Papi and Pardi, 1953), a good deal of research work has been done on different arthropods which inhabit sea shores or river banks, to ascertain what is the diffusion rate of the phenomenon, and to enquire further into the mechanisms of this type of orientation. Particular attention has been paid to the Wolf Spiders (Lycosidae) of the genus *Arctosa*. These are found to be very suitable for analysis from the point of view of phenomena related to astronomical orientation, as a result of their biological and ecological characteristics (Papi, 1955a and b, 1959; Papi and Serretti, 1955; Papi, Serretti and Parrini, 1957; Tongiorgi, 1959).

Some species of *Arctosa* live on river banks or on sandy beaches and demonstrate their capacity to orient themselves in such a way that, on

---

* Presented by F. Papi.

** This research work was supported by the Rockefeller Foundation (GA MNS 5938).

finding themselves on the water, they can maintain a constant orientation towards the azimuth coinciding with the direction straight towards the bank. Every population has its own theoretical escape direction, which corresponds to that inland on the bank it inhabits. When the sky is covered, the spiders orient themselves visually from objects in their surroundings, while under a clear sky they use the direction of the sun or the plane of polarisation of the light from the blue sky, making with it an angle which varies continously in the course of the day. As with many other animals capable of orienting themselves astronomically, the diurnal variation in angle of orientation depends on an endogenous rhythm, or internal clock, which is synchronised with the natural rhythm of illumination.

After it was shown that *Arctosa* is capable of learning a new direction of escape (Papi, Serretti and Parrini, 1957), we engaged in a programme of research to study the components of orientation, both innate and learned, and to look into the learning mechanism of escape direction of *Arctosa*. We wish to report in the following the principal results obtained.

To understand better what follows, we wish to restate the methods used for testing the animals and for evaluating the precision with which they can orient themselves. The spiders are tested under clear sky, putting them singly, usually for 2 minutes, in a cylindrical glass basin containing water, which is subdivided into 16 sectors by means of 8 diameters. The animals are capable of skating rapidly on the surface of the water. In an attempt to escape they dash repeatedly against the walls of the basin. Every collision (an escape attempt) is recorded together with the sector in which this takes place. From the distribution of n escape attempts, it is possible to calculate the direction (r, singular resultant direction) of the mean vector and its length or, which gives a measure of the dispersion of the attempts at escape. If we number the sectors from 0 to 15 starting from that in the N (North) direction and proceeding clockwise, we can then apply the following formulae:

$$\operatorname{tang} r = \frac{\sum_{i=0}^{15} n_i \cdot \sin i \cdot \dfrac{360°}{16}}{\sum_{i=0}^{15} n_i \cdot \cos i \cdot \dfrac{360°}{16}}$$

$$or = \sqrt{\left(\sum_{i=0}^{15} n_i \cdot \sin i \cdot \frac{360°}{16}\right)^2 + \left(\sum_{i=0}^{15} n_i \cdot \cos i \cdot \frac{360°}{16}\right)^2}$$

where n_i is the number of escape attempts registered in the ith sector. The direction of the vector is then expressed in degrees measured in a clockwise direction from N, while its length or varies between 0 and 1 and tends to 0 with the growth in dispersion of the escape attempts.

An example is shown in fig. 1 A, which represents a 2 minute test of a specimen of *Arctosa variana* found on the north bank of Fiume Morto near Pisa (the theoretical direction of escape being 2°). In its displacement from the escape direction expected, the dispersion, and the number of such escape attempts, this specimen shows a very similar behaviour pattern to the spider population, in general, that comes from that river bank.

To evaluate other tests at the same time one can proceed in an analogous fashion considering the overall distribution of escape attempts. However, instead of this method which we have used in the past, we prefer another one in which a count is made of both the direction r and the length or of the vectors, calculated according to the distribution of the escape attempts obtained from the single tests. To calculate the direction R and the length OR of the mean vector of these accumulated distributions the following formulae are used:

$$\tan R = \frac{\Sigma\, or_i \cdot \sin r_i}{\Sigma\, or_i \cdot \cos r_i}$$

$$OR = \sqrt{(\Sigma\, or_i \cdot \sin r_i)^2 + (\Sigma\, or_i \cdot \cos r_i)^2}$$

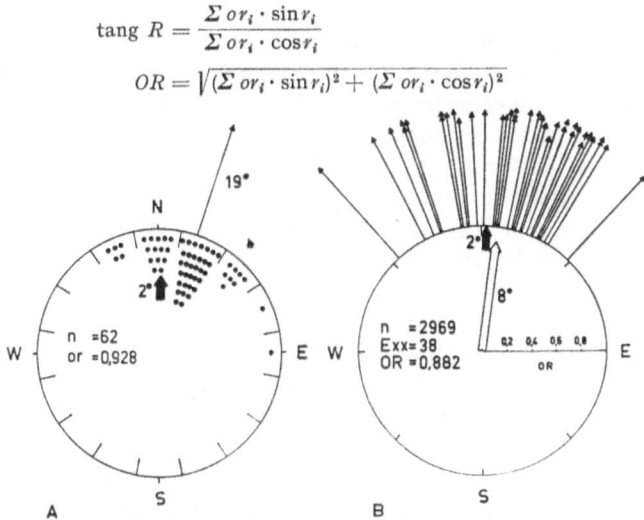

Fig. 1A and B. *A. variana.* A. The result of testing a single specimen having a theoretical escape direction of 2° (black arrow inside circle). The animal made 62 escape attempts (n) in 2 minutes distributed between six sectors and represented by the black dots. The direction of the mean vector (external arrow) is equal to 19°, its length (or) is 0.928. B. The result of testing 38 specimens all having a theoretical escape direction of 2° (black arrow inside). The arrows outside represent the directions and lengths of the mean vectors calculated on the basis of the single tests; the white arrow inside gives the direction and the length of the mean vector of the accumulated distribution. n = total number of escape attempts recorded. $Exx.$ = number of the specimens. OR = length of the mean vector of the accumulated distribution. The scale is valid for measuring the length of all vectors

This method also permits us to evaluate whether an accumulated distribution is statistically different from a random one, by means of the methods elaborated by GREENWOOD and DURAND (1955) and by DURAND and GREENWOOD (1958) (cf. also SCHMIDT-KOENIG, 1961).

An example of this is given in Fig. 1 B, in which a graphical representation of the orientation of a group of specimens of *A. variana* is displayed, all of which possessing the same theoretical direction of flight.

Three species of genus *Arctosa* were used for the experiments: *A. variana* C. L. KOCH, *A. cinerea* (FABR.) and *A. perita* (LATR.). For information on the ecology see PAPI (1959).

II. The orientation of specimens hatched and reared in the laboratory

The female of the *Arctosa*, like those of the other Lycosids, lays the eggs in a cocoon which she carries attached to the spinnerets. After the hatching the young mount on the back of the mother, but subsequently abandon her after a few days. By keeping fertilised females in captivity, or those already possessing an egg-cocoon at the time of capture, we were able to obtain several hatchings in the laboratory. The young were reared singly in glass tubes containing damp sand and were subjected to a natural illumination rhythm. However, only during the periods of the tests were they exposed to direct sunlight, or light coming from the sky.

The larger part of the experiments were performed on *A. variana*. The young came from nine egg-cocoons obtained from different females. Four of them were collected on the N bank of Fiume Morto near Pisa and had a theoretical escape direction equal to 2°, three came from the S (South) bank of the same river and had a theoretical escape direction of 182°, and, finally, two others were captured on the banks of the river Chioma (Province of Leghorn). On this river (for further details see later) the spiders orient themselves in various directions, according to the point of the bank inhabited, or else exhibit an orientation which is quite imprecise.

551 young specimens were used in 2,169 tests of 2 minutes each, during the course of which 59,747 escape attempts were recorded. The tests were performed for different ages of the specimens, varying between 0 and 68 days.

It proved essential to test such a large number of specimens due to the strong variability in their behaviour. In each test the dispersion of escape attempts was not usually much different from that of specimens found in nature. This shows that each young spider, in the course of a single test, generally demonstrates a preference for a well-determined direction, distributing its escape attempts between a few adjacent sectors. This chosen direction varies strongly, however, from one individual to another, especially for the younger specimens for which the resultant directions have a larger dispersion.

We have few relative experiments for animals of a few days old, and so are unable to give a definite judgement on their orientation. Often a negative phototactic tendency seems to accumulate the resultant single directions towards the azimuth opposite to the sun; on other occasions they seem to be distributed in a random fashion. With a growth in age however, a tendency to go towards N increases, independently of the bank from which the mother comes, or of her orientation. The negative phototactic tendency remains and manifests itself with a certain irregularity, with accumulation of r in sectors near W in the morning and in sectors near E in the afternoon. Very occasionally the opposite tendency appears to show up, that is to say, a positive phototactic tendency. However, at least towards the thirtieth day of the spiders' life, sometimes even much before, the tendency to direct itself towards N prevails over every other (Figs. 2—4).

It seems important to us to emphasize that the tendency towards N is manifested throughout the whole of the day and that there does not exist important differences in the behaviour of the offspring of the females of the N and S banks of Fiume Morto and River Chioma.

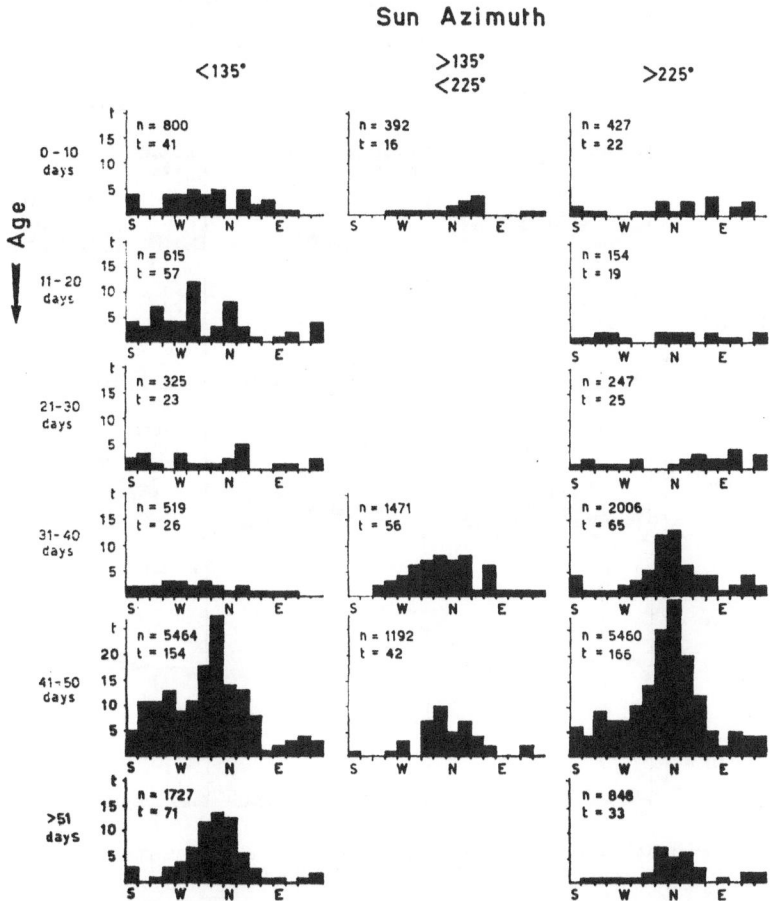

Fig. 2. *A. variana*. Orientation of 193 offspring of females from the N bank of Fiume Morto, hatched and reared in the laboratory. In each diagram: the ordinates give the number of single resultants, the abscissae give the sectors in which they occur. The diagrams are ordered according to the age of the young (horizontal lines) and to the sun's position at the time of the tests (vertical lines). n = number of escape attempts recorded; t = number of tests executed, each test having a duration of 2 minutes

Observations have also been made on the young of *A. cinerea*. The orientation of two hatches, one obtained from a female found on the Tyrrhenian coast near Pisa (with a theoretical direction of escape equal to 90°) and one obtained from a female coming from Henriksberg near Tvärminne in South Finland (theoretical escape direction equal to about 310°) were tested. In both cases a tendency towards N was again observed.

The mechanism by which the young orient themselves towards the N is certainly the same that regulates the orientation of the specimens found in nature, and is based on the compensation of the apparent

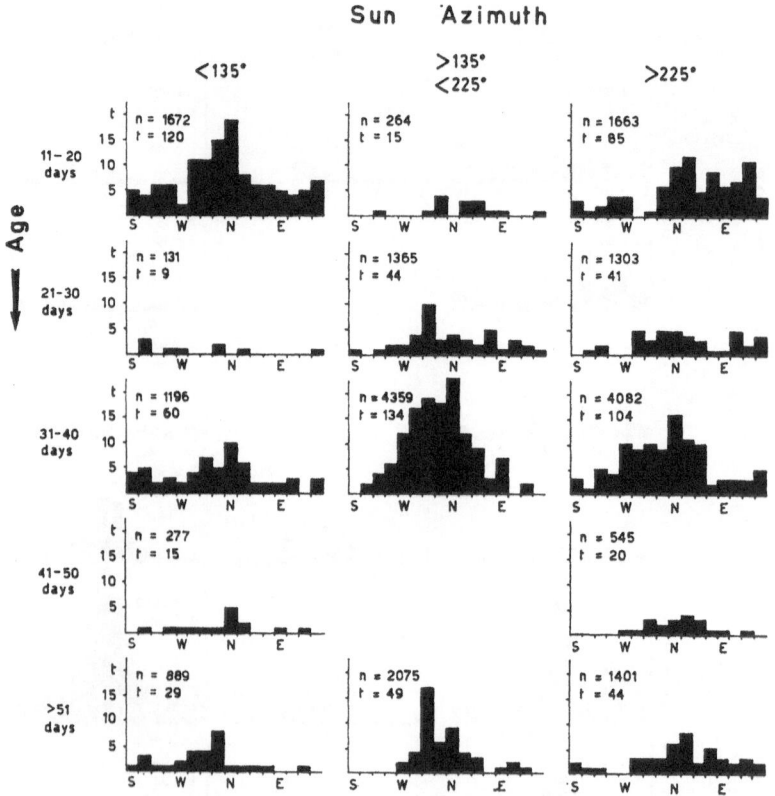

Fig. 3. *A. variana*. Orientation of 180 offspring of females of the S bank of Fiume Morto, hatched and reared in the laboratory. Other explanations as for Fig. 2

movement of the sun. In fact, it has been possible to obtain the anticipated modification of the escape direction also with the young after a phase-shift by six hours of their internal clock, realised by means of a rhythm of artificial illumination.

One can therefore conclude that the mechanism which permits a variation of angle of orientation in the course of the day, in such a manner that the escape direction is maintained relatively constant, is innate and that there does not exist innate differences between the various populations in the directions preferred.

Birds (HOFFMANN, 1953) and fishes (BRAEMER, 1959) also have an innate mechanism which permits compensation for the solar motion. The behaviour of *Arctosa* reminds us, in particular of that of the Hemipter *Velia currens*, which have an innate tendency to direct themselves towards S (BIRUKOW, 1956). For the Talitridae of the sea shores the

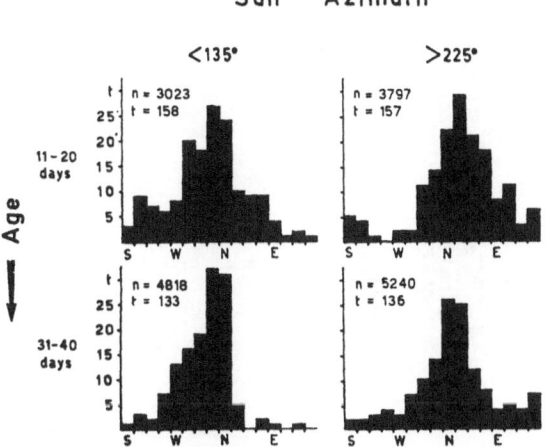

Fig. 4. *A. variana*. Orientation of 178 offspring of females of River Chioma, hatched and reared in the laboratory. Other explanations as for Fig. 2

mechanism of compensation for the solar motion is also innate, but the direction chosen is different from one population to another in relation to the orientation of coast inhabited (PARDI, ERCOLINI, MARCHIONNI and NICOLA, 1958; PARDI, 1960).

With our present state of knowledge, we could only speculate on the reason for which the spiders have an innate tendency to direct themselves towards N and not in other directions. We feel that the development of the tendency to direct themselves towards N, which manifests itself only after a certain time, deserves some attention. It is worthwhile to note that it cannot be a consequence of tests already performed; both because during the test every precaution is taken not to reward the spiders whatever direction they tend to choose, and also because the tendency to direct themselves towards N is manifested also by animals which are tested for the first time, provided that they are not very young. The development of the tendency to direct themselves towards N is therefore an autonomous process which matures independently of experience. Furthermore, it is to be excluded that the younger specimens are unable to orient themselves astronomically towards a determined direction on the grounds of an incomplete development or an imperfect functioning of their receptors or physiological mechanisms (eyes, internal clock, etc.)

given that specimens of the same age, as we shall see, can learn a determined direction of escape very rapidly.

III. The orientation of young spiders found in nature

Since in all populations one finds a similar behaviour of the young hatched and reared in the laboratory, it follows that the animals learn the correct direction of escape in the course of their lives.

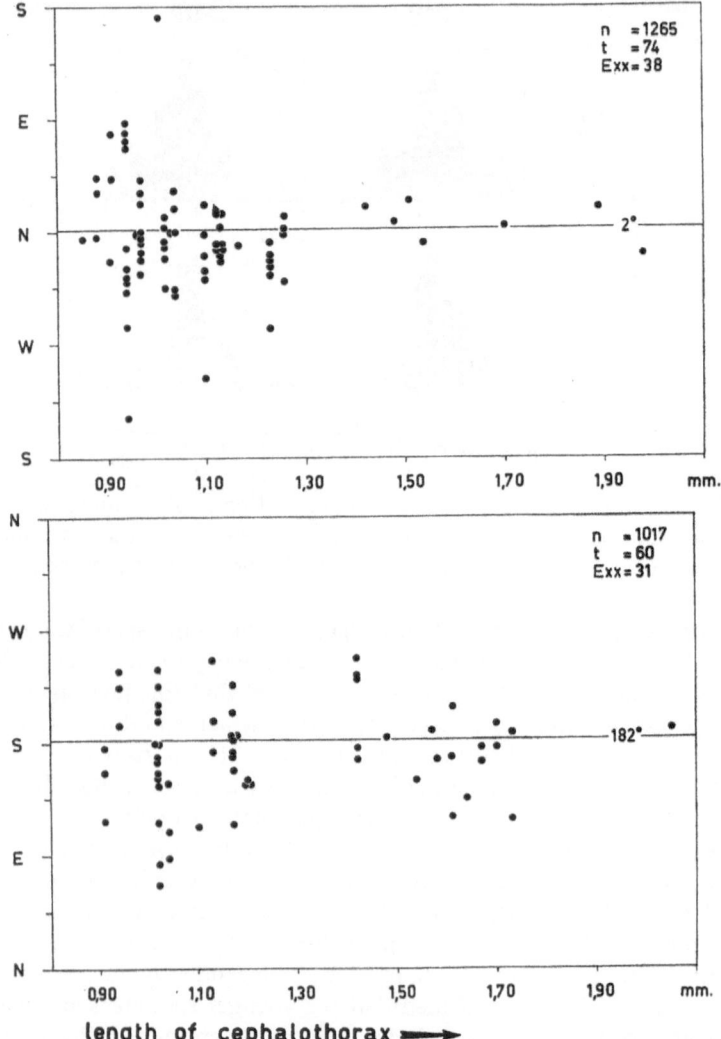

length of cephalothorax ➤

Fig. 5. *A. variana*. Orientation of very young specimens collected in nature. The resultant single directions are represented by black dots and are ordered in accordance with the length of the cephalothorax of the animals. Above, the orientation of the young with theoretical escape direction equal to 2°; below, that of the young with theoretical escape direction equal to 182°. *n* = number of escape attempts recorded; *t* = number of tests effected; *Exx.* = number of specimens involved

To ascertain the manner in which the learning takes place in nature, the smallest specimens that could be found were collected and tested. The young of *A. variana* of the two banks of Fiume Morto (theoretical escape directions equal to 2° and 182°) were studied. However, in so far as the smallest were only a few days old, their orientation was already decidedly different. The young of the N bank already oriented themselves approximately towards the N; the young of the S bank with about the same precision towards the S (Fig. 5). It is not improbable that the precision of orientation improves with age; in this respect, a diminution of the deviation from the expected theoretical escape direction for the larger specimens seems to confirm this hypothesis. In any case, one can conclude that the learning of the direction of escape, at least approximately, is very rapid in nature and occurs during the first few days of life.

IV. The learning of the escape direction in the young

In order to verify what has been deduced in the preceeding arguments, that is to say, that the learning of the escape direction occurs in the first few days of life, we have reared young spiders hatched in the laboratory

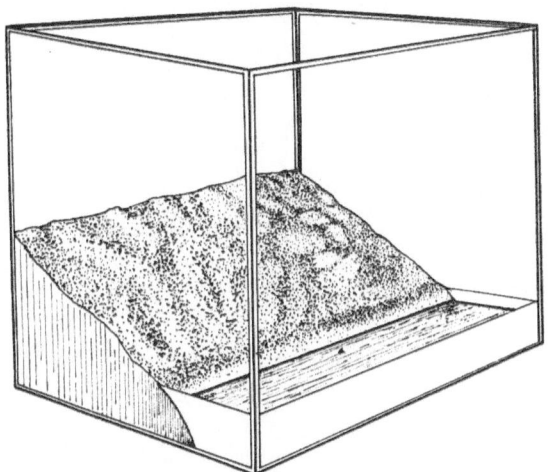

Fig. 6. Tank used in the experiments for the learning of escape directions.
Other explanations are given in the text

in a container in which the conditions of their natural environment were reproduced as far as possible.

For this, glass tanks were used measuring 50 cm × 40 cm and of height 50 cm (Fig. 6). On the bottom, attached to a wall was fixed a bath containing water, while a load of damp sand formed a plane inclined to the horizontal by an amount of about 25°, descending to the water's edge. In order to ascertain whether the presence of water was necessary for the learning of the escape direction, we also used a container without water, containing only damp sand, placed also so as to form an

inclined plane of 25°. It seems worthwhile pointing out that the sand lower down
was kept damper than that higher up, also in this tank.

The tanks were placed and oriented on a terrace where an uninterrupted view
of the sky was available in every direction. It was foreseen that the animals would
learn to direct themselves to that azimuth which, in their tanks, would take them
away from the water and/or take them most rapidly upward.

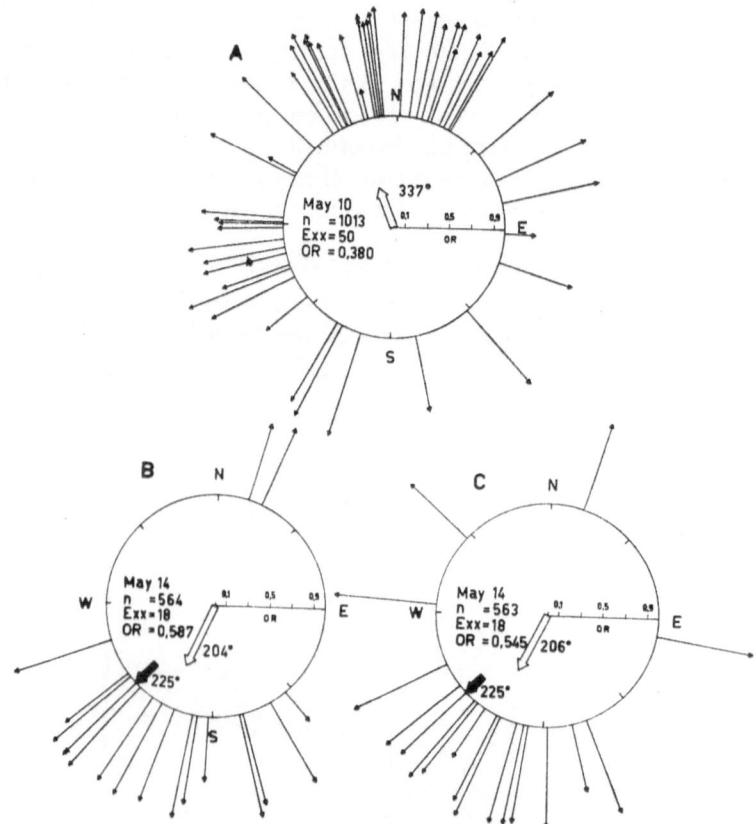

Fig. 7A—C. *A. cinerea*. Experiments on the learning of an escape direction (direction of learning = 225°).
A. Orientation of 50 young one day before being placed in the tanks. B. Orientation of 18 specimens after
remaining about 60 hours in a tank containing a water-bath (Fig. 6). C. Orientation of 18 other specimens
after remaining about 60 hours in a tank without water-bath. Other explanations as for Fig. 1B

The first experiment was performed on *A. cinerea*. On May 3, a
female with her young on her back was taken from the beach at S.
Rossore near Pisa (theoretical direction of escape equal to 90°). The young,
each as they left the mother, were collected and isolated. 50 of them were
tested on May 10; the dispersion of *r* was strong, but with an evident
accumulation towards N (26 *r* out of 50 falling between NW and NE)
(Fig. 7A). As was anticipated, the time spent in nature on the back of

their mother did not influence them in their choice of orientation. In fact, they oriented themselves in the same way as those of the same age hatched in laboratory. On May 11, at 20.00 hours, one half of the animals were introduced into the tank that did not contain water, the other half were placed in the tank containing the water-bath. Both the tanks were oriented in such a way that the direction of learning was SW (225°). On May 14, 18 specimens were still living in both the tanks, and were removed between 7.00 and 8.00 hours and tested shortly afterwards. The orientation of the two groups was very similar, and two thirds of the spiders chose directions approximating to that expected (Fig. 7 B—C). The resultant directions of the accumulated distributions deviated only 19°—21° from that expected; the general distributions were different from random ones with $p < 0.01$.

Other experiments were performed on *A. variana* and showed that the young were able to learn the escape directions expected within two to three days of leaving their mother.

It is therefore seen that the learning of the correct escape direction occurs during the very first few days of autonomous existence.

As regards the mechanism of learning, up to now we have not got much experimental data. The young inexperienced ones, during the tests are strongly attracted to dark objects placed near the testing basin, even when the sky is perfectly clear. It seems to be true that, if those in nature find themselves on water, they direct themselves by sighting the bank, as spiders of all ages do when the sky is covered over. But, as we have seen, direct contact with water is not necessary for the learning of the escape direction. Observations in nature confirm that the young of *A. cinerea* are seldom in contact with water. We believe that the young, as soon as they leave the mother, are able to direct themselves on land following the line of maximum inclination, that is to say perpendicular to the bank. This behaviour could be due to the gradient in humidity or, what is much more likely, to the geotactic orientation.

The manner in which the animals learn to direct themselves also astronomically, in that direction which carries them upward from the bank, remains to be investigated.

Finally, it is interesting to point out that the animals which were in the tank containing only sand, and that had learnt the direction "inland" without ever coming in contact with water, had then chosen this direction when for the first time they were placed on water.

V. Modification of orientation induced by captivity, and disorientation induced experimentally

We can now ask the following: once the spiders have learnt to orient themselves correctly in their natural habitat, whether this orientation remains invariant for the rest of their lives or whether it can be modified due to other experience. Reference has already been made elsewhere to specimens of *A. variana* held in captive for twenty-one days (PAPI, SERRETTI and PARRINI, 1957). It is found that if the animals are kept under a natural rhythm of illumination one observes at the end of the captivity that there still remains a prevalent orientation towards the escape direction which was present at the moment of capture. However, the orientation becomes less precise due to marked differences in tests made at different times of the day and due to a greater dispersion of the resultant single directions. An example of the effect of a brief captivity is given in Fig. 8A—B, where the orientation of 17 specimens of *A. variana* immediately after their capture and after twelve days of imprisonment is represented.

Specimens of *A. variana* were also kept in captivity from autumn to spring for five and half months. In tests made in spring there was still found a prevalent orientation in the original direction, even though imprecise. Occasional observations confirm that even in nature, spiders, at the end of hibernation, possess an orientation less precise which is rapidly improved in the course of the good season.

The workers of the ant *Formica rufa* and the Rove Beetle *Pedaerus rubrothoracicus* were unable to orient themselves astronomically at the end of their hibernation (JANDER, 1957; ERCOLINI and BADINO, 1962). In the case of *Arctosa* it has not been possible to decide whether these modifications in orientation were the result of lack of exercise of the capacity of orientation, or whether it should be attributed to other causes (e.g. an imperfect synchronisation of the internal clock in captivity and at the end of hibernation).

The alteration in capacity of orientation, which is found when the animals are kept repeatedly and for long periods on water trying in vain to reach the "bank", is certainly a result of experience. Specimens of *A. variana*, with escape direction towards N and towards S, were placed six times a day for 10 minutes each in vessels containing water. The experiments were performed under a clear sky and the animals, as in the normal tests, tried to escape orienting themselves astronomically, repeatedly battering themselves against the sides of the vessel. In general the spiders alternated series of escape attempts with ever increasing pauses, so that the activity decreased in the course of a 10 minute experiment. After several days the orientation of these animals differed considerably from that of the control animals. Either they tended in some other direction or had on the whole a random distribution. Fig. 8 summarises the results obtained with specimens taken from the S bank of Fiume Morto (theoretical escape direction 182°). The group of animals subjected to treatment appeared to be completely disoriented at the end of the experiment (Fig. 8D); also the length of the mean individual vectors was strongly reduced in many cases.

In a similar experiment performed on animals with theoretical escape direction of 2° such disorientation was obtained after four days, an din the following days notwithstanding the continuation of the treatment, there was a marked tendency to direct themselves towards S and W.

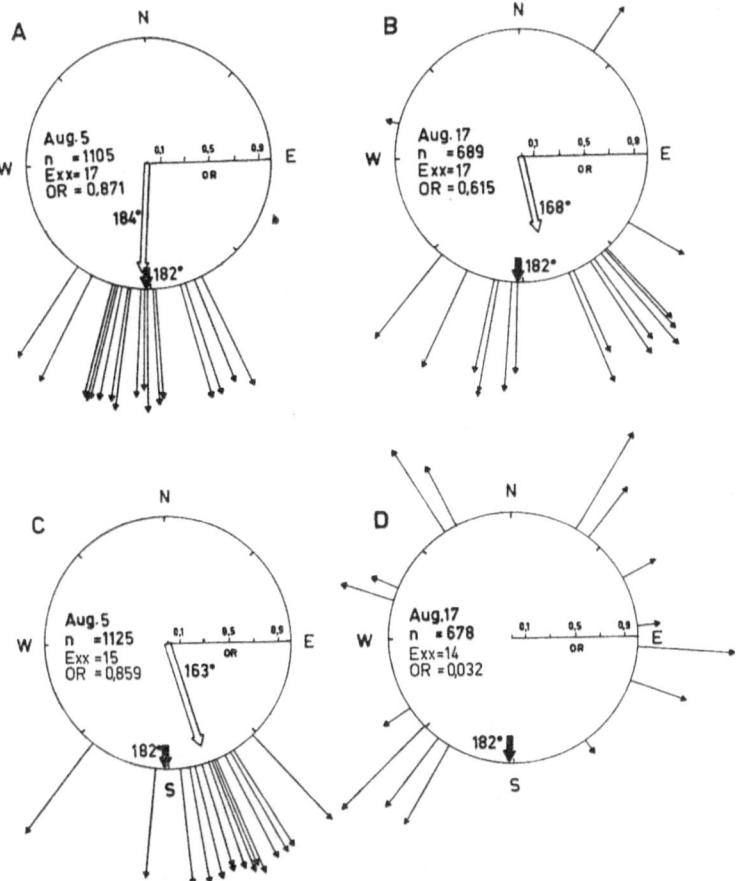

Fig. 8A—D. *A. variana*. Disorientation induced experimentally. A. and B. Orientation of control animals immediately after capture (A) and after 12 days of captivity in laboratory (B). No test was executed during the captivity. C. and D. Orientation of the animals subjected to treatment immediately after capture, and after 12 days. The animals were forced to remain on the water for 10 minutes, six times a day. In D the distribution of *r* is not statistically different from a random one (*p* > 0.05); the length of the mean vector of the accumulated distribution is so short that it cannot be represented on the graph. Other explanations as for Fig. 1 B

VI. The learning of a new escape direction

The ability of *Arctosa* to learn a new direction of escape was demonstrated some few years ago (PAPI, SERRETTI and PARRINI, 1957). Further experiments performed with a greater number of specimens have confirmed this fact.

Part of this research was performed using the same tanks (Fig. 6) as those used in the study of the learning of young spiders. In particular, we shall discuss an experiment performed on *A. cinerea*.

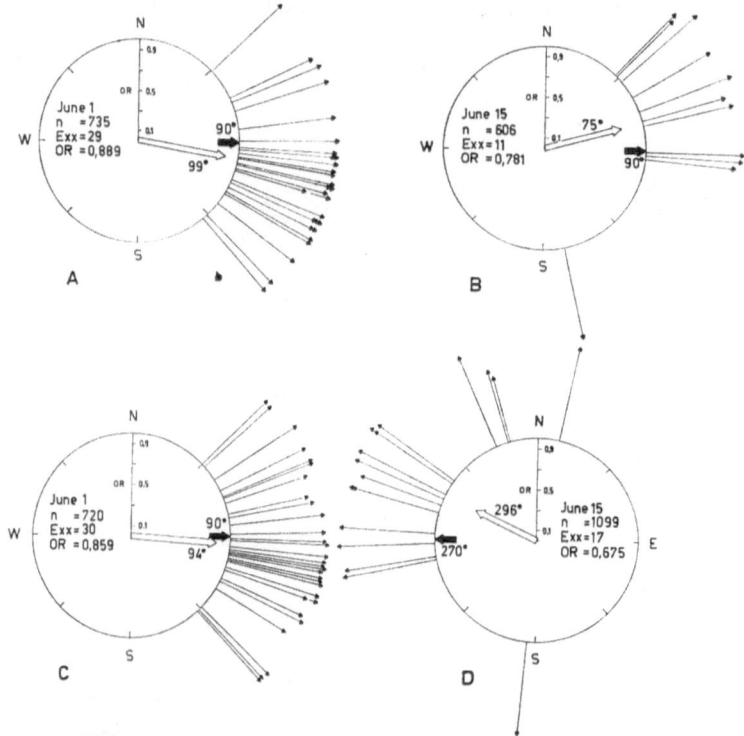

Fig. 9A—D. *A. cinerea*. Specimens with theoretical escape direction towards E. The learning of a new escape direction. Above: orientation of the control animals before (A) and after (B) being in a tank whose theoretical direction of escape was E. Below: orientation of a group of animals before (C) and after (D) being in a tank whose theoretical escape direction was W. Other explanations as for Fig. 1 B

Three lots of about 30 specimens, each having an escape direction towards E were placed, on June 1, in three tanks containing water-baths. In one tank, the orientation of the little artificial bank was such that the theoretical escape direction was equal to W, in the second it was S, while in the third, used for control purpose, the artificial bank was oriented in the same direction as the natural one from which the animals were taken. Due to the frequent cannibalism, the stay in the tanks was not continued after June 15. On this date, all those animals still living were taken from the tanks and tested.

The 11 control animals still oriented themselves towards directions approximating to E, but seemingly with reduced precision (Fig. 9A—B).

The experiment on the animals of the tank with escape direction towards W appears to have been very successful (Fig. 9C—D). At least 12 specimens out of 18 are well oriented and none of these oriented themselves towards the E quadrant. The direction of the mean vector of the accumulated distribution is also very close to the expected one. The learning of the S direction seems to have presented greater difficulties, and we are unable to explain the reason for this. As shown in Fig. 10, only about half the animals chose directions about that of S or in any case, nearer

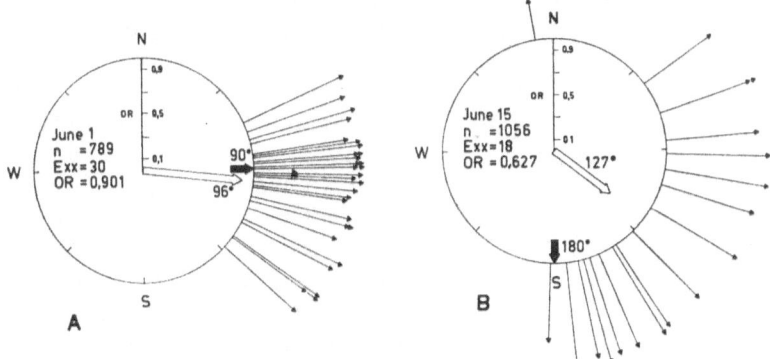

Fig. 10A and B. *A. cinerea*. Specimens with theoretical escape direction towards E. The learning of a new escape direction. Orientation of a group of animals before (A) and after (B) being in a tank whose theoretical escape direction was S. Other explanations as for Fig. 1B

S than E. The direction R deviates only 31° towards S, but one should observe that it was 52° distant from that which the control animals (Fig. 9B) had at the end of the experiment.

We could now enquire whether this capacity to learn a new escape direction, observed under experimental conditions, has a biological significance for the species; that is to say, if the spiders in nature use such ability to adapt themselves if they find themselves transferred to a bank oriented in different direction.

Two equal experiments were effected transfering specimens of *A. variana* from the S bank of Fiume Morto to the N bank. All the spiders had been previously tested in order to exclude specimens whose escape directions were more than 45° from the theoretical direction. Before the animals were released they were marked with a spot of black shellac on the cephalothorax. On October 2, 1956 the first experiment was performed with the transference of 120 specimens. Unfortunately, there was a sudden drop in temperature in the days immediately following and probably many spiders went underground to hibernate. Only two specimens could be recaptured, after ten and fourteen days respectively. On July 13, 1957 the second experiment was performed with the transference of 117 other specimens. 12 of them were found again between two and nine days after their release.

The Fig. 11 summarises the results obtained testing the orientation of all the specimens recovered. Of the three animals recaptured after two days, two oriented themselves towards S and one approximately NE, but with a strong dispersion of escape attempts ($or = 0.469$). Of the four specimens found again after four days, one was still oriented towards

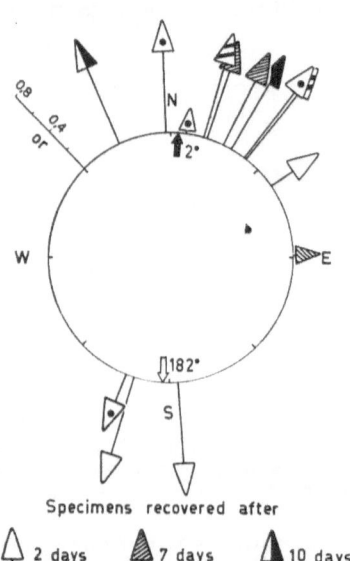

S but with a strong dispersion of escape attempts ($or = 0.490$), two were well oriented in the N quadrant and one showed such a strong dispersion of escape attempts ($or = 0.215$) that one cannot attribute any significance to the resultant direction, very close to N. Of the three specimens recaptured after seven days, two are well oriented in the N quadrant while the third shows a strong dispersion ($or = 0.220$) and has a value of r close to E. All the four specimens recovered between nine and fourteen days show that they have learnt well the new direction of escape.

We can therefore conclude that *Arctosa* is able to learn a new escape direction both under experimental conditions as under natural ones. The animals usually learn a second direction of escape more slowly than do the young inexperienced when learning the first direction. Besides this, there exists individual differences in the time necessary to learn a direction.

Specimens recovered after

△ 2 days ◮ 7 days ⬆ 10 days
△ 4 days ◭ 9 days ⬛ 14 days

Fig. 11. *A. variana*. Orientation of 14 specimens of the S bank of Fiume Morto transferred to the N bank and recaptured after 2—14 days. Each of the external arrows represents the mean vector calculated on the basis of the test performed on each specimen; their length gives the measure of the dispersion (*or*). The white internal arrow represents the theoretical escape direction for spiders of the S bank, the black internal arrow for those of the N bank

Some further points must be added on the behaviour of the spiders in the course of the learning, as also on those specimens, which, after having correctly learnt the new direction, are held in captive in the laboratory. During the course of the experiment conducted in the tanks, on each day that the weather was good, five specimens chosen at random were removed from the tanks, tested and replaced again. The animals that should have learnt the S direction after three days already manifested a net tendency towards SE and they oscillated about this direction for the rest of the experiment. The behaviour of the animals in the tank, with theoretical escape direction towards W, proves more interesting. The single resultant directions gradually move from the old to the new direction (Fig. 12). Nearly all the animals seem to have chosen the intermediate direction in the N hemisphere, shifting their escape direction anticlockwise. This preference was also observed in another analogous experiment for about four-fifths of the animals.

If besides the resultant directions, we consider the distribution of escape attempts recorded in the test of animals that are learning, a conflict of tendency between the old and new directions can be observed. Several specimens attempt to escape directly in an intermediate direction, others however direct themselves towards the old as well as the new direction. Even in this second case the mean vector falls in an intermediate direction. In Fig. 13 two typical examples are represented, between which is to be found many behaviour patterns of intermediate type.

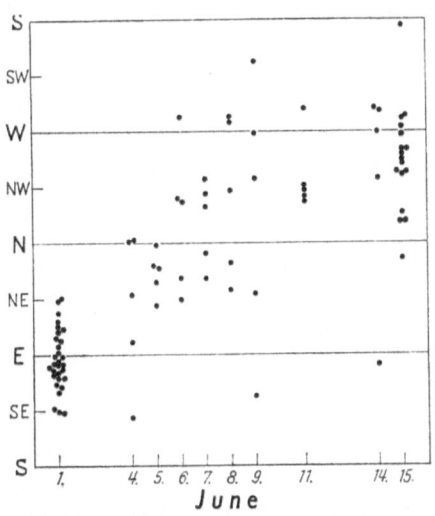

Fig. 12. *A. cinerea*. Specimens with theoretical escape direction towards E. The learning of the W direction in a tank. The distribution of the resultant single directions for all the specimens before being introduced into the tank (1. VI.), and also on successive days for 5 specimens taken at random from the tank. At the end of the experiment (15. VI.) all specimens still living were tested. Every dot represents a single resultant direction. The dots are, in some places, not placed exactly on the vertical corresponding to the date, due to lack of space

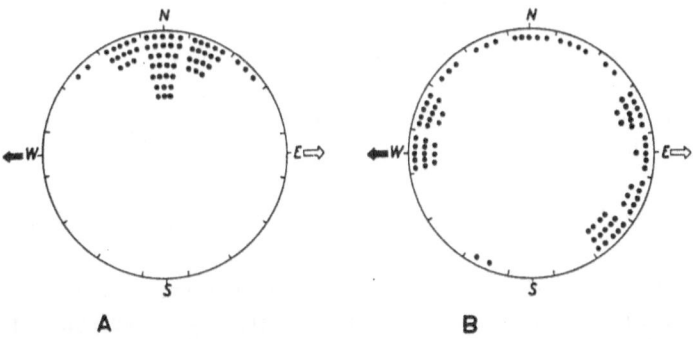

Fig. 13A and B. *A. cinerea*. Result of testing two specimens in a state of conflict between the old escape direction (white arrow) and the new one (black arrow). In A the animal always directs itself towards an intermediate direction, in B it distributes its escape attempts prevalently in directions approximating to the old and new escape directions. Each dot represents an escape attempt

When an animal seeks to escape in both directions, the escape attempts are not regularly alternated between one another. More usual, is that it begins a series of attempts towards one of two directions. The change to the other direction takes place gradually while the attempts continue, or else suddenly after a pause in activity. It may be assumed that the direction in which the animal directed itself initially was the stronger of the two. However since the experimental situation is such that each collision with the vessel represents a frustrated attempt to reach the bank in a determined direction, it is understandable that after an initial series of futile attempts in a given direction the weaker tendency towards the other direction may at least prevail.

Sometimes the tendency in a direction is just noticeable, in so far as during a pause between two successive series of escape attempts in the same direction, the animal turns in the opposite direction, begins as if to make an escape attempt in that direction, but immediately stops, and turns back in the old one instead. More unusual, between two series of escape attempts in opposite directions, the animals go round in circles in the test vessel. This behaviour is frequent for animals disoriented, for example during tests when the sky is covered over.

The behaviour of the spiders that are subjected to the learning of a new direction of escape and that are then held in captive in the laboratory varies considerably. If the training is interrupted when they still show an intermediate orientation, they return after a few days to the old escape direction. The majority of the animals that learn well a new direction maintain it and behave as do animals that have not been subjected to such learning; a percentage can return to an intermediate orientation and sometimes reassume also, either rapidly or gradually, their old direction. Those animals that return to an intermediate direction are observed to manifest the same state of conflict that was observed in the course of learning.

One can therefore conclude that the learning of a new direction of escape does not cancel immediately the tendency to direct themselves towards the old direction. This tendency seems to reinforce itself if the animal, during the captivity, is deprived of the possibility to orient itself.

VII. Variability of the orientation in nature in relation to environmental conditions

After we ascertained that the escape direction is learned and, in general, that the orientation changes as a result of experience it is to be expected that the populations subjected to different environmental conditions can show differences in capacity to orient themselves. In so far as this is concerned, we can consider populations with varying habitat, and populations that, though having similar habitat, are subjected to astronomical conditions differing according to the varying latitude in which they live.

The populations of *A. variana* which live on the banks of Fiume Morto demonstrate, as we have seen, a precise orientation. This can be related to the fact that the part of the Fiume Morto from which the

a

b

Fig. 14a and b. Two habitats of *A. variana*: Fiume Morto in S. Rossore near Pisa (top), and River Chioma south of Leghorn (bottom)

spiders come is actually an artificial canal with straight banks (Fig. 14, top), and with an approximately constant level of water. The animals live close to the water in those lengths of the bank which have little herbaceous vegetation. (Unfortunately in the last few years the banks have not been kept clear of vegetation and *A. variana* have become extremely rare there.) At the same time this species can be found on the exposed beds of streams, especially in the sun-beaten tracts which have sparse vegetation (Fig. 14, bottom). The theoretical escape direction is here very different from one point to another, and in the course of time changes with the variation in level of the water. All the same, the spiders in this environment are even more capable of orienting themselves. A good proportion of their escape attempts is in fact towards directions, which in the areas of the bank where the animals are captured, correspond to that inland. The orientation is much more imprecise than for the populations of Fiume Morto; there is in fact a greater dispersion of escape attempts and also a lower constancy in choice of direction for animals kept in captivity. In certain extreme cases, as for example for animals coming from little islets in the exposed beds of streams, one could not recognise any definite orientation (Tongiorgi, 1962).

Also it was observed that in the case of *Lycosa fluviatilis* Blackw. only those individuals which live on bare banks or on banks with sparse vegetation are capable of precise astronomical orientation (Papi and Syrjämäki).

The nearness of the water and/or a more or less decided slope towards the water of inhabited terrain has a clear influence on the precision of orientation, as results from the comparison of diverse populations of *A. perita* (Tongiorgi, 1962). As a rule this species lives on dunes that are only occasionaly approached by waves and so, according to the nature of the coast, is found at varying distances from the sea. Less frequently one can find it in the immediate neighbourhood of the shore. In Fig. 15 one can observe the orientation of specimens of three populations, living: A. on the dunes of the Gombo on the Tyrrhenian coast near Pisa at 100—120 m. from the sea; B. on the beach of Buca del Mare, also on the Tyrrhenian coast near Pisa, but at 20—30 m. from the shore; C. on the beach of Henriksberg in the Southern Finland, where the animals were collected at a distance of 0.5—4 m. from the sea. In the first case (Fig. 15 A) the prevalent orientation of the animals is erroneous, in so far as the direction of the mean vector is displaced 108° from that expected. Furthermore, the dispersion of r is so strong that the accumulated distribution is hardly different from a random one ($OR = 0.321$, $0.01 < < p < 0.05$). The animals of Buca del Mare (Fig. 15 B) are on the other hand rather well oriented and the accumulated distribution of r ($OR = 0.523$, $0.001 < p < 0.01$), is different from a random one. The animals of Henriksberg (Fig. 15 C), finally, are oriented with great precision and extremely small dispersion of r ($OR = 0.892$, $p < 0.001$). There therefore exists a strong correlation between habitat and precision of orientation, that must be ascribed to experience.

Finally one encounters differences in the mechanism of astronomical orientation between populations living in different latitudes (PAPI and SYRJÄMÄKI). Specimens of Italian *A. cinerea* are incapable of definite solar orientation for all twenty-four hours, in so far as at night they

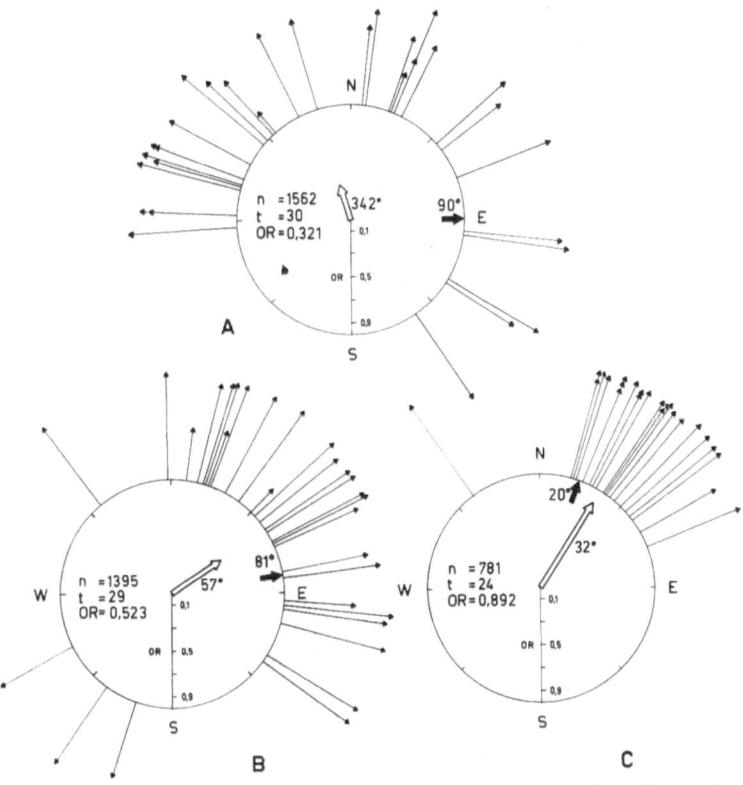

Fig. 15A—C. *A. perita*. Different precision of orientation in populations living at different distances from the sea. A. 15 specimens from Gombo captured between 100—200 m. from the shore: strong dispersion of the single resultant directions, and erroneous general orientation. B. 15 specimens from Buca del Mare, captured between 20—30 m. from the shore: moderate dispersion of the single resultant directions and correct general orientation. C. 12 specimens from Henriksberg, captured between 0.5—4 m. from the shore: very precise orientation. *t* = number of tests performed. Other explanations as for Fig. 1 B

assume in an irregular manner angles of orientation valid during the day or show themselves to be completely disoriented. Finnish populations of the same species are instead correctly oriented throughout the twenty-four hours in the period around the time of the Summer solstice. For further details we refer to PAPI and SYRJÄMÄKI; we just limit ourselves to stressing that in this case it is yet to be established whether the differences found must be attributed to learned or innate components.

References

Birukow, G.: Lichtkompaßorientierung beim Wasserläufer *Velia currens* F. (Heteroptera) am Tage und zur Nachtzeit. I. Herbst- und Winterversuche. Z. Tierpsychol. **13**, 463—484 (1956).

Braemer, W.: Versuche zu der im Richtungsgehen der Fische enthaltenen Zeitschätzung. Verh. dtsch. zool. Ges., Zool. Anz. **23** Supplementband, 276—288 (1959).

Durand, D., and J. A. Greenwood: Modifications of the Rayleigh test for uniformity in analysis of two-dimensional orientation data. J. Geol. **66**, 229—238 (1958).

Ercolini, A., e G. Badino: L'orientamento astronomico di *Paederus rubrothoracicus* Goeze. (Coleoptera Staphylinidae.) Boll. Zool. **28**, 421—432 (1962).

Greenwood, J. A., and D. Durand: The distribution of length and components of the sum of *n* random unit vectors. Ann. Math. Stat. **26**, 233—246 (1955).

Hoffmann, K.: Die Einrechnung der Sonnenwanderung bei der Richtungsweisung des sonnenlos aufgezogenen Stares. Naturwissenschaften **40**, 148 (1953).

Jander, R.: Die optische Richtungsorientierung der Roten Waldameise *(Formica rufa L.)*. Z. vergl. Physiol. **40**, 162—238 (1957).

Papi, F.: Astronomische Orientierung bei der Wolfspinne *Arctosa perita* (Latr.) Z. vergl. Physiol. **37**, 230—233 (1955a).

— Ricerche sull'orientamento di *Arctosa perita* (Latr.) (Araneae-Lycosidae). Pubbl. Staz. zool. Napoli **27**, 80—107 (1955b).

— Sull'orientamento astronomico in specie del gen. *Arctosa* (Araneae-Lycosidae). Z. vergl. Physiol. **41**, 481—489 (1959).

— e L. Pardi: Ricerche sull'orientamento di *Talitrus saltator* (Montagu) (Crustacea-Amphipoda). II. Sui fattori che regolano la variazione dell'angolo di orientamento nel corso del giorno. L'orientamento di notte. L'orientamento diurno di altre popolazioni. Z. vergl. Physiol. **35**, 490—518 (1953).

— e L. Serretti: Sull'esistenza di un senso del tempo in *Arctosa perita* (Latr.) (Araneae-Lycosidae). Atti Soc. tosc. Sci. nat., Mem. (B) **62**, 98—104 (1955).

— — e S. Parrini: Nuove ricerche sull'orientamento e il senso del tempo di *Arctosa perita* (Latr.) (Araneae-Lycosidae). Z. vergl. Physiol. **39**, 531—561 (1957).

— and J. Syrjämäki: The sun-orientation rhythm of Wolf Spiders at different latitudes. Arch. ital. Biol. **101** (in press).

Pardi, L.: Innate components in the solar orientation of littoral Amphipods. Cold Spr. Harb. Symp. quant. Biol. **25**, 395—401 (1960).

— A. Ercolini, V. Marchionni e C. Nicola: Ricerche sull'orientamento degli Anfipodi del litorale: il comportamento degli individui allevati in laboratorio sino dall'abbandono del marsupio. Atti Accad. Sci. Torino, Cl. Sci. fis. mat. nat. **92**, 1—8 (1958).

— u. F. Papi: Die Sonne als Kompaß bei *Talitrus saltator* (Montagu) (Amphipoda, Talitridae). Naturwissenschaften **39**, 262—263 (1952).

— — Ricerche sull'orientamento di *Talitrus saltator* (Montagu) (Crustacea-Amphipoda). I. L'orientamento durante il giorno in una popolazione del litorale tirrenico. Z. vergl. Physiol. **35**, 459—489 (1953).

Schmidt-Koenig, K.: Die Sonne als Kompaß im Heim-Orientierungssystem der Brieftauben. Z. Tierpsychol. **18**, 221—244 (1961).

Tongiorgi, P.: Effects of the reversal of the rhythm of nycthemeral illumination on astronomical orientation and diurnal activity in *Arctosa variana* C. L. Koch (Araneae-Lycosidae). Arch. ital. Biol. **97**, 251—265 (1959).

— Sulle relazioni tra habitat ed orientamento astronomico in alcune specie del gen. *Arctosa*. (Araneae-Lycosidae.) Boll. Zool. **28**, 683—689 (1962).

Geographische Prägung, Tag- und Nachtorientierung trans-ozeanisch wandernder Pazifischer Goldregenpfeifer (Pluvialis dominica fulva)*

Von E. G. Franz Sauer

University of Florida, Department of Biology, Gainesville, Florida (U.S.A.)

Mit 1 Abbildung

Im Sommer 1960 zogen wir zehn Pazifische Goldregenpfeifer nahe der Boxer Bucht auf der St. Lorenz-Insel, Beringsee, von Hand auf. Die Vögel konnten dort den natürlichen Himmel sehen, und es war zu erwarten, daß ihre organischen Zeitgeber auf Lokalzeit abgestimmt waren. Ehe die zehn Goldregenpfeifer in ihre erste Zugstimmung kamen, wurden sie Ende August 1960 in verhangenen Käfigen mit Flugzeugen von ihrer Geburtsinsel nach Madison/Wisconsin und im Februar 1961 nach San Francisco verfrachtet.

Sechs Vögel (vier in San Francisco) lebten in Klimakammern, in denen die Tag-Nacht- und Jahresrhythmen jenen entsprachen, denen die Vögel im Ablauf ihres normalen Jahreszyklus an und zwischen den beiden Orten Boxer Bucht (St. Lorenz-Insel) und Honolulu (Hawaii) ausgesetzt gewesen wären. Die vier übrigen Vögel (sechs in San Francisco) sahen in Madison und später in San Francisco wenigstens einen Teil des natürlichen Himmels bei Tag und Nacht.

Während ihrer frühjahrlichen und herbstlichen Zugperioden wurden die Richtungswahlen der zugaktiven Vögel anhand ihrer Abflüge und Landungen in einem für den Gebrauch für Watvögel konstruierten Rundkäfig automatisch registriert. Die Tonbandwiedergabe von Flugrufen Pazifischer Goldregenpfeifer über einen im Zentrum unter dem Käfig gelagerten Lautsprecher steigerte die Zugaktivität der Vögel erheblich.

Die Versuche im Frühjahr und Herbst 1961 wurden in San Francisco an der *California Academy of Sciences* unter dem natürlichen Himmel und im *Morrison Planetarium* der Akademie durchgeführt. Sie erstreckten sich

* Kurzfassung des Vortrages; Veröffentlichung der zugrundeliegenden Daten in Proc. 13th Internat. Ornith. Congress, Ithaca, New York, 1962. Mit Unterstützung der *National Science Foundation*, Forschungsstipendium NSF G-10724.

auf **159** Versuche mit einer Registrierzeit von **249** Std **52** min. Die Zugaktivität der zehn Goldregenpfeifer kulminierte in der zweiten Hälfte
des Mai und September. Als Bezugssystem für die Beurteilung der Zug-
diagramme wurden willkürlich die Großkreisrouten zwischen den in
Frage kommenden Orten herangezogen (Abb. 1) und erwiesen sich als
überaus brauchbar. Die bisherigen Auswertungen lassen sich folgendermaßen kurz zusammenfassen:

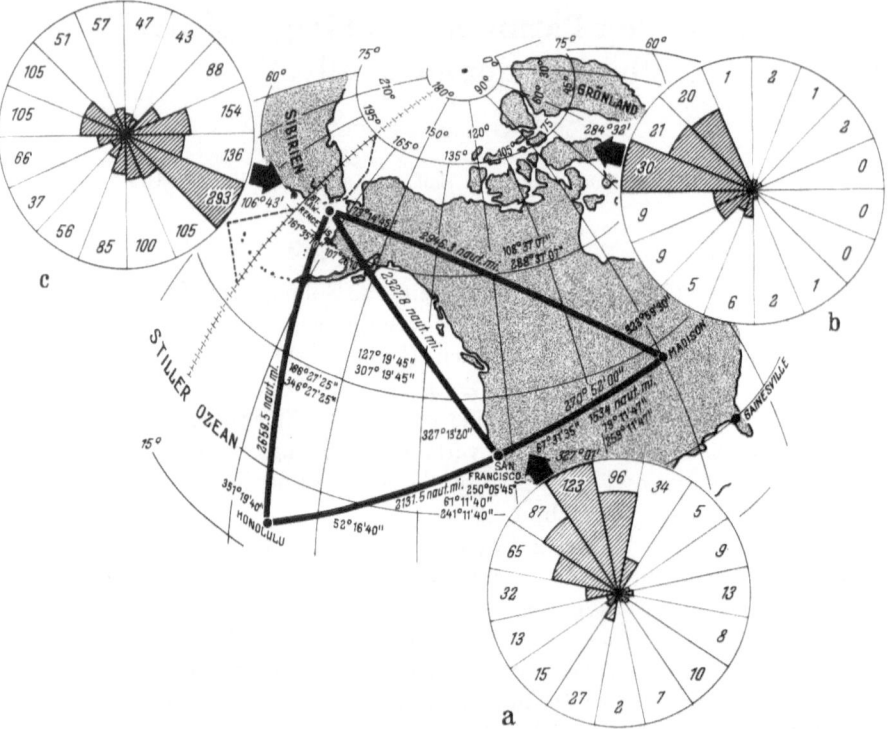

Abb. 1a—c. Bogen der Großkreise zwischen Boxer Bucht, Madison, San Francisco und Honolulu. Drei
Beispiele von Zugversuchen im Frühjahr (a, b) und Herbst (c) 1961 in San Francisco.

a) LD ♂,   7—6—61, 0815—0945, C:112, klar, 4/10 StCu, W 5—8, Spitzen 10—20.
         8   1040—1217, C:411, klar, 4/10 Cu, später CiCu, W-NW 10—14, Spitzen 25.
         9   0825—0955, C:23, klar, SW 2—3.
       TC: 546, dt—2h 50m (42° 30').

b) RB ♂, 13—5—61, 1535—1745, C:56, klar, hohe Ci, W 10.
      15   1440—1650, C:53, klar, W 2—10. TC: 109.

c) LD ♂, 14—9—61, 1324—1453, C:504, klar, 1/10—3/10 St, SW 10—14, 18, W 2—5.
     19   1155—1328, C:357, klar, W 1—3, S, gelegentlich SW 2—4.
     25   0920—1050, C:189, klar, W 1—3.
     26   0845—1015, C:93, klar, N 2—4.
     29   1036—1206, C:190, klar, W 3—6.
     30   1348—1524, C:58, klar, W 2—3.
     2—10—61, 1521—1625, C:137, 1/10—5/10 Ci und niedere St, W—SW 4—6.
     TC: 1528, dt — 3 h 16 m 40 s (49° 10').

Erklärungen: Zeit in San Francisco Lokalzeit. C Registriersumme des Einzelversuches, TC Registriersumme
der Zusammenfassung; dt- Zeitdifferenz zwischen innerer Uhr und Lokalzeit; Bewölkung: Ci Cirrus, CiCu
Cirrocumulus, Cu Cumulus, St Stratus, StCu Stratocumulus; Grad der Bewölkung von 1/10—10/10; Kompaßrichtungen kennzeichnen Windrichtungen; Windgeschwindigkeit in Knoten

Drei typische Beispiele der Reaktionen der Vögel aus den beiden Versuchsgruppen bei klarer Sicht sind in Abb. 1a—c aufgezeigt. In den Diagrammen entsprechen die 16 Sektoren den 16 Registriereinheiten; die Zahlen geben die Anzahl der Abflüge und Landungen in den entsprechenden Richtungssektoren an. Die Lage der Mittelwerte, durch Vektoren-Addition errechnet, ist mittels der schwarzen Pfeile dargestellt.

Die inneren Uhren der Vögel der ersten Versuchsgruppe gingen bis zu 3 Std 16 min 40 sec (einem Winkelbetrag von 49° 10' entsprechend) gegenüber der Lokalzeit von San Francisco nach. Unabhängig von diesen verschiedenen Zeitdifferenzen näherten sich die Mittelwerte der frühjahrlichen Zugversuche mit diesen Vögeln dem Anfangskurs des Großkreisbogens, der San Francisco mit der Boxer Bucht verbindet (Abb. 1a). Die Vögel bevorzugten diese Richtung auch noch unter völlig bewölktem Himmel, doch nur solange der Stand der Sonne durch die Wolken deutlich wahrzunehmen war.

Die Vögel der zweiten Versuchsgruppe, die im Herbst 1960 dem Himmel Madisons ausgesetzt waren und deren innere Uhren jetzt in Phase mit der Lokalzeit von San Francisco waren, wählten in San Francisco einen mittleren Frühjahrszugkurs, der einer guten Annäherung an den Großkreisbogen entspricht, der nicht von San Francisco, sondern vielmehr von Madison nach der Boxer Bucht führt (Abb. 1b).

Der qualitative Unterschied zwischen den beiden Mittelwerten der Frühjahrsbefunde mit den Vögeln der beiden Versuchsgruppen stellte sich in Versuchen unter zunehmend dicht bewölktem Himmel heraus. Wenn ein menschlicher Beobachter den Sonnenstand nur noch angenähert richtig angeben konnte und häufig mißdeutete, verloren allein die Vögel der Versuchsgruppe 1 ihre ursprüngliche Orientierung. Sie wichen von ihrem frühjahrlichen Zugziel entsprechend ihrer inneren Zeitumstimmung gegenüber der Ortszeit um einen entsprechenden Winkelbetrag nach rechts ab. Man kann diesen Befund dahin deuten, daß die umgestimmten Vögel in dieser Situation nicht länger zu einer Navigation, also einer Orientierung nach zwei Himmelskoordinaten befähigt waren. Sie gaben diese zugunsten einer einfachen Azimutorientierung auf, bei der sie unter Vernachlässigung der Sonnenhöhe erwartungsgemäß eine der Phase ihrer inneren 24stündigen Uhr entsprechenden Azimutverrechnung der Sonne ausführten. Bei zurückgestellter Uhr ergibt das eine Rechtsabweichung. Die Befunde bestätigen zusätzlich die Deutung, daß sich die Goldregenpfeifer optisch nach der Sonne orientieren. Bei Verfrachtungen der Vögel ist ihre Navigationsleistung solange unabhängig von der Gangphase ihrer inneren Uhr geblieben, solange diese eine Zeitdifferenz zur Lokalzeit aufwies. Zudem war die Entscheidung der Vögel für Navigation oder Kompaßorientierung sichtabhängig (vgl. auch unten).

Die Ergebnisse lassen sich weiterhin dahin deuten, daß die zunächst auf der St. Lorenz-Insel aufgewachsenen Vögel vor ihrer ersten Zugphase einer sensitiven Phase ausgesetzt waren, während der sie sich die geographische Lage ihres prospektiven Brutgebietes (in diesem Falle mit dem Geburtsort identisch) einprägten. In einer zweiten sensitiven Phase zu Ende der ersten Zugperiode (wahrscheinlich auch schon während dieser) erlernten die Vögel der zweiten Versuchsgruppe die geographische Lage Madisons. Die Deutung beruht vornehmlich darauf, daß diese Vögel nach ihrer Verfrachtung nach San Francisco und der Angleichung ihrer inneren Uhr an dessen Lokalzeit weder eine artspezifische Zugrichtung noch zielgerichtet zum Brutplatz, sondern vielmehr den Parallelkurs zu jener geprägten Route flogen. Die Frage nach einer statistischen Sicherung der Unterschiede in den Mittelwerten, die von den Zugversuchen mit den Vögeln der beiden Versuchsgruppen resultierten, ist in diesem Zusammenhang gegenstandslos. Die Versuche unter stark bewölktem Himmel, bei denen nur die Vögel der ersten Versuchsgruppe die erwartungsgemäße Rechtsabweichung zeigten, weisen auf die qualitative Verschiedenheit der beiden in Frage stehenden Mittelwerte hin. Wir prüfen derzeit die naheliegende Arbeitshypothese, nach der diese Goldregenpfeifer der zweiten Versuchsklasse bei einem wirklichen Abflug von San Francisco nicht auf dem fehlweisenden Parallelkurs bleiben, sondern vielmehr doch ihr artgemäßes Zugziel erreichen würden. Nach dieser Annahme fliegen die Vögel nur so lange parallel zum Prägungskurs, solange sie keine Zeitdifferenz zwischen ihrer inneren Uhr und der für ihren jeweiligen Flugort gültigen Lokalzeit wahrnehmen können. Bei jedem von der N-S-Richtung abweichenden Flug wird ein Vogel jedoch schon nach einer relativ kurzen Flugzeit eine Zeitdifferenz wahrnehmen können und dann mittels seines sich wieder einspielenden Navigationsmechanismus richtig kompensieren können. Nach dieser Vorstellung wären Arten, die genau in nördlichen oder südlichen Richtungen ziehen, bei Ost- oder Westverfrachtungen mit nachfolgender Uhrangleichung an die entsprechende Lokalzeit schwerlich zu einem Korrektionsflug zum ursprünglichen Heimatgebiet befähigt. Das Fußfassen und die N-S-Ausbreitung des Kuhreihers in der Neuen Welt könnte einem solchen Mechanismus zugrunde liegen.

Die Herbstversuche in San Francisco erzielten im wesentlichen Mittelwerte, die den Frühjahrskursen entgegen lagen (Abb. 1c), doch zeigte sich dabei auch, daß die Prägung der Vögel auf eine experimentell bestimmte Winterherberge durch ihren Aufenthalt in San Francisco beeinflußt werden konnte. So kann man bei diesen Vögeln nicht von einer irreversiblen geographischen Prägung sprechen.

Die Nachtversuche und die ersten Kontrollversuche im Planetarium resultierten mit gleichen Richtungswahlen wie bei Tage und weisen darauf hin, daß sich die Vögel nachts nach den Sternen orientieren.

Nach den Befunden ist anzunehmen, daß die Goldregenpfeifer keine angeborene Kenntnis ihres Geburts- oder Brutplatzes haben, noch bekundeten sie eine mystische Fähigkeit, von jedem Ort unter jeder Bedingung nach „Hause" zu finden. Es wäre unangebracht zu denken, daß die Vögel eine starre, erblich streng genormte Zugrichtung zwischen Brut- und Überwinterungsgebiet verfügbar hätten. Es ist vielmehr wahrscheinlich, daß genetisch ein bestimmter Bereich für mögliche Modifikationen der Zugrichtung festgelegt ist. Innerhalb des Modifikationsbereiches würde sich dann der individuelle Kurs unter dem Einfluß der Lernvorgänge während der beiden sensitiven Phasen und der während des Fluges auf den Vogel wirkenden Umweltfaktoren ausprägen. Man hat darin eine Leistungsstufe zu sehen, die über das reine Instinktniveau hinausreicht und die Vögel ausgezeichnet für ihren trans-ozeanischen Flug von mehreren Tausend Kilometern anpaßt. Ausgerüstet mit diesem Mechanismus kann sich der bezüglich seines Winterquartieres noch ungeprägte, unerfahrene Erstzieher mit großer Erfolgswahrscheinlichkeit dem langen Überwasserflug aussetzen. Für ihn liegt an dessen Ende die Hawaiische Inselkette als über 3000 km breite Zielscheibe.

Neuere Aspekte über die Orientierungsleistungen von Brieftauben

Von Klaus Schmidt-Koenig

Max Planck-Institut für Verhaltensphysiologie (Abt. H. Mittelstaedt) und Dept. of Zoology, Duke University Durham N. C. (USA)

Mit 9 Abbildungen

Inhaltsverzeichnis

A. Vergleich zweier Stämme und Versuchsgebiete

1. Einleitung

Der weitaus größte Teil unserer Kenntnisse über die Orientierungsleistungen von Brieftauben ist von Kramer u. Mitarb. mit Tieren der Wilhelmshavener Kolonie gewonnen worden (vgl. Kramer, 1959). Der Wilhelmshavener Raum steht jedoch im Verdacht, womöglich wegen seiner geographischen Gliederung kein günstiges Versuchsgelände abzugeben: Aus nördlichen Richtungen können die Tauben nur in begrenztem Umfang aufgelassen werden. Selbst der vorhandene Raum ist kaum ausgenutzt worden. Vielleicht ist die nord-polarisierte Abflugstendenz der Tauben, die aus nahezu allen veröffentlichten Diagrammen sofort ins Auge springt (z. B. Wallraff, 1959b, Abb. 8), ein Dressureffekt als

Folge überwiegender Auflassungen aus Süden. Die Nord-Überlagerung ist zwar erkannt worden, ihr volles Gewicht kann jedoch erst an Ergebnissen, die unter symmetrischen Verhältnissen zustande gekommen sind, gemessen werden.

Nachdem schon J. G. PRATT, G. KRAMER und U. v. ST. PAUL (KRAMER et al., 1956, 1957, 1958; PRATT, 1955) in den fünfziger Jahren den Anfang gemacht hatten, werfen nun neuere Untersuchungen in North Carolina Licht auf das, was im Wilhelmshavener Raum tatsächlich an den üblichen Kriterien der Orientierungsleistung verzerrt und überdeckt worden ist. Die bisherigen Vorstellungen ändern sich dadurch in einigen Punkten.

Bevor man aus unterschiedlichen Versuchsergebnissen von verschiedenen geographischen Orten auf Unterschiede in Art und Stärke der orientierenden Faktoren schließen kann, muß u. a. der „genetische Faktor" — Unterschiede, die auf unterschiedlicher Leistungsfähigkeit der benutzten Taubenstämme beruhen — ausgeschaltet sein.

Genetisch bedingte Unterschiede der Orientierungsleistung sind in einigen Befunden von PRATT (1955) angedeutet, HOFFMANN (1959) fand sie zwischen englischen und Wilhelmshavener Tauben. Andere Befunde an weit entfernt voneinander angesiedelten Tauben ein und desselben Stammes wiesen auf Unterschiede hin, die mit der geographischen Lage der Schläge oder Versuchsgebiete bzw. den assoziierten orientierenden Faktoren in (nicht näher bekanntem) Zusammenhang zu stehen schienen (HOFFMANN, 1959).

Zunächst mußten weitere Anhaltspunkte über den „genetischen Faktor" gewonnen werden.

Die Arbeiten an der Duke Universität werden von der National Science Foundation (Vertrag G-9816 und G-19849 mit P. H. KLOPFER und K. SCHMIDT-KOENIG) und vom Office of Naval Research (Vertrag 301-681 mit der Duke Universität) finanziert. Ich danke vor allem Dr. P. H. KLOPFER für seine stete Hilfsbereitschaft und anregende Diskussionen.

2. Material und Methodik

Im Jahre 1959 verpflanzten wir Tauben des im Max Planck-Institut Wilhelmshaven gezüchteten Stammes (MP-Tauben) an die Duke University in Durham, North Carolina. Die Nachkommen dieser Tauben dienten zu vergleichenden Versuchen mit Tauben des Stammes der Duke University (DU-Tauben). — Das experimentelle und statistische Vorgehen ist an anderer Stelle ausführlich beschrieben (SCHMIDT-KOENIG, 1963a). Hier sei nur folgendes erwähnt: Zwei identische Serien von Auflassungen wurden durchgeführt, die erste im Juli und August 1960, die zweite im März und April 1962. Jede Serie beinhaltete zwei Runden, jede Runde bestand aus 4 Auflassungen; je eine aus 16 km N, 12 km S, 31 km W und 66 km E in dieser Reihenfolge. An jeder Auflassung nahmen je 15—20 MP- und DU-Tauben teil. Jede Serie wurde mit ungeflogenen Tieren begonnen, die letzte Auflassung jeder Serie war also die achte Auflassung für jeden überlebenden Teilnehmer. Verluste wurden durch Tauben ersetzt, deren Flugerfahrung in ergänzenden Auflassungen auf den Stand der Versuchsgruppe gebracht worden war.

Zu weiteren Vergleichen wurden entsprechende Daten von vier weiteren Runden herangezogen, die mit DU-Tauben im Anschluß an die 1960er Serie in anderem Zusammenhang durchgeführt worden waren (SCHMIDT-KOENIG, 1963), und die Orientierungsleistung über die achte Auflassung hinaus erfaßt. Das experimentelle Vorgehen unterschied sich von den dem direkten Vergleich von DU- und MP-Tauben gewidmeten Serien nur insofern, als der Auflaßplatz 12 km S durch einen Platz 31 km S ersetzt und die Auslaßfolge N, W, S, E war.

Die Orientierungsleistungen der Tauben wurden an verschiedenen Kriterien, die im folgenden getrennt besprochen werden, verglichen und zu andernorts gewonnenen Ergebnissen in Beziehung gesetzt.

3. Ergebnisse und Diskussion

a) **Heimkehrleistung.** In der 1960er Serie (Abb. 1) waren die MP-Tauben den DU-Tauben in beiden Runden überlegen ($p < 0.001$ bzw. $p = 0.004$[1]), der Unterschied verminderte sich jedoch im Verlauf der

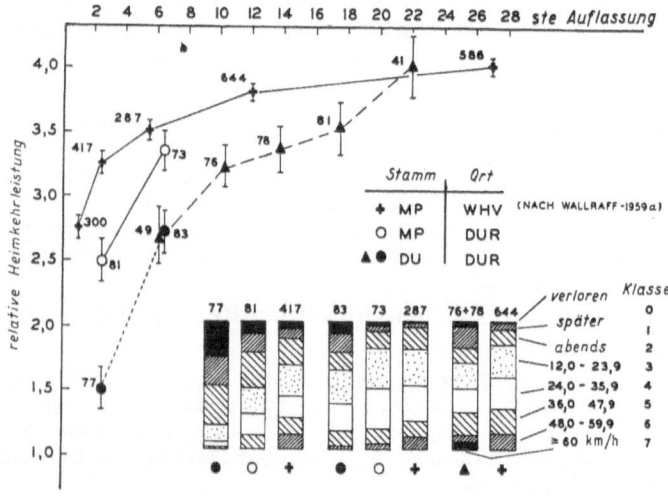

Abb. 1. Heimkehrleistung (Ordinate) der beiden Runden der 1960er Serie (MP Tauben weiße Kreise, DU-Tauben schwarze Kreise) und der anschließenden Runden (DU-Tauben Dreiecke) als Funktion der Zahl absolvierter Flüge (Abszisse). Einige Werte sind in Säulendiagramme aufgeschlüsselt; sie können an der Stichprobengröße und den beigegebenen Symbolen identifiziert werden. Neben den Säulendiagrammen steht der Schlüssel für die Klassen. Die Ordinate ist in Klasseneinheiten unterteilt. — Die Leistungskurve von MP-Tauben von Wilhelmshaven (WALLRAFF, 1959a) ist mit Kreuzen dargestellt. Jedem Symbol ist die Stichprobengröße und der mittlere Fehler des Mittelwertes ($\sigma_{\overline{x}}$) beigegeben

Serie. Die Heimkehrleistung beider Kollektive verbesserte sich mit steigender Zahl absolvierter Flüge, vor allem in den unteren Bereichen durch Auslese wenig begabter Versuchsteilnehmer. Die Heimkehrschnelligkeit stieg weiter über die achte Auflassung hinaus an, schließlich erreichten die DU-Tauben etwa das Niveau der MP-Tauben in Wilhelmshaven, deren Leistungskurve ebenfalls in Abb. 1 wiedergegeben ist. Die Kurven sind allerdings unter verschiedenen Versuchsbedingungen zu-

---

[1] Die Heimkehrleistung wurde stets mit dem Mann-Whitney-U-Test (SIEGEL, 1956) geprüft; Wahrscheinlichkeitswerte sind für den zweiseitigen Test angegeben.

stande gekommen und eignen sich nur zu annähernden Vergleichen. — Nachdem der DU-Stamm durch zahlreiche Experimentalauflassungen (SCHMIDT-KOENIG, 1963, 1963b und unveröff.) ebenso scharf ausgelesen worden war wie der MP-Stamm seit Jahren ausgelesen wird (die meisten DU-Tauben verbuchten mehr als 20 Flüge), gab es in der 1962er Serie (Abb. 2) zwischen DU- und MP-Tauben keinen nachweisbaren Unterschied mehr ($p = 0{,}32$ bzw. $p = 0{,}64$).

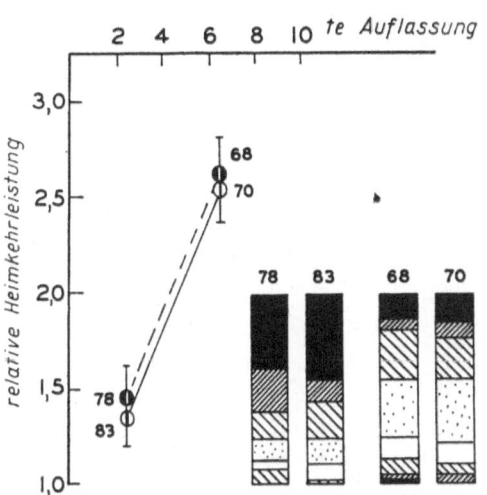

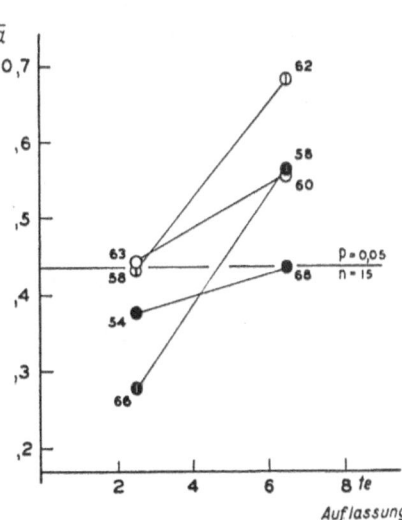

Abb. 2. Heimkehrleistung der beiden Runden der 1962er Serie. Die Symbole der 1962er Serie sind durch einen Vertikalstrich gekennzeichnet. Darstellung wie in Abb. 1

Abb. 3. Die Streuung der Abflüge beider Serien dargestellt als $\bar{a}$ (Ordinate), das arithmetische Mittel der Längen der 4 mittleren Vektoren je Runde. Abszisse und Symbole wie in Abb. 1 bzw. Abb. 2. Die Signifikanzgrenze von 5% für Zufallsverteilung ist für die mittlere Stichprobengröße (15) angegeben

b) Anfangsorientierung. Leider ist es im Moment noch unmöglich, Unterschiede zweier Stichproben anhand vektorieller Größen statistisch zu prüfen. Wo Unterschiede nicht sehr ins Auge fallen, muß es bei vorläufiger Beurteilung bleiben.

In den 16 Auflassungen beider Serien flogen die MP-Tauben 5mal, die DU-Tauben 10mal zufallsverteilt[1] ab. Aus Abb. 3 kann man ebenfalls den Eindruck gewinnen, daß die DU-Tauben stärker streuten als die MP-Tauben. Abb. 3 zeigt außerdem an, daß sich die Streuung der Abflüge mit zunehmender Flugerfahrung des Kollektivs verringerte.

Die Beziehung zwischen Abflug- und Heimrichtung kann mittels der Heimkomponente h beurteilt werden (Abb. 4). Ihre Berechnung ist von H. MITTELSTAEDT eingeführt worden[2] (vgl. Abb. 7).

---

[1] $p > 0{,}05$ für Zufallsverteilung nach dem von D. DURAND und J. A. GREENWOOD modifizierten Rayleigh-Test (s. SCHMIDT-KOENIG, 1961, Anhang).

[2] $h = a \cos(\alpha - \beta)$, wobei a die Länge des mittleren Vektors α eines Abflugdiagramms und β die Heimrichtung ist.

Ob Unterschiede zwischen den Stämmen und Serien bestehen, ist nicht sicher; eine Überlegenheit der MP-Tauben scheint jedoch nicht ausgeschlossen. Mit einiger Sicherheit kann jedoch gefolgert werden, daß

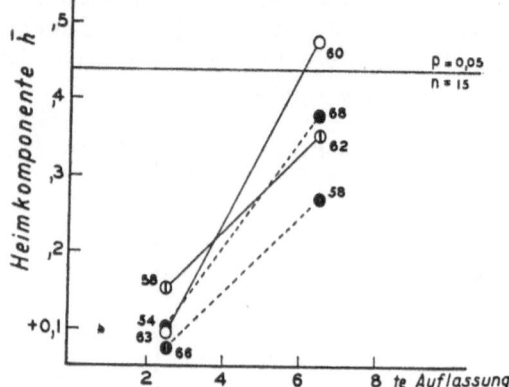

Abb. 4. Die Heimkomponente $\bar{h}$ (Ordinate) beider Serien, arithmetisch gemittelt aus den h der 4 Auflassungen je Runde. Abszisse und Symbole wie in Abb. 3. Die Signifikanzgrenze von 5% für Zufallsverteilung dient als grober Anhaltspunkt. Werte, die darunter liegen, sind sicher nicht heimgerichtet

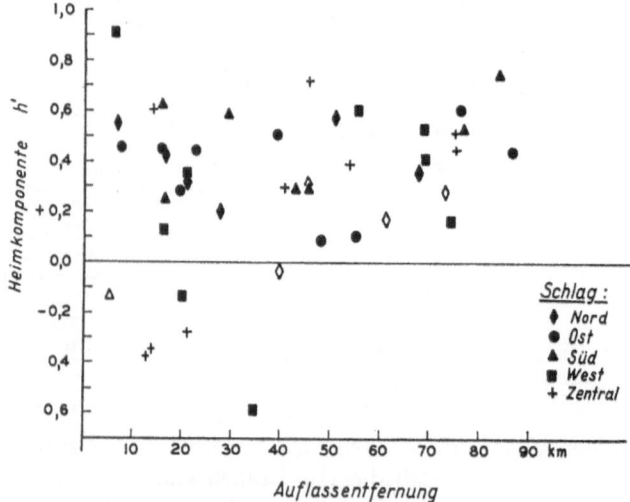

Abb. 5. Die Heimkomponente h (Ordinate) aller Auflassungen deren Heimkehrleistung in Abb. 1 zu-sammengefaßt ist (die Wilhelmshavener Kurve ausgenommen) nach Auflassungen aufgeschlüsselt. Die ver-schiedenen Symbole stehen für die verschiedenen Auflaßplätze (vgl. Schlüssel; der Ost-Auflaßplatz ist 66 km nicht 64 km entfernt). Weiße Symbole bezeichnen MP-Tauben, schwarze Symbole DU-Tauben. Abszisse wie in Abb. 1. Die Wahrscheinlichkeitsgrenzen von 5% für Zufallsverteilung sind für die kleinste (16), die mittlere (14) und die größte (21) Stichprobe als grober Anhaltspunkt (vgl. Abb. 4) angegeben. Symbolen im kritischen Bereich sind die Stichprobengrößen beigegeben

sich die Abflüge mit zunehmendem Erfahrungsgrad der Tauben zur Heimrichtung hin verlagerten. Die Werte bleiben jedoch im wesentlichen unterhalb der Wahrscheinlichkeitsgrenze von 5% für Zufallsverteilung.

Die Abflüge sind also (im Durchschnitt dieser vier Auflaßplätze) sicher nicht heimbezogen.

In Abb. 5 sind die Heimkomponenten von allen Auflassungen, deren Heimkehrleistung in Abb. 1 zusammengefaßt war (die Wilhelmshavener Daten ausgenommen), einzeln aufgetragen. Die Analyse ist hier noch durch die sehr unterschiedliche Größe der Stichproben erschwert. Es sieht jedoch so aus, als ob oberhalb der etwa zehnten Auflassung die Heimtendenz nur noch ausnahmsweise eine Verbesserung erfahren kann,

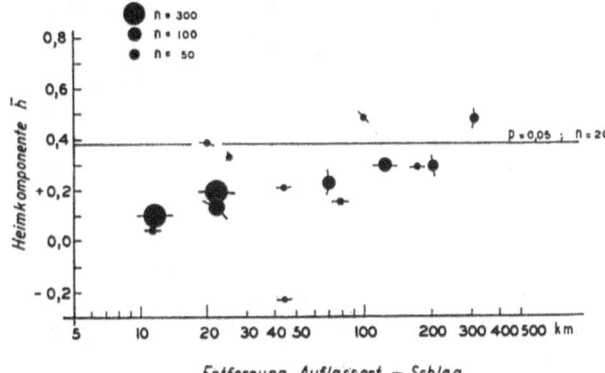

Abb. 6. Dr. H. MITTELSTAEDT (nach Daten von G. KRAMER und H. G. WALLRAFF): Die Heimkomponente $\bar{h}$ (Ordinate) gemittelt von symmetrisch zum Schlag aufgelassenen unerfahrenen MP-Tauben in Westdeutschland als Funktion der Auflaßentfernung (Abszisse). Die Striche an der Peripherie der Symbole geben die Auflaßrichtung an. Die Wahrscheinlichkeitsgrenze von 5% für Zufallsverteilung ist als grober Anhaltspunkt (vgl. Abb. 4) für etwa die mittlere Stichprobengröße (20) eingezeichnet

nämlich an dem Platz (42 km W) bzw. an solchen Plätzen, an denen die Tauben notorisch fehlgerichtet und weit streuend abfliegen. An den übrigen Plätzen scheinen sich die Abflüge innerhalb eines gewissen Bereichs — vielleicht ± 60° von der Heimrichtung — nicht mehr verbessern zu können. Vielleicht ist mit diesem Befund ein wesentliches Charakteristikum des Orientierungsmechanismus bzw. der Brauchbarkeit von Verschwindediagrammen als Kriterium der Orientierungsleistung erfaßt.

H. MITTELSTAEDT hat die Heimkomponente in unsere Auswertungsmethoden eingeführt mit der Berechnung von $\bar{h}$ für eine große Zahl von Abflügen unerfahrener MP-Tauben, die im wesentlichen von G. KRAMER und H. G. WALLRAFF in Deutschland in den vergangenen Jahren aufgezeichnet worden sind (Abb. 6[1]). Daß unerfahrene Tauben heimgerichtet abfliegen, ist demnach, wenn überhaupt, nur sehr knapp wahrscheinlich. Nach den bisherigen Feststellungen verbessert sich weder die

---

[1] Ich möchte auch an dieser Stelle Herrn Dr. MITTELSTAEDT für die Überlassung des Diagramms danken.

Streuung noch die Heimbezogenheit der Anfangsorientierung mit zunehmender Erfahrung der Tauben im westdeutschen Raum (WALLRAFF, 1959a; PRATT und WALLRAFF, 1958). In diesem Befund mag sich die Ungunst der lokalen Verhältnisse niederschlagen. Es besteht zwar kein Zweifel daran, daß schon „die unerfahrene Durchschnittstaube" ein primäres Heimfindevermögen besitzt, an den Abflugkursen läßt sich das jedoch nicht ablesen.

Zusammenfassend kann gesagt werden: Unterschiedliche Leistungsfähigkeit verschiedener Taubenstämme braucht lediglich eine Funktion unterschiedlicher Haltung bzw. verschieden intensiver Übung und Auslese zu sein. Bevor Vergleiche zwischen zwei Versuchsgebieten stichhaltig werden, muß also auch dieser Störfaktor sorgfältig ausgeschaltet sein. — Unter symmetrischen Auflaßbedingungen treten andere Grundzüge der Orientierungsleistung zutage als z. B. unter der polarisierten Situation Wilhelmhavens. Es verbessert sich nicht nur die Heimkehrleistung, sondern — in begrenztem Umfang — auch die Streuung und Heimbezogenheit der Abflüge mit zunehmender Orientierungserfahrung der Tauben. Mit der Starrheit der Anfangsorientierung erscheint der Wilhelmshavener Raum in besonders ungünstigem Licht als Experimentierfeld. Die Hauptkriterien der Anfangsorientierung eignen sich dort wahrscheinlich nicht zur Bearbeitung der Kernfragen des Orientierungsproblems. — Ferner scheinen wenig oder unerfahrene Tauben unter keinen Umständen geeignetes Versuchsmaterial abzugeben. Zur Analyse der Orientierungsfaktoren wird mehr, wenn nicht ausschließliches Gewicht auf die Leistungen gut erfahrener Tauben gelegt werden müssen.

B. Die Rolle der Auflaßentfernung

Mit den voranstehenden Ergebnissen vor Augen können wir uns der Frage zuwenden, ob und in welcher Weise die Verfrachtungsentfernung in den Orientierungsprozeß eingeht. Kenntnis darüber ist von grundlegender Bedeutung für theoretische und praktische Überlegungen zur Analyse der Orientierungsfaktoren und -mechanismen. Bei den Erörterungen des „Karte-Kompaß-Konzepts", das von G. KRAMER 1953 formuliert wurde, ist der Parameter Entfernung erst neuerdings diskutiert worden (SCHMIDT-KOENIG, 1960, 1961). In einer weiteren Versuchsserie, dem „Kreuzschlagexperiment" ist ein erster Versuch gemacht, diese Frage auch experimentell anzugehen.

4. Methodik

Bei Durham haben wir fünf Schläge in Form eines großen, annähernd symmetrischen Kreuzes erstellt (Abb. 7) und mit DU-Tauben besiedelt. Entlang den Verlängerungen der Achsen dieses Kreuzes wurden je zwei Auflaßplätze gewählt. Von den „inneren" Plätzen — so nah wie möglich am Kreuz — hatten die Tauben der verschiedenen Schläge entweder gleiche Heimrichtungen aber unterschiedliche Entfernungen, oder etwa gleiche Entfernungen aber verschiedene Heimrichtungen vor sich. Von den „äußeren" Auflaßplätzen — in etwa dreifacher Entfernung — waren diese Unterschiede auf wenige Prozent vermindert. Entlang der Westachse wurde ein dritter Platz gewählt.

Das experimentelle und statistische Vorgehen ist an anderer Stelle eingehend beschrieben (SCHMIDT-KOENIG, 1963b), hier sei soviel gesagt: Von jedem der neun Auflaßorte wurden zunächst zwei Auflassungen mit je 15—20 etwa erfahrungsgleichen Tauben aus allen Schlägen durchgeführt, die verschiedenen Kriterien der Orientierungsleistung jeweils nach Schlagzugehörigkeit zusammengefaßt. Stimmten die Abflüge beider Auflassungen überein, indem sie entweder beide Male a) nicht zufallsverteilt[1] waren und die resultierenden Vektoren um $\leqq 35°$[2] voneinander abwichen, ober b) beide zufallsverteilt waren, dann wurden die Abflugdaten beider Auflassungen vereinigt und in Abb. 8 mit einem doppelten Vektorsymbol eingetragen (z. B. alle Schläge von Caldwell School). Wenn jedoch die Abflüge entweder c) nicht zufallsverteilt waren, aber um > 35° voneinander abwichen, oder d) einmal zufallsverteilt, das andere Mal nicht zufallsverteilt waren, wurde eine dritte Auflassung durchgeführt. Von den drei Auflassungen wurden die zwei zusammengefaßt und in Abb. 8 mit einem Doppelsymbol wiedergegeben, welche die Voraussetzungen a) oder

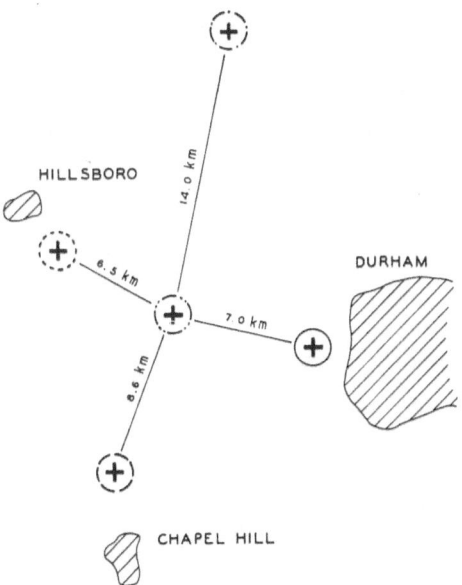

Abb. 7. Die 5 Schläge des „Kreuzschlagexperiments" und ihre Symbole (vgl. Abb. 8) in der Nähe von Durham N. C. USA

b) erfüllten, die dritte mit einem einfachen Symbol (z. B. Ostschlag und Südschlag von Hillsboro, Nordschlag von South Boston). Wenn alle drei nicht übereinstimmten, sind drei einfache Symbole gegeben (z. B. Ostschlag von Reidsville). Ausnahmsweise wurden vier Auflassungen durchgeführt (z. B. Südschlag und Zentralschlag von Burlington).

5. Ergebnisse

a) Anfangsorientierung. In der überwiegenden Mehrzahl, nämlich in 89 von insgesamt 111 Tagesdiagrammen[3] bevorzugten die Tauben eine

---

[1] $p \leqq 0,05$.

[2] Zur Zeit der Versuche gab es noch keinen kreisspezifischen Test zur Trennung zweier Stichproben. 35° schien eine gute Grenze.

[3] Unter Tagesdiagramm sind im folgenden die nach Schlagzugehörigkeit zusammengefaßten, während eines Auflaßtages gewonnenen Werte verstanden.

bestimmte Richtung (im Gegensatz zu zufallsverteilten Abflügen). Die
Tendenz zu weit streuenden Abflügen war jedoch schlagspezifisch verschieden. Am wenigsten neigten die Tauben des Zentralschlages (1 von
21 Tagesdiagrammen zufallsverteilt), am häufigsten die des Nordschlages
(7 von 20 Tagesdiagrammen) dazu. Für einige Schläge sind die Raten
signifikant unterschiedlich.

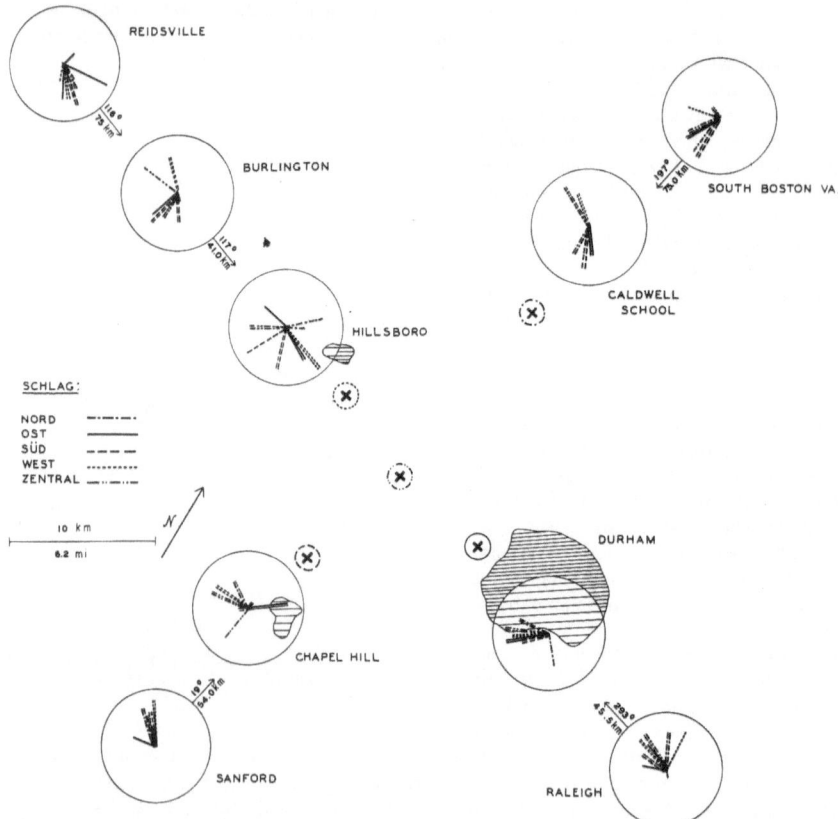

Abb. 8. Abflugdiagramme der Kreuzschlagtauben. Die Vektoren sind in den Symbolen der Schläge gezeichnet
(vgl. Abb. 7 und Schlüssel). Zur Verrechnungsweise s. S. 287. Die 4 inneren Auflaßplätze sind maßstabsgetreu
eingetragen, die 5 äußeren mit Pfeilen und Angaben über Richtung und Entfernung auf den Zentralschlag
bezogen

Weit auseinander strebende Abflugrichtungen kennzeichnen vor
allem die inneren Auflaßplätze (Abb. 8). Tauben, deren Schläge in
gleicher oder nahezu gleicher Richtung lagen, flogen mitunter einander
entgegengesetzt ab. Es besteht jedoch offenbar keine Relation zwischen
den Unterschieden der Abflugkurse und den Unterschieden der Heimrichtungen oder -entfernungen. Genau heimgerichtete Anfangsorientierung gab es selten. Signifikant verschiedene Sichtzeiten wurden eben-

falls für die Angehörigen verschiedener Schläge gefunden. — Obwohl der Streubereich der Vorzugsrichtungen klar abgenommen hat, gab es auch an den äußeren Auflaßplätzen noch signifikant[1] verschiedene Vorzugsrichtungen sowie Zufallsverteilungen. Auch unterschiedliche Sichtzeiten kamen noch vor. Sowohl der Streubereich der Vorzugsrichtungen selbst, als auch die Abnahme der Streutendenz mit zunehmender Entfernung scheint in den verschiedenen Richtungen verschieden zu sein.

b) Heimkehrerfolg. Die Verlustrate[2] war knapp ($p = 0,07$; χ^2-Test) inhomogen unter den Schlägen, doch waren die Verlustraten einiger Schläge (z. B. Westschlag 7,5% und Ostschlag 14%) voneinander signifikant verschieden ($p \leqq 0,01$).

6. Folgerungen

a) Der „Schlageffekt". Mehrere Autoren haben Anhaltspunkte dafür gefunden oder diskutiert, daß gleichstämmige Tauben, die weit entfernt voneinander angesiedelt sind, unterschiedlich gut heimkehren (S. 287). Wie die Kreuzschlagversuche zeigen, gilt das a) tatsächlich auch dann, wenn Haltung und Flugtraining gleich sind, b) nicht nur für die Heimkehrleistung, sondern auch für andere Kriterien der Orientierungsleistung und c) für Kolonien, die nur wenige km voneinander entfernt liegen. Die erhaltenen Ergebnisse lassen sich infolgedessen kaum mit großräumigen topographischen, meteorologischen oder geophysikalischen Faktoren und Variablen erklären, die zur Interpretation der Variabilität des Heimkehrverhaltens herangezogen worden sind (WALLRAFF, 1960). — Erscheinungen, die zunächst ein „Richtungseffekt", dann ein „Ortseffekt" zu sein schienen, mögen in erster Linie schlagspezifische Ursachen haben, die allerdings in verschiedenen Himmelsrichtungen und von verschiedenen Plätzen unterschiedlich reflektiert werden. Ihre Natur ist nicht bekannt.

b) Die Entfernung. Dieser Punkt führt auf die ursprüngliche Fragestellung zurück. Die Heimkomponente h', sämtliche Abflüge je Auflaßplatz und Schlag zusammenfassend, ist in Abb. 9 gegen die Auflaßentfernung aufgetragen. Der mehr qualitative Eindruck von Abb. 8 erfährt damit eine gewisse Quantifizierung, wenn auch die Beurteilung wiederum durch die uneinheitlichen Stichprobengrößen erschwert wird. Ein gewisser Einfluß der Entfernung scheint sich abzuzeichnen: es sieht so aus, als ob grob fehlgerichtete Anfangsorientierung auf kürzere Entfernungen beschränkt sei. Andererseits scheint es auch in größeren

---

[1] Sowohl mit dem nicht kreisspezifischen, aber für viele Fälle ausreichenden Mann-Whitney-U-Test (SIEGEL, 1956), als auch mit dem neuen Test nach WATSON (1962).

[2] Aus personaltechnischen Gründen konnte nur die Zahl überhaupt heimgekehrter und die Zahl verlorener Tauben ermittelt werden.

Entfernungen (bis 90 km) innerhalb der heimzugewandten Hälfte keine
wesentliche Verbesserung mehr zu geben, und Zufallsverteilungen kamen
auch an den äußeren Auflaßplätzen noch vor. Die Streuung der Abflüge
verbessert sich dagegen nicht mit zunehmender Entfernung (SCHMIDT-
KOENIG, 1963b). Dieser vorläufige Hinweis auf eine mögliche Rolle der
Entfernung erfährt noch insofern eine Einschränkung, als sich die
Tauben, bedingt durch die experimentelle Planung des Versuchs, bei

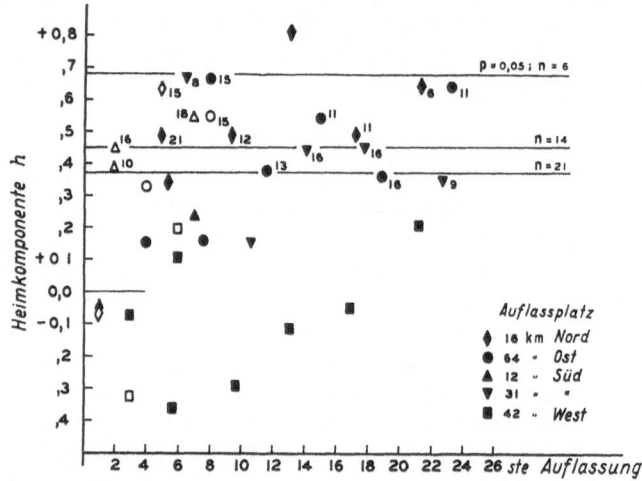

Abb. 9. Die Heimkomponente *h'*, sämtliche Abflüge je Auflaßplatz nach Schlägen vektoriell
zusammenfassend. Leere Symbole kennzeichnen Zufallsverteilungen (*p* > 0,05)

einem Teil der Kurzstreckenflüge noch auf einer Erfahrungsstufe befanden, in der sich die Anfangsorientierung noch verbessern kann. Dieser
Umstand kann jedoch nur einen kleinen Teil der beobachteten Vielfalt
an Vorzugsrichtungen erklären. — Womöglich haben wir einen Orientierungsmechanismus vor uns, der auf kürzere Entfernungen nicht genau
arbeitet. Weitere Versuche, deren Planung die Erkenntnisse der voranstehenden Experimente berücksichtigt, sollen die Rolle der Entfernung
in größerem Maßstab sicherstellen und womöglich die Grenzen der
Leistungsfähigkeit sowie systematische Fehler des Orientierungsmechanismus als Basis für weitere gezielte Forschung herausarbeiten.

Literatur

HOFFMANN, K.: Über den Einfluß verschiedener Faktoren auf die Heimkehrleistung
　　von Brieftauben. J. Ornithol. **100**, 90—102 (1959).
KRAMER, G.: Wird die Sonnenhöhe bei der Heimfindeorientierung verwertet?
　　J. Ornithol. **94**, 201—219 (1953).
— Recent experiments on bird orientation. Ibis **101**, 399—416 (1959).

KRAMER, G., J. G. PRATT and U. V. ST. PAUL: Directional differences in pigeon homing. Science 123, 229—230 (1956).
— — — Two-directional experiments with homing pigeons and their bearing on the problem of goal orientation. Amer. Naturalist 91, 37—48 (1957).
— — — Neue Untersuchungen über den Richtungseffekt. J. Ornithol. 99, 178 bis 191 (1958).
PRATT, J. G.: An investigation of homing ability in pigeons without previous homing experience. J. exp. Biol. 32, 70—83 (1955).
— u. H. G. WALLRAFF: Zwei-Richtungsversuche mit Brieftauben: Langstreckenflüge auf der Nord-Süd-Achse in Westdeutschland. Z. Tierpsychol. 15, 332—339 (1958).
SCHMIDT-KOENIG, K.: Die Sonne als Kompaß im Heim-Orientierungssystem der Brieftauben. Z. Tierpsychol. 18, 221—244 (1961).
— Hormones and homing in pigeons. Physiol. Zool. 1963 (im Druck).
— On the role of stock, selection and experience in pigeon homing. (1963a) Ms.
— On the role of the loft, the distance and site of release in pigeon homing. (1963b) Ms.
SIEGEL, S.: Nonparametric Statistics. London, New York, Toronto: McGraw-Hill 1956.
WALLRAFF, H. G.: Über den Einfluß der Erfahrung auf das Heimfindevermögen von Brieftauben. Z. Tierpsychol. 16, 424—444 (1959a).
— Örtlich und zeitlich bedingte Variabilität des Heimkehrverhaltens von Brieftauben. Z. Tierpsychol. 16, 513—544 (1959b).
— Über Zusammenhänge des Heimkehrverhaltens von Brieftauben mit meteorologischen und geophysikalischen Faktoren. Z. Tierpsychol. 17, 82—113 (1960).
WATSON, G. S.: Goodness-of-fit tests on a circle II. Biometrika 49, 57—63 (1962).

Panspecific Reproductive Convergence in *Lepidochelys kempi*

By Archie Carr

Department of Biology, University of Florida, Gainesville, Florida (U.S.A.)

With 3 Figures

Recent studies indicate that marine turtles may be capable of oriented open-sea travel comparable to that of the classic animal navigators. Because of the large size and slow travel speeds of sea turtles, and of their easy manipulability at young stages, they lend themselves to various kinds of orientation testing, both in the field and in the laboratory. Much, however, remains to be learned of the ecological geography of all the five genera, and especially of the routes and scheduling of their high-seas travel. Obviously a thorough understanding of the character of the guidance feat of the animal in nature must precede any analysis of the orientation responses that operate during each stage of a long migratory journey. Marine turtles seem peculiarly promising as subjects for field tests in which whole migratory courses are traced out under natural conditions.

Discovery of the migration of green turtles from Brazil to Ascension Island, a minimum overwater distance of 1400 miles against contrary currents of varying direction, provided a valuable natural situation in which much can be deduced by direct zoogeographic appraisal and from results of tagging. A recent development that seems to show site fixity and long distance homing capacity in another species of marine turtle is the *arribada* — the mass nesting emergence — of the Atlantic ridley, *Lepidochelys kempi* (Garman), during which virtually the entire breeding population aggregates during a day along a short section of the Gulf coast north of Tampico, Mexico.

For 75 years after the Atlantic ridley (see Fig. 1) was described as a new genus and species from Key West, Florida, its breeding place and habits remained unknown. Some eighteen years of interest on my own part, and repeated reconnaissance throughout the known range of the species, failed to locate a single authentic nesting emergence. Beginning with the finding of two hatchlings by Fugler and Webb (1957), a slow accumulation of clues delimited a strip of the coast between Corpus

Christi, Texas, and southern Veracruz, Mexico, as a region in which very desultory nesting occurs. The meagreness of the reproduction there, however, made it seem impossible that the world population of the Atlantic ridley could be derived from this area. The zoogeographic situation was discussed by CARR (1957, 1961, and 1961a) and by CARR and CALDWELL (1958).

The solution to the stubborn problem recently came, in the shape of an extraordinary film made by Sr. ANDRÉS HERRERA, an engineer of

Fig. 1. The Atlantic ridley (*Lepidochelys kempi* GARMAN). Mature female from Pinellas County, Florida

Tampico, Mexico. The film was made June 18, 1947, and was in Ing. HERRERA's desk during many of the years that the search for the breeding place of the ridley was going on. The film shows an incredible aggregation of ridleys nesting on the almost uninhabited shore between Tampico and Soto la Marina, at a place known as Rancho Nuevo in the Municipio de Aldana. The film was shown by Dr. HENRY HILDEBRAND in 1961, at the Austin, Texas, meetings of the American Society of Ichthyologists and Herpetologists. The circumstances under which it was made, and the results of Dr. HILDEBRAND's subsequent inquiries and observations in the area are reported in a paper in CIENCIA (HILDEBRAND, in press).

The salient features of the *arribadas* are these. Each year, at an unpredictable time between April and June a number of ridleys estimated as — on the occasion of the filmed emergence — about 40,000 appears off

shore and awaits a strong wind, whereupon it moves to the beach. During
a six-hour period there may be 10,000 turtles on the sand at one time.
The traffic is so dense through the mile-long site of maximum emergence
that the eggs laid by a turtle are often dug up by one nesting a little later.
Both Ing. HERRERA and other people consulted by Dr. HILDEBRAND,
and two eye-witnesses of other *arribadas* with whom I have recently
talked, say that during the nesting times coyotes congregate at the place
in numbers seen nowhere else. At least four informants have independently said that the coyotes congregate in advance of the arrival of the turtle
aggregations, suggesting that whatever concordance of factors it is that
determines the time and exact place of the emergence also controls the
massing of the predators. The emergence is repeated three times each
season, at ten day intervals (the average interval for *Chelonia* is 12.5 days),
and never occurs at precisely the same place in a given season.

Ethologically the film is of the greatest interest. It shows that as the
turtles move inland they test the sand by pushing their snouts into it —
not just on the wave-washed lower beach, where the green turtle does the
same thing, but all the way up into the loose sand where a nest site is
finally selected. The digging, covering and concealing mannerisms in
general suggest those of *Eretmochelys*, rather than those of *Caretta* or
Chelonia. A unique stereotype revealed by the film is the finishing of the
nest-covering operation by energetic pounding of the site with the
plastron. That males migrate with the females and that copulation
occurs off the nesting beach, as in the case of the green turtle, is suggested
by a section of the film showing a male coming ashore and attempting
to mount one of the nesting females on the upper beach. Both the film
and word-of-mouth information corroborate scattered previous indica-
tions that, unlike most sea-turtles, *Lepidochelys* usually nests in the
daytime. All reported *arribadas* have been diurnal.

The most extraordinary aspect of this great massing is that it con-
stitutes essentially the whole reproductive effort of an entire species of
animal. The distribution of the genus in Atlantic waters is shown in
Fig. 2. It is not known from what extremes of the range turtles travel to
the Mexican rendezvous. It is possible, for example, that the European
ridley records represent waifs permanently lost to the breeding population. Nevertheless the Mexican assemblage is quite clearly the only
breeding colony of the *kempi* in existence, and so must certainly be
recruited from a tremendously extensive range.

The migratory journey cannot be thought of as passive transportation
by surface currents, which are mainly contrary during the season of
migration. If the movement were confined to costal waters, with the
aggregation simply snowballing as it picked up recruits along the
approach to Mexico, no navigation would be required in making the

landfall. During the many years that the breeding locality of the ridley remained unknown, however, I repeatedly canvassed shrimp trawlers, snapper fishermen and other coastal residents along the way between the more northerly centers of abundance and Mexico, and found no knowledge of seasonal massing of turtles anywhere. This suggests that the Rancho Nuevo aggregation moves in from open water and is a con-

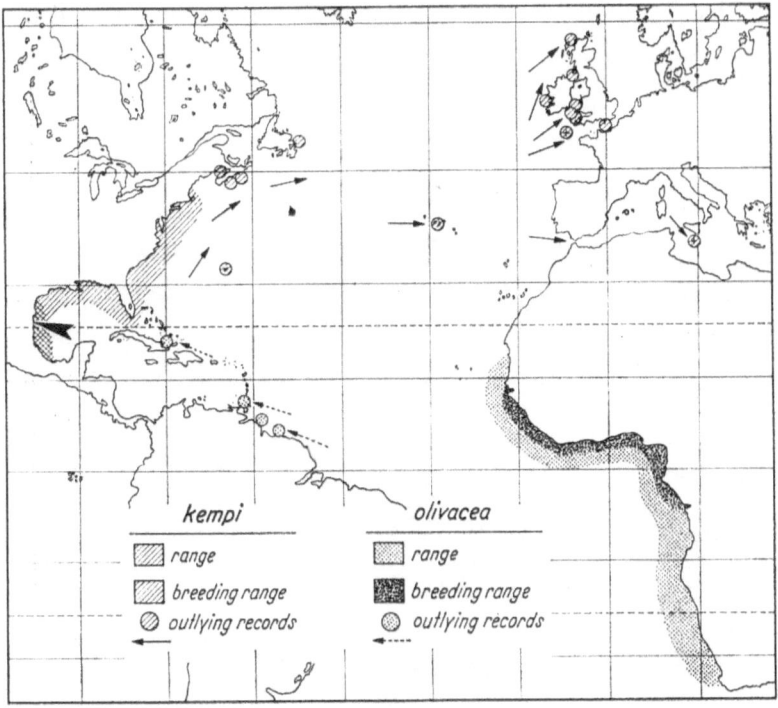

Fig. 2. Distribution of *Lepidochelys* in the Atlantic. The African ridley, *L. olivacea* is indistinguishable from populations in the Pacific, but different from the American form, *L. kempi*. The two Atlantic stocks are genetically isolated from each other, although as the small circles on the map show, waifs from each straggle far downstream in the global currents. The thin arrows near the circles show current trend there (Gulf Stream, Florida Current or Equatorial Current). The big arrow point indicates the location of the mass *arribada* at Rancho Nuevo in the State of Tamaulipas, Mexico

vergence of separately migrating individuals or small bands. If so, then, as in the case of green turtles going to Ascension Island from Brazil, the ridley migration depends upon a refined, composite orientation process that copes with varied slants of current and with drastic changes in heading. While parts of the migratory trip would seem to demand a bicoordinate navigatory process, the final massing off the nesting beach must be controlled by local sources of information — either landmarks by means of which a final site selection is made, or signals that emanate from the turtles themselves, and draw them more tightly together once they have arrived in the right general region. Olfaction may be involved. If the exceptionally active nosing of the sand mentioned earlier is, as it seems, a smelling maneuvre it suggests that the ridley is more dependent on this sense than any of the other sea turtles. Moreover, the ridleys are the only sea turtles that have conspicuous secretory pores at each seam between the inframarginal scales. The function of the secretion from these pores is not known, but the most reasonable assumption is that it is an

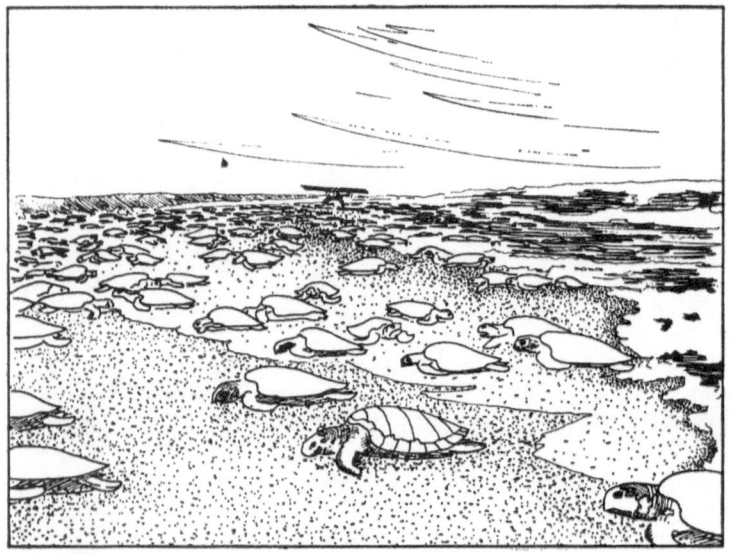

Fig. 3. Nesting aggregation of the Atlantic Ridley at Rancho Nuevo, Tamaulipas, Mexico. The drawing is a tracing taken from an enlarged frame of a 16 mm. color film made by Ing. ANDRÉS HERRERA on June 18, 1947

olfactory aid to sex — or species — recognition. If so, then it may be a signal for the final massing off the nesting beach, and possibly even a scent beacon that marks the sand of the shore for *arribadas* of later years.

Progress in animal navigation research is being hindered by temporary cleavages among the special fields of investigation involved. Although conspicuous advances are being made in studies of both physiological rhythms and orientation senses and behavior, there is as yet no model to show how the internal clock may be meshed with orientation responses in the guidance problems that animals face in nature. Another gap separates experimental research and the results of field observations of navigating animals, and these obviously must furnish the grounds for judging what cues and senses are in use at each stage of any given journey.

HASLER (1956) postulated that many cues may enter the guidance process in migrating fishes. LINDAUER (in press) has shown that even in the relatively short trips that bees make, both terrestrial and solar guideposts are used. SCHMIDT-KOENIG found low correlation between "vanishing-direction" and homing success in pigeons. There can thus be no doubt of the need for careful field studies of entire journeys — routes and schedules — of navigation and piloting animals. That *Lepidochelys kempi*, a species that masses at a single site for reproduction, offers extraordinary opportunity for such study seems evident.

Acknowledgments: I am indebted to Dr. HENRY HILDEBRAND of the University of Corpus Christi for telling me of the HERRERA film of the Tamaulipas arribada and for helping me obtain a copy of it. To Ing. ANDRÉS HERRERA I owe thanks for his permission to copy the film. My own field work on the *Lepidochelys* problem has been supported by grants from the National Science Foundation.

References

CARR, A.: Notes on the zoogeography of the Atlantic sea turtles of the genus *Lepidochelys*. Rev. Biol. trop. (S. José) 5, 45—61 (1957).
— The ridley mystery today. Anim. Kingdom 74, 7—12 (1961).
— Rätsel der Ridley-Schildkröten. Tier, no. 12, 32—36 (1961a).
— and D. K. CALDWELL: The problem of the Atlantic ridley turtle *(Lepidochelys kempi)* in 1958. Rev. Biol. trop. (S. José) 6, 245—262 (1958).
HASLER, A. D.: Perception of pathways by fishes in migration. Quart. Rev. Biol. 31, 200—209 (1956).
HILDEBRAND, H.: A ridley sea turtle, *Lepidochelys kempi* (GARMAN) rookery in the western Gulf of Mexico. Ciencia (in press).
LINDAUER, M.: Kompaßorientierung. Ergebn. Biol. 26, 158—181.
SCHMIDT-KOENIG, K.: Neue Aspekte über den Orientierungsmechanismus der Brieftaube. Ergebn. Biol. 26, 286—297.

Namenverzeichnis — Author Index

Die gewöhnlich gesetzten Ziffern weisen auf die entsprechenden Stellen im Text und die *kursiven* Seitenzahlen auf das Literaturverzeichnis hin

Numbers in *italics* refer to the page-numbers in the bibliography, ordinary numbers refer to the page-numbers in the text

20*

Sachverzeichnis — Subject Index